THE ECOLOGY OF
TEMPERATE CEREAL FIELDS

THE ECOLOGY OF
TEMPERATE CEREAL FIELDS

THE 32ND SYMPOSIUM OF

THE BRITISH ECOLOGICAL SOCIETY

WITH THE ASSOCIATION OF

APPLIED BIOLOGISTS

UNIVERSITY OF CAMBRIDGE

1990

EDITED BY

L.G. FIRBANK
Anglia Polytechnic,
Cambridge

N. CARTER
Rothamsted Experimental Station,
Harpenden

J.F. DARBYSHIRE
Macaulay Land Use Research Institute,
Aberdeen

G.R. POTTS
The Game Conservancy Trust,
Fordingbridge

OXFORD

BLACKWELL SCIENTIFIC PUBLICATIONS

LONDON EDINBURGH BOSTON

MELBOURNE PARIS BERLIN VIENNA

© 1991 the British Ecological Society
and published for them by
Blackwell Scientific Publications
Editorial offices:
Osney Mead, Oxford OX2 0EL
25 John Street, London WC1N 2BL
23 Ainslie Place, Edinburgh EH3 6AJ
3 Cambridge Center, Cambridge
    Massachusetts 02142, USA
54 University Street, Carlton
    Victoria 3053, Australia

Other editorial offices:
Arnette SA
2, rue Casimir-Delavigne
75006 Paris
France

Blackwell Wissenschaft
Meinekestrasse 4
D-1000 Berlin 15
Germany

Blackwell MZV
Feldgasse 13
A-1238 Wien
Austria

First published 1991

Set by Excel Typesetters Co., Hong Kong
Printed and bound in Great Britain
by Hartnolls Ltd, Bodmin, Cornwall

DISTRIBUTORS

    Marston Book Services Ltd
    PO Box 87
    Oxford OX2 0DT
    (*Orders*: Tel: 0865 791155
            Fax: 0865 791927
            Telex: 837515)

USA
    Blackwell Scientific Publications, Inc.
    3 Cambridge Center
    Cambridge, MA 02142
    (*Orders*: Tel: 800 759–6102)

Canada
    Oxford University Press
    70 Wynford Drive
    Don Mills
    Ontario M3C 1J9
    (*Orders*: Tel: 416 441–2941)

Australia
    Blackwell Scientific Publications
    (Australia) Pty Ltd
    54 University Street
    Carlton, Victoria 3053
    (*Orders*: Tel: 03 347–0300)

British Library
Cataloguing in Publication Data

British Ecological Society
    (Symposium: 32nd: 1990:
    University of Cambridge)
    The ecology of temperate cereal fields.
    1. Temperate regions. Cereals
    I. Title   II. Firbank, L.G.
    III. Association of Applied Biologists
    633.109123

ISBN 0–632–03147–6

Library of Congress
Cataloging in Publication Data

British Ecological Society.
    Symposium (32nd: 1990:
    University of Cambridge)
    The ecology of temperate cereal fields/
    The 32nd Symposium of the
    British Ecological Society
    and the Association of Applied Biologists,
    University of Cambridge, 1990;
    edited by L.G. Firbank . . . [*et al.*].
            p.   cm.
    ISBN 0–632–03147–6
    1. Grain — Congresses.
    2. Grain — Ecology — Congresses.
    I. Firbank, L.G.   II. Association of
    Applied Biologists.   III. Title.
    SB188.2.B75   1990
    633.1 — dc20

CONTENTS

v

PART 2: THE SOIL

PART 3: SPECIES INTERACTIONS

PREFACE

Agriculture was big news in Britain in the spring of 1990. Headlines about mad cow disease had followed swiftly on the heels of revelations about the dangers of lysteria in soft cheeses and salmonella in chickens. On the inside pages were articles concerned with the levels of nitrates in drinking water, codes of practice for the use of pesticides and the introduction of genetically engineered organisms. A more careful look at the scientific and agricultural press would have revealed yet other concerns. The long-awaited research programme into Agriculture and the Environment had finally begun, along with the first experiments designed to investigate the effects of setting aside land from production. The perceived over-production of European agriculture suddenly had to be set against the possible effects of global warming and the changing politics of eastern Europe. And over all of these factors, there was the increasing desire of people to live in or visit an attractive countryside and to eat food with fewer or no artificial additives. The policies and practices which had done so much to improve the productivity of food were being questioned over their emphasis on quantity rather than on quality — of produce, of environment, of life.

Central to concerns of quality in agriculture are issues of ecology. However, these issues go deeper than many non-specialists appreciate. Ecology is not simply to be equated with conservation; the principles of ecology are central to the study of crop growth, the reduction of pest species and the efficient use of soil and other resources. Indeed, the cereal field may be the most intensively studied ecosystem in the world. The problem is not that there is insufficient knowledge about cereal ecology, but rather that the information is fragmented into many subdisciplines.

It was entirely appropriate, therefore, that the British Ecological Society should choose the ecology of temperate cereal fields as the subject of its 1990 symposium. The objective of the symposium was to give an integrated view of the ecology of temperate cereal fields. We hoped to demonstrate to ecologists that the cereal field is worth studying; to conservationists that there are elements of the cereal field that are worth conserving and to agriculturalists that an appreciation of ecology is essential if the twin goals of quality and profitability are to be met.

The task of selecting topics and authors was not easy, but we were helped greatly by members of the co-organizing body, the Association of Applied Biologists, which includes many scientists looking at the ecology of cereals. We would like to thank in particular Roger Austin, Allan Langton, Gareth Jones and David Tottman for their contributions at this stage.

We adopted a largely north-western European perspective, biased towards large-scale integrated studies. Several non-ecological papers were included to put the symposium in a wider context. We also tried to include a variety of viewpoints, from industry to conservation, and a variety of styles and techniques. Unfortunately we were forced to exclude much excellent and relevant research. Naturally, the programme had to be modified slightly; here we thank Nick Mills for reading Professor Altieri's paper; sadly, Richard Southwood, John Lawton and Oliver Rackham all had to decline offers to participate; and Richard Brown's paper on agrochemicals and soil organisms could not be finished in time for inclusion in this volume.

The conference was held at Newnham College and the Lady Mitchell Hall, Cambridge, on 3–5 April. Much of the administration was carried out by Sharon Flint at Newnham and by Roger Rooke and Carol Millman of the Association of Applied Biologists. As well as the main lectures and a poster display, Mark Tatchell organized a visit to Rothamsted Experimental Station, and Nigel Boatman and Nick Sotherton organized a visit to the farm of Mr John Wilson at Ixworth Thorpe, Suffolk, who actually practises many of the techniques which encourage the successful marriage of farming and conservation. Jonathon Porritt, then chairman of the Friends of the Earth, gave an excellent after-dinner talk.

The papers in this volume have been edited by ourselves and a number of referees. Susan Sternberg, Emmie Williamson and Karen Biggs looked after things for us at Blackwells. We would like to thank all these people for their contribution, but most of all we thank the authors who have produced what we believe is a unique and timely synthesis of cereal ecology.

L.G.F., N.C., J.F.D., G.R.P.
May 1991

PART 1
THE WIDER CONTEXT

Session Chairman: EARL OF CRANBROOK
Session Organizer: L.G. FIRBANK

There is no more a standard cereal field than there is a standard woodland or a standard garden. The species present and their abundances reflect the characteristics of soil, climate, drainage and land management. Large differences are apparent to the most casual observer between vast prairies and small fields, surrounded by hedgerows and close to other habitats. A visit to an arable nature reserve shows the colourful weeds which must have exasperated farmers before the use of herbicides removed them from our more intensive farms. Changing crop rotations, cultivation techniques, fertilizer usages and crop varieties all have an impact upon the cereal ecosystem.

One of the themes running through this volume is the development of management strategies that minimize damage to the ecosystem while still producing economically acceptable yields. However, we should be deluding ourselves if we ignore the many other factors which farmers and their advisors must take into account.

In this section there are several papers which provide a wider context for the more ecological papers to follow. Dick Potts sets the scene for the volume with a discussion of the ecological value of cereal fields. He outlines how it has been diminished by intensification, notably as a result of increased pesticide usage. He makes the first of several pleas for a less intensive approach to cereal farming. There are, however, problems in offering ecological arguments to policy makers. These are well described by the Earl of Cranbrook, here drawing on his experience in the UK House of Lords.

The prospects of changes to cereal farming depend, as ever, on the development and availability of suitable crop varieties, a topic addressed by Professor Fischbeck. He also looks at the prospects for biotechnology in altering the type of crops that will be grown. Also of central importance in cereal farming are economic issues. Peter Costigan and Paul Biscoe show how the yields of cereals depend upon the weather and upon the choice of levels of inputs. They show that the choice of inputs is fundamentally an economic one, which is inevitably influenced by the prospects for world trade in cereals. Michael Murphy provides this essential economic view-point of cereal production and consumption.

Most papers were written between late spring and early autumn 1990, when the outcome of the Uruguay round of the GATT talks (General Agreement on Tariffs and Trade) was not known. As this introduction is written, the EC has just agreed on reductions of subsidies to farmers, but not enough to satisfy other participants at GATT. However, the potential importance of these talks on cereal farming, and hence on the cereal ecosystem and the wider landscape should be clear from the papers by Potts and Murphy.

There have also been new developments in Europe concerning the way in which cereals are to be produced. On 3 August 1990, the European Commission submitted a proposal for a Council Regulation (EC) COM (90) 366 on 'the introduction and the maintenance of agricultural production methods compatible with the requirements of the protection of the environment and the maintenance of the countryside, which has become known as the Regulation on Extensification.' This Regulation was revised on 4 April 1991, and is currently unofficial and subject to further revision.

If this Regulation is approved, its policies must be implemented by the twelve nations of the EC. These measures will meet the many calls made in this volume for a less intensive agriculture.

1. THE ENVIRONMENTAL AND ECOLOGICAL IMPORTANCE OF CEREAL FIELDS

G. R. POTTS

The Game Conservancy Trust, Fordingbridge, Hampshire SP6 1EF, UK

INTRODUCTION

The many factors, global and local, in the mix of problems centering on cereal-field ecosystems make for a potent brew. Dealing with the issues of farm surpluses, set-aside, declining farm incomes, agrochemicals, polluted or depleted aquifers and similar sensitive issues probably requires collective effort and diplomacy of a kind that has hitherto been reserved for national defence. Problems at the cereal-growing/environment interface are beginning to be dealt with at an international level, for example in the current 'Uruguay Round' of talks on the General Agreement on Tariffs and Trade (GATT), and in the plans for central and eastern Europe within the European Agricultural Forum of the Council of Europe. The general aim is for more consistency between nations in measures relating to environmental protection, and reduced farm subsidies, but progress so far has been minimal.

This paper outlines the ecological significance of modern cereal crops, identifies a major drawback of intensification and its associated environmental damage — both spatially and temporally — and introduces some possible solutions. In particular it is argued that the policy known in Europe as 'extensification', whereby inputs are reduced, is, for environmental reasons, preferable to that of 'set-aside', whereby the size of the area used to grow crops is reduced.

In this review paper there are many references to The Game Conservancy's study of the cereal ecosystem on the South Downs in Sussex. This continuing long-term study, which began in 1968, is described in detail elsewhere (Potts & Vickerman 1974; Potts 1986; Aebischer & Potts 1990a; Potts & Aebischer 1991; Aebischer, this volume, pp. 305–331).

SCALE OF THE CEREAL-FIELD ECOSYSTEM

The area planted to cereals in any one year is vast: 4 million ha in the UK, 34 million ha in the European Community and a world total estimated at

more than 700 million ha, giving a global yield of 1729 million tonnes
(Home Grown Cereals Authority (HGCA) London, estimates for 1989).
Because cereals are often rotated with other crops, the total area involved
in cereal production directly or indirectly in any one year must be about
1000 million ha. Data for 1989 compiled by the consultancy County
National Westminster for the British Agrochemicals Association show that
a little over 35% of the world usage of pesticides was on cereal crops. On a
more meaningful ecological scale the area given over to cereals in the UK,
17% of the total land area, is sixteen times the combined area given over to
all Nature Reserves in the UK (calculated from Nature Conservancy
Council data). It follows that major changes in the cereal-field ecosystem
will have repercussions for wildlife in general. Early and well-known
examples include the use of the cereal seed-dressings dieldrin and
carbophenothion which, respectively, reduced the size of the sparrowhawk
Accipiter nisus population (Newton 1986) and caused considerable
mortality to geese wintering in the UK (Stanley & Bunyan 1979).

INTENSIFICATION

Intensification in the cereal ecosystem takes many forms both temporally
and spatially, but almost all of them take the system towards that of a
monoculture. Hedges are removed so that fields become bigger, rotations
become less complex and even disappear (see Potts & Vickerman 1974 for
the Sussex study area), mixed farms give way to specialized farms and
mixed-farming regions give way to specialized regions (Southwood 1972;
O'Connor & Shrubb 1986). For example cereals give way to grass in the
west of Britain, grass gives way to cereals in the east. Farm livestock are
separated from the crops grown to feed them. Farmyard manure (a
mixture of dung and straw) gives way to slurry (a mixture of dung and
water) in the west, and to surplus straw and burning in the east. Cropping
of winter wheat in southern England can extend to 11 months of the year;
irrigation evens out the seasonal and annual variations in crop growth with
well-known depletion of aquifers (especially in the USA) and some damage
to wildlife (Serre *et al.* 1989). Like the question of nitrate or pesticide
pollution of aquifers and watercourses, these subjects are all relatively
tangible and understood. A useful introduction is the Natural Environment
Research Council Publication *Agriculture and Environment* (Jenkins 1984)
and a recent comprehensive review with worldwide coverage is given by
Conway & Pretty (1991). This paper concentrates on developments within
the cereal crop consequent on the use of pesticides.

YIELD OF CEREALS

Throughout the world, cereal yields have been increasing. Figures for the UK show an increase of approximately 2% per annum in recent decades. The principal reasons are the use of inorganic fertilizers, the breeding of higher-yielding varieties and of course pesticides (Costigan & Biscoe, this volume, pp. 56–68). Together in the UK these factors have resulted in an exponential increase in the average yield of wheat from 0·5 t/ha in medieval times to 6 t/ha today (Cooke 1967, updated by HGCA/MAFF statistics).

Before one can consider reasonably the recent adverse ecological effects of this twelve-fold increase in cereal yields, it is important to realize that at least 95% of the total yield today is necessary to meet the demand for food. Famine has long been banished from western Europe and, where it still exists elsewhere, tends to be the result of war rather than of poor crops. Consequently, the necessary adjustments to production required to bring supply and demand into reasonable balance, i.e. with adequate strategic reserves, are relatively small, though with wide significance for farm profitability and the environment.

In the UK, cereal fields have recently been regarded by notable UK conservation bodies as 'man-made and artifical; deserts in all but name' and as 'devoid of interesting wildlife'. Perceptions are less extreme on the mainland of Europe but obviously these expressions are more true now than formerly, and it is self-evident that wildlife will have been squeezed out as intensification has proceeded. Here we need to consider how far the process has gone and, indeed, whether or not it matters.

POTENTIAL RICHNESS OF THE CEREAL ECOSYSTEM

The temperate cereal-field flora and fauna are basically a mixture of vestiges of the communities of the steppe or prairie ecosystems that cereals have almost entirely replaced, and of ruderal species which exploit the cereal ecosystem. In Europe most of the latter originated, like cereals, in the mediterranean region. Although the cereal ecosystem is not stable in the sense of sustainability without anthropogenic inputs, distinct cereal flora and fauna have evolved over thousands of years. Compared to the 'Fertile Crescent' of the Middle East where cereal cultivation began over 10 000 years ago, Britain is not ideally suited to the growing of grain, yet cereals have been grown here for at least 7000 years (Edwards & Hirons

1984), rather earlier than usually given (Barker 1985). The cereal ecosystem thus pre-dates some other ecosystems such as heather moorland, coppice woodland and perhaps even chalk grassland, all three of which many conservationists regard as important in conservation terms. On the other hand the cereal ecosystem has changed markedly to its present form.

Plants

Excluding cereals and grasses a total of 131 species of flowering plants have been encountered in cereal crops in the Sussex study area, which has calcareous soils. Widening the geography to the whole of Britain would bring in many calcifuge species (e.g. corn marigold *Chrysanthemum segetum* and weasel snout *Misopates orontium*) or eastern species of sandy soils (such as flixweed *Descurania sophia*) so that the total rises to over 200 (Smith 1986), or even approaches 300 (Wilson 1990). Given that all arable weeds would have been found in cereals at some time or other, then the total for western and central Europe is over 700 (Hanf 1983).

Many cereal weeds are now very rare. Some specialists, originating mostly in crops which have themselves become rare, such as flax, are virtually extinct (Kornas 1988). The abundance of most others has been dramatically reduced by herbicides. Some became extinct in our Sussex study area before our monitoring began, e.g. cornflower *Centaurea cyanus* and pheasant eye *Adonis annua* (Hall 1980), and many others have declined over the past 20 years. At least five species have become extinct in the past 20 years in Britain, and several others like the two above are in imminent danger of extinction; twenty-nine have been of British Red Data Book status for some time (Perring & Farrell 1983) and many others have suffered catastrophic declines. In Germany, seventy-five species from the cereal flora are 'endangered', fifteen already extinct (Eggers 1984). In Britain the cereal ecosystem flora prior to the use of herbicides and inorganic fertilizers probably contained about 17% of all British flowering species, compared to 10% for woodland (Wilson 1990).

In the Sussex study area in the mid 1950s, as in the rest of Britain, the use of herbicides in cereals became routine. It was, however, 15 years before the Sussex study was in place to measure the ecological impact. Individual species responded differently depending on the activity spectra of the herbicides. In addition, as Firbank (this volume, pp. 217–218) clearly demonstrates, for the corncockle *Agrostemma githago*, other factors were leading to declines, in this case the cause was improved methods of seed cleaning. Nevertheless, many long-term effects of herbicides on the cereal-field flora have now been quantified despite the buffering effect of the seed-bank (Wilson 1990), and they have been catastrophic. Of the 131 dicotyledonous weed species recorded in the Sussex study, only cleavers *Galium aparine* is certainly more numerous now than it was in the years prior to the use of herbicides (Aebischer, this volume, pp. 312–313), though in this case factors such as more winter cereals, ploughing of hedge bottoms, and increased fertilizer use have also contributed (Boatman 1989). Swine cress *Coronopus squamatus* may however have benefited from the 'tramline system' introduced in the late 1970s to facilitate regular spraying without wheel damage, and hedge mustard *Sisymbrium officinale* has been encouraged with the incorporation of oilseed rape in the arable rotation.

Some species such as corn gromwell *Lithospermum arvense* and narrow-fruited corn salad *Valerianella dentata*, once abundant in most cereal fields in the Sussex study area, survive now in only five and three field edges respectively out of a total of about 500, whilst corn parsley *Petroselinum segetum* survives in only one field corner, within half a metre of a regularly sprayed sterile strip around cereal crops. Such species as these are expected to disappear soon from this study area of 62 km^2: corn spurrey *Spergula arvensis*, always local to the more acid clay-cap areas, has not been found since 1981, perhaps as a result of liming. The monocotyledons have fared better with very large increases in *Bromus sterilis* and *B. mollis* (Aebischer, this volume, pp. 311–313).

Arthropods

A total of about 550 species of insects have been identified in the cereal ecosystem of the Sussex study area, over eighty species of spiders and three species of harvestmen. Many of these species spend a good deal of the year outside the cereal field, for example the Coleoptera and Heteroptera which overwinter in hedgerows (Sotherton 1985; S.J. Moreby, personal communication). Other species not included here, because they breed elsewhere, visit cereal-field plants for nectar, e.g. butterflies (Dover 1989). Yet others, also not included here, use cereal crops for resting. For example it is not unusual to find that more than half the species of winged aphids in cereal crops belong to trees rather than cereals (Potts & Vickerman 1974). Several important groups have been only incompletely identified to species, particularly parasitic Hymenoptera, Collembola and Acari so that the true total of arthropods must approach 700 species for the period 1970–89 as a whole. Pending identification of all species sampled, some idea of the total can also be obtained independently, as follows.

Meyer (in Heydemann 1983) found an average of sixteen species of herbivore per plant species (b) in the arable ecosystem in central Germany. This compares with twenty-four species per tree species in a forest in the USA (Futuyuma & Gould 1979). In all habitats herbivores share host plants, in one study estimated at 2·22 plant hosts per herbivore (c), (Thomas 1990). The number of plant species in cereals on the Sussex study area (d) was probably near 140 species prior to the use of herbicides compared to fifty-five today (d'). Cohen's review (summarized in Pimm 1982, p. 167) gives similar ratios of predator and parasitoid species to prey species for arthropods in a variety of habitats, on average 0·76 (e).

We can use the estimates b to e in the above to estimate the number of insects and spiders (a) in the Sussex study as follows:

1 before the use of herbicides

$$a = (b/c \times d) + e(b/c \times d)$$
$$a = 1776 \text{ species}$$

2 after the use of herbicides

$$a' = (b/c \times d') + e(b/c \times d')$$
$$a' = 698 \text{ species.}$$

An independent estimate (known to be an underestimate) of $e(b/c \times d')$ is 416 (Sunderland, Chambers & Stacey 1984, updated; K.O. Sunderland (personal communication) and adding parasitoids) giving $a' > 963$ species for UK cereal crops. Heydemann (1983) considered that before the use of herbicides 8% of the total insect fauna of Central Europe was found in cereal crops. In the UK with 22 500 species in total this (a) would be 1800 species. This could be an underestimate because, for example, the 180 species of spider recorded from UK cereals (Sunderland 1987) is 26% of total UK species of spider, spiders being better known than many of the insects. Of course these species totals for the cereal ecosystem are gross approximations but they do suggest a loss of several hundred species of arthropod from the cereal ecosystem in the past 50 years.

The indirect impact of herbicides on arthropod species is difficult to assess because many of them are polyphagous, and the relationship between the density of a host-plant density and that of its herbivores is not straightforward. For example, below a threshold density of hosts the herbivores may disappear. Knotgrass *Polygonum aviculare* does not attract ovipositing females of the strictly host-specific knotgrass beetle *Gastrophysa polygoni* below about eight plants per metre (Sotherton 1982).

Most trials with herbicides show reductions in the overall number of field-layer and crop arthropods of about 50% (Potts 1986, Table 6.5), increasing to 80% in the case of plant bugs: Heteroptera (Sotherton, this volume, p. 386, Table 17.9). Most of this reduction is caused by the removal of the food source of the arthropods but in the case of the plant bugs more subtle effects may also be occurring. For example species such as *Calocoris norvegicus* can feed and breed on the cereal itself, yet are more abundant where herbicides are not used. In this case it may be that the effect of weeds on the microclimate is crucial at some stage, possibly just after hatching (S.J. Moreby, personal communication). The adverse effect of the reduction in arthropods by herbicide use can be overcome, at least to a considerable extent, by the selective use of herbicides at cereal crop margins (i.e. Conservation Headlands — Sotherton, this volume, pp. 381–392).

The decline of undersowing, that is, using cereal crops as a nurse to establish grass/clover mixed leys, has exacerbated the herbicide effect on arthropods (Potts 1970; Aebischer 1990). Cultivations disrupt the life cycle of the sawfly, in the pre-pupal and pupal stages, whereas undersown areas are not cultivated until after the adults have emerged (Potts 1970, 1986).

Many species have, without doubt, declined in cereal crops as an indirect result of the use of herbicides and, more recently and to a lesser extent, fungicides and insecticides. In Germany, Heydemann (1983) has documented a loss of 50–80% of species of Coleoptera and Hymenoptera: Formicidae from arable land in the period 1951–81; the same could have happened in the UK in many groups. Later in this volume Woiwod (pp. 299–301) describes a 55% loss in total numbers of nocturnal Lepidoptera as sampled by light trapping, starting about 1950. We do not know how many arthropod species disappear from the cereal ecosystem each year but over 70% of the taxa recorded in the Sussex study are declining significantly, equivalent to a 50% decline in overall numbers every 10 years (Aebischer & Potts 1990b). Some of the data for Sussex for the period 1970–89 inclusive are given in Fig. 1.1 — but most of these species — all of which had survived the use of herbicides for at least 15 years — are still relatively common. Detailed analysis of the Sussex data follows in the paper by Aebischer (this volume, pp. 305–332); here the concern is about the wider relevance of the results to conservation rather than to agriculture.

None of the thirty species of Heteroptera we have identified in cereals in Sussex are to be found amongst the seventy-nine species of Heteroptera listed in the British Red Data Book on insects. Indeed only one of the Red Data Book species *Eurygaster austriaca* lived in cereals and this is the only one, out of a total of 540 species of Heteroptera, which has almost certainly become extinct in the UK. It is still common, however, in cereals overseas, though not so numerous as its close relative *E. integriceps*, a major pest in southern and eastern Europe.

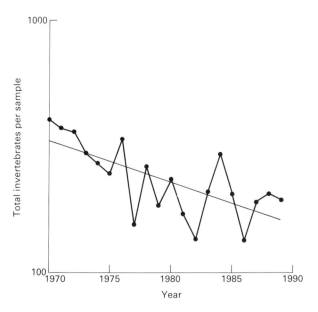

FIG. 1.1. Reduction in the mean number (logarithmic scale) of invertebrates, excluding Acari, Collembola, Thysanoptera and Aphididae, recorded per sample in the Sussex study from 1970 to 1989. The average annual rate of decline was $3\cdot8 \pm 0\cdot9\%$ ($t_{18} = 4\cdot11$, $P < 0\cdot001$).

Ten species of *Dolerus* sawflies (36% of the genus of twenty-eight species for the UK) have declined in Sussex since the early 1970s, plus forty species (5%) of the 800 UK rove beetles (Staphylinidae), and twenty-five species (7%) of 347 UK leaf and mould beetles. None of these declining species are in the Red Data Book. However, the collective decrease, averaging around 5–7% per annum per species (Aebischer & Potts 1990b) is, none the less, a major change in one of our largest ecosystems.

The sawflies inhabiting cereal crops illustrate a general point about the cereal ecosystem rather well, that the ecology and biology of the species which are not of economic importance is remarkably poorly known. In Sussex (this study), France (Chevin & Chambon 1984) and Poland (Miszulski & Lipinska 1988) it appears that about twenty species of sawflies belong to the cereal ecosystem in each case, a total of over thirty species. Some studies of fluctuations in densities of *Dolerus* sp. have been made in Sussex (Aebischer 1990) and in Germany (Freier & Wetzel 1984). The larvae of several species, however, have not been properly described. Knowledge about their Ichneumon parasitoids — nine species of which were found in the Sussex study — has scarcely moved forward for the major part of the century, indeed not since Morley (1911), although the taxonomy of some of them is better understood (Fitton 1974). Yet, as we shall see, sawfly larvae are important foods of at least one group of birds, the gamebirds.

Fungicides

From the late 1930s almost all cereal seed has been dressed with fungicides. With the exception of those mercury-based fungicides which were shown to be toxic to granivorous birds (Potts 1986) and have now been phased out, the ecological effects of such dressings are not known.

In the Sussex study area, as elsewhere in temperate zones, foliar fungicides began to be used on cereals in the early 1970s with 'saturation use' by the early 1980s. Only one of these compounds, the organophosphate pyrazophos, is directly insecticidal: in trials it reduced the abundance of gamebird chick food items by about 60% (Sotherton, Moreby & Langley 1987; Sotherton & Moreby 1988). Such effects, though, are not discernable in June if the fungicide is applied before 1 March.

The ecological side-effects of fungicides on the pathogenic and sapro-phytic fungi are not understood but the progressive decline in abundance of the fungus-feeding fauna following the introduction of fungicides is one of the outstanding features of the Sussex monitoring study (Aebischer, this volume, pp. 305–331). Whether cause and effect are involved is not yet known, but seems likely (Aebischer & Potts 1990b).

To complete the picture we would have to include the astonishingly complex assemblage of micro-organisms found on all plants: the bacteria, yeasts and filamentous fungi (Blakeman, this volume, pp. 171–191). In addition are the soil flora and fauna and the viruses, all of which are dealt with in subsequent papers in this volume (Jagnow, pp. 113–137; Zwart & Brussaard, pp. 139–168; Carter & Plumb, pp. 193–208).

Insecticides

Prior to 1968 there was no use of insecticides on cereals in the Sussex study area excepting seed dressings (γ-HCH usually; dieldrin rarely) and minor uses of DDT to control leatherjackets (larval Tipulidae: Diptera).

In Sussex the 1968 and 1970 outbreaks of cereal aphids were very widespread but, as elsewhere, most farmers were not concerned by the presence of the aphids and aphicides were not widely used in Britain until the next outbreak, in 1975.

As a result of trials with aphicides in 1975 we warned farmers of the adverse effect of insecticides, especially of the organophosphate dimethoate, and began to recommend the more aphid-specific carbamate pirimi-carb (Vickerman, Potts & Sunderland 1976; Vickerman & Sunderland 1977). National surveys of aphicide use were carried out in 1975 and 1976, and repeated in 1980 and 1981, covering hundreds of farms each year. It was concluded that adverse effects on the survival rates of partridge chicks occurred when the insecticides were used in June, and that these effects were broadly proportional to the percentage of cereals treated and to the toxicity of the insecticides used to the preferred insect food of partridge chicks (Potts 1981).

In 1984 the first adverse effects on partridge and pheasant chicks were encountered in the Sussex study area. Survival rates of the chicks of both species were reduced by almost 40% where demeton-S-methyl — also like dimethoate, an organophosphate — was used (Potts 1986). The effects on partridge density were not measurable, however, because the two areas treated were relatively small.

Ministry of Agriculture, Fisheries and Food data show 60 000 ha to have been treated with dimethoate in 1982. Surveys of Cereals and Gamebirds Research Project subscribers showed dimethoate to be the most frequently used aphicide on UK cereals in 1984 and 1988 (Wratten *et al*. 1990). A 1989 extension of these surveys suggested that the area sprayed with this chemical could now be as much as 400 000 ha per year.

The first widespread farm-scale use of dimethoate in the Sussex study area took place in mid June 1989, when it was used on all the cereals in a contiguous block of 543 ha just at the time of peak partridge-chick hatch; the rest of the study area, 2482 ha, was not treated.

Stubble counts established that although adult gamebirds had not died as a result of the spraying, the chick survival rate was halved (Potts & Aebischer, unpublished). Recent cases in the USA where significant numbers of adult gamebirds, in particular sage grouse *Centrocercus urophasianus*, died as a result of the use of dimethoate on lucerne (Blus *et al*. 1989) might be explained by the fact that 560 g of active ingredient were applied per hectare, whereas in Sussex the figure was only 336 g (Vickerman & Sunderland 1977), the recommended rate of application in the UK. Alternatively it may be that insects are a vital source of food for adult sage grouse.

In Sussex the removal of chick food by dimethoate reduced chick survival rates of both partridge and pheasant to a level well below the minimum necessary for population maintenance. The effect on partridge chicks was almost exactly that predicted from a model published prior to the spraying (see Potts & Aebischer 1990). The absence of a direct effect on the chicks was in line with our expectation and with work in France in the early 1980s (Saint-André de la Roche & Douville de Granssu 1982).

Full recovery of the sawfly population (the most important chick food item) might take up to 7 years (Aebischer 1990), even without the further spraying which took place in 1990. Simulation models of partridge populations suggest that frequent use of dimethoate could seriously reduce the population density of the resident wild gamebirds in the areas treated (Potts 1990). The implications for other less well-studied species would seem equally grim.

Effects like these are presumably being repeated, but unobserved, on a vast scale. An indication of their magnitude in Britain is given by the honey bee *Apis mellifera*. In 1989 at least 344 colonies, about 5% of the total, were very seriously affected by insecticide use on farmcrops (E. Fenner, British Beekeepers Association), almost 70% of these cases being caused by dimethoate (ADAS, Harpenden). This is of course despite the legal necessity on the part of farmers to warn hive owners so that the bees can be kept safe, an option not available for the protection of wild species.

As is well shown by Wratten & Powell later in this volume (pp. 233–257), the natural enemy complex of cereals has at least the potential to keep cereal aphids below the level where it becomes necessary to spray them. Whether the upsurge in aphicide use is in part a response to pesticide-induced declines in some of the predator species remains to be demonstrated, but seems likely. Whether such effects will seriously limit future attempts at Integrated Crop Protection in cereal crops remains to be seen (see Cavalloro & Sunderland 1988).

More widely, several cereal pests are increasing at the present time, resulting in yet more use of insecticide. The Russian wheat aphid *Diuraphis noxia* has reached Australia, South Africa, Chile and the USA. In the last case its first recorded presence was in Texas in 1986, but by the end of that year it had already reached six other states (Stoetzal 1987). Hughes (1988) reviewed the international problem, and judged that it was likely to worsen. Another predominantly Russian pest of cereals, the grain ground beetle *Zabrus tenebrioides*, was first identified in the Sussex study area on 8 March 1990. It has been increasing through most of Europe for many years (e.g. Basset 1978; Cate 1980).

Outside the UK the situation may generally be worse. Several insecticides not allowed in the UK, such as methyl parathion, can have very severe effects on the cereal fauna. For example in Germany, Basedow (1987) recorded an 81% loss in the density of ground beetles, Carabidae, from 1971 to 1984. During this period the use of insecticide in the area studied increased from 6% in 1971, to 100% in 1983; methyl parathion was particularly damaging. Another example is monocrotophos, widely used in eastern Europe on maize to control the weevil, *Tanymechus dilaticollis*, and which has also caused notable mortalities amongst gamebirds (A. Zajak & D. Farkas, personal communication. May 1981, Wildlife Protection Station of the Plant Protection and Agrochemistry Centre, Hungary, and Nikodemusz (1983)).

In North America the very widespread use of insecticides in cereals, especially of the carbamate carbofuran principally to control grasshoppers, may be having adverse effects on many species. So far this has only been demonstrated clearly where the chemical has been used on grasslands (McEwen 1981; Fox *et al.* 1989), or in rice (Flickinger *et al.* 1980).

The implication of all this is clear. The insecticides, which the Sussex study has shown to have important ecological side-effects, are being used on a global scale. So the ecological impact is also likely to be measurable globally.

Birds and mammals

Most birds of dry grasslands, moor and heath, and of hedgerows and woodland edge will feed in cereal crops but, as with the arthropods, few of them can reside there exclusively throughout the year. Perhaps fifteen species in Europe (compared to forty-four for dry grassland: P. Goriup, ICBP, personal communication) could be said to be members of the cereal ecosystem throughout the year although very many more feed in cereal fields. In Britain perhaps only the skylark *Alauda arvensis* and corn bunting *Miliaria calandra* can exist more or less entirely in cereal fields. On The Game Conservancy's main experimental farm, Manydown in Hampshire, twenty-seven species were regularly found to feed in cereal crops during the summer (British Trust for Ornithology, unpublished). The total for Britain is much higher; we need to add several cereal-field species that do not breed on this farm, e.g. tree sparrow *Passer montanus*, quail *Coturnix coturnix* and the species clearly associated primarily with other habitats such as reed bunting *Emberiza schoeniclus* and oyster-catcher *Haematopus ostralegus* which nevertheless frequently breed in cereal crops or which, like the swallow *Hirundo rustica*, feed over them. Finally we need to add those species which feed on cereal fields only in winter such as the golden plover *Pluvialis apricaria*. In all, my estimate would be that at least seventy-eight species of birds are affected (like the stone curlew *Burhinus oedicnemus*, last seen in the Sussex study area in 1981) or could be affected by the ecological state of cereal crops. Although many species are involved in the UK, there are far more in Europe: the two species of bustard *Otis tarda* and *O. tetrax*, Montagu's harrier *Circus pygargus*, and several species of lark and bunting. A recent BTO estimate is that 130 species of bird regularly breed on farmland in the UK, with at least sixteen of them declining significantly (Marchant *et al.* 1990). At least nine of the fifteen species mentioned earlier as the main bird inhabitants of cereal fields are declining at the present time (Marchant *et al.* 1990), reminiscent of the situation with arthropods.

The difficulty of estimating the number of species feeding to a notable extent in cereal fields is underlined by the fact that there are, according to my calculations, approximately half a million kilometres of cereal-field margins in Britain alone, adjacent to an enormous variety of other habitats.

The position with mammals is more difficult to summarize. Few mammals live entirely in cereal fields, even the hare *Lepus europaeus*, field mouse *Apodemus sylvaticus* and harvest mouse *Micromys minutus* often venture outside them. Most, however, feed on cereals, e.g. rabbit *Oryctolagus cuniculus*, and roe deer *Capriolus capriolus*, or over them, e.g. several species of bats (Chiroptera). In mainland Europe one could add the field vole *Microtus arvalis* and hamster *Cricetus cricetus*. An even more diverse suite of species occupies the cereal ecosystem of North America, as is the case with birds too.

Thus it is clear that the cereal ecosystem is far from a 'man-made desert' though most of the species it contains have greatly declined and are still declining. The grey partridge in its world range down from 120 million individuals to 10 million (Potts 1986, updated) is merely a well-quantified example; it seems not to be atypical.

DISCUSSION

Having described the inexorable progression of intensification in cereal growing we can return to the two outstanding problems we began with, cereal surpluses and the environmental damage caused in producing the surpluses.

In Europe cereal surpluses are being dealt with by lower production-coupled subsidies, downward adjustment of price-support mechanisms and by set-aside. The first two may, in the long term, help to reduce the inputs which cause the environmental damage but there is no sign of that happening yet; with the latter, a measure transferred (inappropriately in my opinion) from low input systems in North America, this cannot happen. By definition, set-aside removes amounts of *land* cultivated (but in Europe, so far, less than 2%), rather than inputs. By keeping the price of grain high relative to world prices, set-aside in the USA is bound to encourage intensification, possibly even in Europe (but see Murphy, this volume, pp. 75–76). As a policy it is surely a further case showing that the input of science into important European Community (EC) legislation can be negligible. The Earl of Cranbrook takes up this problem later in the volume.

At this point it may be useful to summarize the main measures taken within the EC to deal with cereal surpluses and the environmental problems relating to agriculture. EC Agricultural Structures Regulation 797 of 1985, especially Article 19, introduced the Environmentally Sensitive Areas (ESAs). In them extensification is possible but states are not empowered under the Regulation to deal with the vast majority of the countryside. The ESAs occupy less than 2% of farmland in the UK and although all the Sussex study area is in one, less than 12% of the area within this ESA is involved in the scheme. The main impact in the study area has been to encourage farmers to switch five cereal fields to pasture. Only in one of the nineteen UK ESAs has any extensification of cereals been approved: Conservation Headlands in Breckland, where, however, it has been very successful (Sotherton, this volume, pp. 373–397).

EC Regulations 797 of 1985, 1760 of 1987 and 1094 and 1272 of 1988 introduced the supply-side control system of set-aside, designed to aid farmers in the transition to lower overall output by removing land at the margins of production where inputs are already low. Although 4·2% of 261 cereal growers involved with The Game Conservancy's Cereals and Gamebirds Research Scheme have opted for set-aside, the total area set aside is only 0·6% of the total area of cereals grown, compared to the national 2%: set-aside appears to have been adopted mostly by those with a small area given over to cereals.

Even if set-aside land has any ecological value — then, as at present planned, it is strictly temporary. This has been a main feature of set-aside in the USA throughout its erratic 50-year history there (Berner 1988); the areas set-aside are determined by relatively short-term near-market forces. In view of the small proportion of land involved in set-aside, the value of it would have to be enormous to offset the environmental damage of intensification; our studies, like those earlier in the USA (Berner, 1988), show that it is usually of no value for gamebirds. Moreover, because the cereals which are set-aside tend to be those with low inputs, the remaining cereals will be more intensively grown than cereals as a whole prior to the set-aside. Unbelievably, this is explicit government policy in the UK!

The European Parliament has recently begun to recognize some of the problems arising from set-aside and demanded that it be used not as an end in itself but as a 'transition to ecologically sound farming', i.e. extensification (see COM (89) 353 final of Doc C3 — 139/89). The EC policy known as extensification (defined in Article 4 of EC Regulation 4115 of 1988) could solve much of the problem of environmental damage and surpluses at a stroke, though it would not — as at present envisaged — cover the Conservation Headland system because it is not yet applicable to cereals and because it demands a 20% reduction in *production* across the farm (set-aside is a minimum 20% reduction in *area*). Most attention is at present being given to the so-called organic cereals option, or zero agro-chemicals, but that is a more radical approach than is necessary for most environmental reasons. Moreover, forecasts are that organic cereals are unlikely to account for more than about 2% of total cereals — at present they account for less than a tenth of this (i.e. 0·2% of the total) (Woodward & Lampkin 1990; Organic Farmers & Growers Association, personal communication).

Extensification, because it works through lower agrochemical inputs, is very attractive from the environmental viewpoint and could potentially affect a much greater area of land than set-aside. Though at present it is too demanding for most modern cereal growers unless government aided, there is the added attraction that it would aid integrated pest management, helping to control such pests as cereal aphids (Wratten & Powell, pp. 233–257; Altieri, pp. 259–272, this volume). In fact a new policy could be generated which would easily solve all the existing problems by rearranging existing policies. For example, set-aside and extensification need not be mutually exclusive as at present. Some set-aside could be established by undersowing, the cereal nurse being allowed under extensification. This set-aside would be used as extensively grazed pasture and provide an entry for cereals in the autumn of the fifth year. This is traditional ley farming sustained from the late 19th century through to the mid 1960s. Conservation Headlands (a limited form of extensification) would then be allowed around the cereal fields which were not undersown, thus restoring the flora and fauna in its essentials to that preceding the pesticide-driven intensification. There would be few other restrictions, excepting foliar insecticides in summer where only pirimicarb (Aphox) would be allowed, except near field margins and only when really necessary. It should be noted that the new synthetic pyrethroids cleared for use on cereals during this conference are similar in their ecological effects to dimethoate and other broad-spectrum insecticides; they are certainly as harmful to the insect species preferred as food by gamebird chicks (Sotherton 1990).

A major advantage of extensification is that the farming infrastructure remains to expand acreage again if necessary (e.g. if the present genetic resistance to fungicides becomes more widespread) whereas this is impossible when a farm adopts 100% set-aside. It is therefore to be hoped that hybrid or derivative schemes, such as those outlined above, will one day find favour. At the present time (April 1990) the news is that such policies are at best, at least in the UK, on the 'back burner'. Meanwhile in Germany, Conservation Headlands — subsidized by the Länder (states) — are a feature of the landscape with great benefit to wildlife (Helfrich 1989) as a free-standing scheme in their own right. In parts of the USA and Sweden, much the same effect can be achieved with 'undisturbed small grains' allowable on set-aside (Potts 1990).

For once there may not be a shortage of finance for such schemes. The current GATT round, mentioned at the start of this account, demands that subsidies to farmers be reduced. The obvious objection, that farmers need financial support, could be overcome if essentially the same monies were in part de-coupled from production and reoriented towards environmental objectives.

I believe the main problem is that at present too few perceive the cereal ecosystem as an ecosystem as opposed to an artificial system. If this could be overcome extensification would be given greater priority and a richer environment would appear in the wider countryside. This conference volume should help.

ACKNOWLEDGMENTS

I would like to thank my colleagues, past and present, at The Game Conservancy for a great deal of help in the preparation of this essay.

REFERENCES

Aebischer, N.J. (1990). Assessing pesticide effects on non-target invertebrates using long-term monitoring and time-series modelling. *Functional Ecology*, **4**, 369–373.

Aebischer, N.J. & Potts, G.R. (1990a). Sample size and area: implications based on long-term monitoring of partridges. *Pesticide Effects on Terrestrial Wildlife* (Ed. by L. Somerville & C.H. Walker), pp. 257–270. Taylor & Francis, London.

Aebischer, N.J. & Potts, G.R. (1990b). Long-term changes in numbers of cereal invertebrates assessed by monitoring. *Proceedings of Brighton Crop Protection Conference — Pests & Diseases*, pp. 163–172. British Crop Protection Council, Farnham.

Barker, G. (1985). *Prehistoric Farming in Europe*. Cambridge University Press, Cambridge.

Basedow, T. (1987). Der Einfluss gesteigerter Bewirtschaftungsintensität im Getreidebau auf die Laufkäfer (Coleoptera, Carabidae). Auswertung vierzehnjähriger Untersuchungen (1971–1984). *Mitteilungen aus der Biologischen Bundesanstalt für Land und Forstwirtschaft Berlin-Dahlem*, **235**, 1–123.

Basset, P. (1978). Damage to winter cereals by *Zabrus tenebrioides* (Goeze) (Coleoptera: Carabidae). *Plant Pathology*, **27**, 48.

Berner, A.H. (1988). Federal land retirement program: a land management albatross. *Transactions of the North American Wildlife and Natural Resources Conference*, **49**, 118–131.

Blus, L.J., Staley C.S., Henny C.J., Pendleton, G.W., Craig, T.H., Craig E.H. & Halford, D.K. (1989). Effects of organophosphorus insecticides on sage grouse in southeastern Idaho. *Journal of Wildlife Management*, **53**, 1139–1146.

Boatman, N.D. (1989). Selective weed control in field margins. *1989 Brighton Crop Protection Conference — Weeds*, pp. 785–795. British Crop Protection Council, Farnham.

Cate, P. (1980). The cereal ground beetle (*Zabrus tenebrioides* Goeze), an important cereal pest in Austria. *Pflanzenarzt*, **33**, 115–117.

Cavalloro, R. & Sunderland, K.D. (Eds) **(1988).** *Integrated Crop Protection in Cereals*. A.A. Balkema, Rotterdam.

Chevin, H. & Chambon, J.-P. (1984). Recherches sur les biocénoses céréalieres: inventaire des hyménoptères Symphytes. *La Défense des Végétaux*, **227**, 156–162.

Conway, G.R. & Pretty J.N. (1991). *Unwelcome Harvest: Agriculture and Pollution*. Earthscan Publications Ltd, London.

Cooke, G.W. (1967). *The Control of Soil Fertility*. Crosby Lockwood Ltd, London.

Dover, J.W. (1989). A method for recording and transcribing behavioural observation of butterflies. *Entomologist's Gazette*, **40**, 95–100.

Edwards, K.J. & Hirons, K.R. (1984). Cereal pollen grains in pre-elm decline deposits: implications for the earliest agriculture in Britain and Ireland. *Journal of Archaeological Science*, **11**, 71–80.

Eggers, T. (1984). Some remarks on endangered weed species in Germany. *Proceedings on Weed Biology, Ecology and Systematics*, **7**, 395–402. Paris, France.

Fitton, M.G. (1974). A review of the British species *Tryphon* Fallen (Hym. Ichneumonidae). *Entomologist's Monthly Magazine*, 153–171.

Flickinger, E.L., King, K.A., Stout, W.F. & Mohn, M.M. (1980). Wildlife hazards from Furadan 3G applications to rice in Texas. *Journal of Wildlife Management*, **44**, 190–197.

Fox, G.A., Mineau, P., Collins, B. & James, P.C. (1989). The impact of the insecticide carbofuran (Furadan 480F) on the burrowing owl in Canada. *Technical Report Series Canadian Wildlife Service*, 72.

Freier, Von B., & Wetzel, Th. (1984). Abundanzdynamik von Schadinsekten im Winterweizen. *Zeitschrift für Angewandte Entomologie*, **98**, 483–494.

Futuyuma, D.J. & Gould, F. (1979). Associations of plants and insects in a deciduous forest. *Ecology*, **49**, 33–50.

Hall, P.C. (1980). *Sussex Plant Atlas: an Atlas of the Distribution of Wild Plants in Sussex*. Booth Museum of Natural History, Brighton.

Hanf, M. (1983). *The Arable Weeds of Europe with their Seedlings and Seeds*. BASF UK Ltd.

Helfrich R. (1989). Das 'Acker-und Wiesenrandstreifenprogramm' in Bayern — ein Programm zur Verbesserung der gesamtökologischen Situation in der Feldflur. Schriftenreihe Bayer. *Landesamt für Umweltschutz*, **89**, 155–160.

Heydemann, B. (1983). Die Beurteilung von Zielkonflikten zwischen Landwirtschaft, Landschaftspflege und Naturschutz aus Sicht der Landschaftspflege und des Naturschutzes. *Schriftenreihe für ländliche Sozialfragen*, **88**, 51–78.

Hughes, R.D. (1988). *A Synopsis of Information on the Russian Grain Aphid Diuraphis noxia.* CSIRO Division of Entomology, Technical paper, 28.

Jenkins. D. (Ed.) **(1984).** Agriculture and the Environment. *Proceedings of ITE Symposium No.13*, Natural Environment Research Council.

Kornas, J. (1988). Speirochoric weed in arable fields: from ecological specialisation to extinction. *Flora*, **180**, 83–91.

Marchant, J.H., Hudson, R.J., Carter, S.P. & Whittington, P.A. (1990). *Population Trends in British Breeding Birds*. British Trust for Ornithology, Tring.

McEwen, L.C. (1981). Review of grasshopper pesticides vs. rangeland wildlife and habitat. *Proceedings of the Wildlife–Livestock Relationships Symposium*, pp. 362–382. Forest, Wildlife and Range Experiment Station, University of Idaho, Moscow, Idaho.

Miczulski, B. & Lipinska, T. (1988). Wystepowanie osliniarek (Hymenoptera, Symphyta) w lanach pszenicy ozimej i jeczmienia jarego. *Polskie Pismo Entomologiczne*, **58**, 673–684.

Morley, C. (1911). *Ichneumnologia Britannica iv. The Ichneumons of Great Britain: Tryphoninae.* H.W. Brown, London.

Newton, I. (1986). *The Sparrowhawk.* T. & A.D. Poyser, Calton.

Nikodemusz, E. (1983). Toxic effects of pesticides to small game. *Proceedings of the International Conference on Environmental Hazards of Agrochemicals in Developing Countries, Alexandria, Egypt*, **1**, 448–460.

O'Connor, R.J. & Shrubb, M. (1986). *Farming and Birds.* Cambridge University Press, Cambridge.

Perring, F.H. & Farrell, L. (1983). *British Red Data Book 1. Vascular Plants.* Royal Society for Nature Conservation, Nettleham, Lincs.

Pimm, S.L. (1982). *Food Webs.* Chapman & Hall, London.

Potts, G.R. (1970). Recent changes in the farmland fauna with special reference to the decline of the grey partridge (*Perdix perdix*). *Bird Study*, **17**, 145–166.

Potts, G.R. (1981). Insecticide sprays and the survival of partridge chicks. *The Game Conservancy Annual Review*, **12**, 39–48.

Potts, G.R. (1986). *The Partridge: Pesticides, Predation and Conservation.* Collins, London.

Potts, G.R. (1990). Agricultural programs: the European perspective. *Perdix V, Gray Partridge and Ring-necked Pheasant Workshop* (Ed. by K.E. Church, R.E. Warner & S.J Brady). Kansas Department of Wildlife and Parks Emporia.

Potts, G.R. & Aebischer, N.J. (1991). Modelling the population dynamics of the grey partridge: conservation and management. *Bird Population Studies: Their Relevance to Conservation and Management* (Ed. by C.M. Perrins, J.-D. Lebreton & G.J.M. Hirons), pp. 373–390. Oxford University Press, Oxford.

Potts, G.R. & Vickerman, G.P. (1974). Studies on the cereal ecosystem. *Advances in Ecological Research*, **8**, 107–197.

Saint André de la Roche, G. & Douville de Granssu, P. (1982). Influence of insecticide treatment of cereals on young partridges. *La Défence des Végétaux*, **214**, 64–70.

Serre, D., Birkan, M., Pelard, E. & Skibniewski, S. (1989). Mortalité, nidification et réussite de la reproduction des perdrix grises (*Perdix perdix belesiae*) dans le contexte agricole de la Beauce. *Gibier Faune Sauvage*, **6**, 97–124.

Smith, A. (1986). *Endangered Species of Disturbed Habitats.* Nature Conservancy Council Internal Report.

Sotherton, N.W. (1982). Effects of herbicides on the chrysomelid beetle *Gastrophysa polygoni* in laboratory and field. *Zeitschrift für angewandte Entomologie*, **94**, 446–451.

Sotherton, N.W. (1985). The distribution and abundance of predatory arthropods overwintering in field boundaries. *Annals of Applied Biology*, **106**, 17–21.

Sotherton, N.W. (1990). The effects of six insecticides used in U.K. cereal fields on sawfly larvae (Hymenoptera: Tenthredinidae). *Brighton Crop Protection Conference — Pests and Diseases*, pp. 999–1004. British Crop Protection Council, Farnham.

Sotherton, N.W. & Moreby, S.J. (1988). The effects of foliar fungicides on beneficial arthropods in wheat fields. *Entomphaga*, **33**, 87–99.

Sotherton, N.W., Moreby, S.J. & Langley, M.G. (1987). The effects of the foliar fungicide pyrazophos on beneficial arthropods in barley fields. *Annals of Applied Biology*, **111**, 75–87.

Southwood, T.R.E. (1972). Farm management in Britain and its effects on animal populations. *Proceedings of the Tall Timbers Conference on Ecological Animal Control by Habitat Management*, **3**, 29–51.

Stanley, P.I. & Bunyan, P.J. (1979). Hazards to wintering geese and other wildlife from the use of dieldrin, chlorfenvinphos and carbophenothion as wheat seed treatments. *Proceedings of the Royal Society of London*, B, **205**, 31–45.

Stoetzal, M.B. (1987). Information on and identification of *Diuraphis noxia* (Homoptera: Aphididae) and other aphid species colonizing leaves of wheat and barley in the United States. *Journal of Economic Entomology*, **80**, 696–704.

Sunderland, K.D. (1987). Spiders and cereal aphids in Europe. *International Organisation for Biological Control West Palaearctic Regional Section* X/1, 82–102.

Sunderland, K.D., Chambers, R.J. & Stacey, D.L. (1984). Polyphagous predators and cereal aphids. *Glasshouse Crops Research Institute Annual Report for 1982*, 94–97.

Thomas, C.D. (1990). Fewer species. *Nature, London*, **347**, 237.

Vickerman. G.P., Potts, G.R. & Sunderland, K.D. (1976). The cereal aphid outbreak 1975. *Annual Review of the Game Conservancy*, **7**, 28–33.

Vickerman, G.P. & Sunderland, K.D. (1977). Some effects of dimethoate on arthropods in winter wheat. *Journal of Applied Ecology*, **14**, 767–777.

Wilson, P.J. (1990). *The ecology and conservation of rare arable weed species and communities*. PhD thesis, University of Southampton.

Woodward, L. & Lampkin, N. (1990). Organic and low input agriculture. *British Crop Protection Council Monograph*, **45**, 19–29.

Wratten, S.D., Watt, A.D., Carter, N. & Entwistle, J.C. (1990). Economic consequences of pesticide use for grain aphid control on winter wheat in 1984 in England. *Crop Protection*, **9**, 73–77.

2. ECOLOGICAL RESEARCH AND GOVERNMENT

EARL OF CRANBROOK

Glemham House, Great Glemham, Saxmundham, Suffolk IP17 1LP, UK

INTRODUCTION

If ecology is the science of living organisms in the environment, government certainly needs to be informed ecologically over a wide range of land-use issues and in most areas of living resource management. Ecology may be typified in popular perception as a discipline most deeply concerned with the natural environment. Apart from the objection that no such thing as 'natural' habitat (in the sense of being wholly undisturbed by man) can any longer be found in Britain — if indeed anywhere in Europe — this image overlooks the importance of the ecological approach to many aspects of policy setting, strategic decision and routine administration in rural and urban affairs.

The routes by which scientific information reaches government and the means by which it is then put to use are topics of compelling interest. If we believe (as, I confess, I do) that ecologists in general and the British Ecological Society (BES) in particular have something important to offer policy makers, in approach and in content, then it is important both to chart these routes and to facilitate the transfer process.

Some time ago, the BES accepted that it could play a role in this political arena. Members have been invited to participate and to declare fields of interest, and an administrative structure has been put in place to deliver the product. I have not personally been involved in these initiatives but, as a parliamentarian in the UK House of Lords, I have seen some of the results. My parliamentary experience has also included 11 years of familiarization with European Community environmental initiatives. In the following paragraphs I comment on some aspects of the assimilation of science into policy-making at government level in the UK and in the EC.

SCIENCE AND POLICY: THE ADMINISTRATIVE VIEW

Over the past few years, under successive Secretaries of State, the glossy publications of the Department of Environment (clearly intended for public consumption and often widely distributed) have repeatedly em-

phasized the place of science in policy formulation. For instance, under
Kenneth Baker, the United Kingdom's response to the *World Conservation Strategy* (Department of the Environment 1986, p. 9) stated that:

> Preventing new damage to the environment, or even cleaning up
> past dereliction, demands knowledge of the environmental systems
> involved. Without it, we cannot choose courses of action that work
> best and use equipment, money and human effort most efficiently.

Under Nicholas Ridley, a publication produced for the European Year of
the Environment restated the case (Department of the Environment 1987,
Introduction):

> The Government's aim in environmental protection policy is to
> prevent undue risks to human life and health now and in the future
> and to maintain the resources necessary to support Man and his
> activities. Important related objectives are the protection of the
> natural environment and the improvement of public amenity.
>
> The principles underlying this approach to pollution control are
> the need to: base decisions on the best available scientific
> foundation; adopt a preventive, precautionary approach; set
> realistic goals for environmental quality; take technical feasibility
> and economic considerations into account in reaching decisions . . .

Under the present Secretary of State the importance of scientific research
continues to be emphasized. For instance, introducing the clean technology scheme, Chris Patten wrote (Department of Environment 1989):
'The main thrust of the scheme is to offer grant aid towards the cost
of research to take innovative ideas within priority areas to "proof of
concept".'

Within the EC, the development of environmental policies has been
given a new dimension by the Single European Act, Articles 130R, S and
T. These articles have given legal force to certain principles of EC environmental policy (as already set out in successive Action Programmes on the
environment) and have added a further principle: that environment protection requirements shall be a component of the Community's other policies
(see Haigh & Baldock 1989 for a discussion of implications). Four factors
are to be taken into account by the Community in developing its policies on
the environment:

1   available scientific and technical data;
2   environmental conditions in the various regions;
3   the potential benefits and costs of action or lack of action;
4   the economic and social development of the Community as a whole and
the balanced development of its regions.

INSTITUTIONAL STRUCTURES

At appropriate levels, the UK government is provided with institutional measures to ensure that science informs policy. Any attempt to provide an exhaustive list would offend by its omissions. Centrally, the chief scientific adviser to the cabinet must wield great influence. The structure within departments of the state works through the chief scientist of each, with any number of specialist advisory committees. However, the duties of the Chief Scientist to the Department of the Environment (to take an example) are not exclusively scientific, as outlined within the publicity material already mentioned (Department of the Environment 1989, p. 5):

> In addition to advising Ministers of general scientific issues, the Chief Scientist has a special responsibility for ensuring value for money for the Department's research budget . . . He also provides the link with the UK's science base through his membership of the Advisory Boards of the Research Councils, the Natural Environment Research Council and the Science and Engineering Research Council. Current priorities include improving international collaboration on the science underpinning our understanding of the global environment.

Some departments have associated research establishments. Probably all also turn to outside institutions to commission research. Here, universities, the research councils (on which senior departmental representatives may sit, as noted above, or act as assessors), companies and individuals all contribute.

The UK government also makes wide use of formal advisory committees for the provision of expert advice (Everest 1990). Prominent among such bodies is the Royal Commission on Environmental Pollution, originally appointed in 1970 'to advise on matters, both national and international, concerning the pollution of the environment; on the adequacy of research in this field; and the future possibilities of danger to the environment.' The reports of RCEP have tackled nuclear power and the environment (1976), agriculture and pollution (1979), oil pollution and the sea (1981), lead in the environment (1983), waste management (1985), the best practicable environmental option (1988) and the release of genetically engineered organisms (1989); it is currently studying pollution in freshwaters.

The parliamentary select committees also provide important indirect routes by which scientific judgements can inform government while at the same time airing issues in a more public manner. The House of Commons

Environment Committee is appointed under Standing Order No. 130 'to examine the expenditure, administration and policy of the Department of the Environment and associated public bodies.' The Committee has the power, *inter alia* to appoint specialist advisers 'either to supply information which is not readily available or to elucidate matters of complexity within the Committee's order of reference.' In the Lords, there are two standing select committees whose work is relevant: Science and Technology, and the Select Committee on the European Communities, notably Sub-committee 'F', Environment. Both have great freedom of action, the first being appointed simply 'to consider science and technology', and the second 'to consider Community proposals, whether in draft or otherwise, to obtain all necessary information about them, and to make reports on those which, in the opinion of the Committee raise important questions of policy or principle, and on other questions to which the Committee consider that the special attention of the House should be drawn.' Both also appoint suitably qualified specialist advisers to assist their deliberations.

Although technically these Select Committees report to their respective Houses of Parliament, much of the substance of their reports is firmly directed to the government. The BES has submitted evidence, most recently to a subcommittee of the House of Lords Science and Technology Committee which, under the chairmanship of Field Marshall the Lord Carver, looked into the scientific needs of the Nature Conservancy Council (Berry 1990). Members of BES have also acted individually as specialist advisers to this or other parliamentary select committees.

EFFECTIVENESS

Despite the multiple routes designed to bring science into policy formation, doubts persist about the role and impact of scientific advice, and about the interpretation of scientific data by politicians. I am clear that difficulties are imposed by the continuing culture gap, the lack of an adequate quantum of science in our general education. All too often among my more resolutely self-proclaimed non-scientific parliamentary colleagues, I observe the men-tal switch-off whenever any element of what is perceived as scientific or technical vocabulary obtrudes — even, for example, a simple metric unit of length or mass.

We live in a time when science may seem to have grabbed the headlines: action on flue gas desulphurization; unleaded petrol; the Mon-treal protocol on CFCs; automobile exhaust catalytic filters. Are these not examples of science leading politicians?

Yet the respective roles of scientific adviser and policy maker remain distinct, and pitfalls in the translation of science to political action must not be hidden. This problem is universal. In June, 1989, Werkgroep Nordzee's 3rd North Sea Seminar was dedicated to an investigation of the interaction between policy makers and scientists in decision processes. The seminar (ten Hallers-Tjabbes & Bijlsma 1989) was opened by Mrs N. Smit-Kroes, then the Netherlands Minister of Transport and Public Works whose responsibilities covered the marine environment. Her speech gave remarkable insight into ministerial expectations and her personal perception of the intellectual chasm between scientist and policy maker. The Minister criticized the scientific community as being 'confused' about environmental signals and

> unable to advise the authorities concerned quickly enough on their significance. At the same time, however, policy makers do not seem fully able to make clear to scientists what information . . . is required. Politicians — who make decisions — are therefore confronted with uncertainties, often at a time when immediate and effective action is required.

She thought it important that

> scientists should carefully consider how to deal with uncertain factors more effectively . . . it is also important for policy makers to let scientists know exactly what type of information they require . . . and what they expect of science in concrete terms. Politicians will have to provide suitable premises on which to proceed.

The next speaker, Professor R.B. Clark of the University of Newcastle on Tyne, provided a scientist's perception which differed little in its overall conclusion, that

> science, whether national or international, has obviously a role to play in decision making, but it is a much more limited role than many people, especially scientists, seem to expect. Science is overshadowed by political considerations, and rightly so.

The essential task of the scientific interpreter is eased if the parties can meet on common ground. This was achieved in the report by the Royal Commission on Environmental Pollution on lead in the environment. The study took place in the heady political context of the campaign for lead-free petrol by a pressure group, CLEAR, which based its case on the (controversial) results of research into the effects on children's behaviour and IQ attributed to blood lead levels below those hitherto considered to threaten health. The Royal Commission concluded (1983, paragraphs 8.19–20):

> At present the average blood lead concentration of the UK population is about one quarter of that at which features of frank

lead poisoning may occasionally occur. We are not aware of any
other toxin which is so widely distributed in human and animal
populations and which is also universally present at levels that
exceed even one tenth of that at which clinical signs and symptoms
may occur . . . In our view it would be prudent to take steps to
increase the safety margin for the population as a whole. We reach
this conclusion without coming to any judgement on the possible
effects of low concentrations of lead on children's behaviour. We
conclude that measures should be taken to reduce the
anthropogenic dispersal of lead wherever possible.

The Government announced its decision to eliminate lead from petrol on
the afternoon of the day that this report was published. Never before or
since has there been so rapid a response to any report of the Royal
Commission! Scientists were speaking the language of politics.

In many environmental measures, the UK is now regulated by EC
directives. Directives on the quality of drinking water and surface waters
have, like many others, provoked dispute, and provide my second
example. For drinking water, nitrate limits were set at 50 mg/l on the basis
of medical research and WHO recommendations, taking into account
suitable safety margins. Limits for pesticides (the toxicology in general
being well known) were set at very low levels — effectively zero — on the
precautionary principle. In the face of subsequent proposals to extend the
same nitrate limits to surface waters, the Select Committee of the Lords
has been severe in its criticism (House of Lords 1989, p. 47), being
concerned that:

in this instance there appears to have been a complete breakdown in
the consultative process and that the Commission may, for whatever
reason, have brought forward its proposals without adequately
taking account of the best scientific advice . . . The Committee
appreciate the constraints under which the Commission must work.
It is however incumbent on the Member States to collaborate with
the Commission to ensure that proposals for Community legislation
are founded upon the best available scientific advice.

THE ROLE OF THE
BRITISH ECOLOGICAL SOCIETY

Where, in this polarity, can or should BES stand? As John Sheail (1987)
has emphasized, the early ecologists were leading influences in establishing
the politics of environmental concern. Today, this is a mass movement.
BES may need to identify its future role with some precision. The Society

is a repository of expertise, but should not dissipate its efforts. There is an important role for direct intervention where ecological science can define the issues.

Ecologists may also be able to apply their professional skills, and learn something of political imperatives, through personal participation. This need not be at a national level. Local government still provides the essential means by which this nation is administered. Ecologists prepared to stand for election to parish, district or county councils could be in a position to import their expertise into the democratic processes.

Above all, ecologists should press for the incorporation of their discipline into the educational curriculum. Only if there is a basic grounding in ecological science among the public at large (including the politicians) can we hope for informed decisions, in keeping with environmental needs.

REFERENCES

Berry, R.J. (1990). Reorganisation of the Nature Conservancy Council. *British Ecological Society Bulletin*, **21**, 2–6.

Department of the Environment (1986). *Conservation and Development: the British Approach.* May 1986. D38NJ. HMSO, London.

Department of the Environment (1987). *Protecting Your Environment: a Guide.* 'European Year of the Environment'. Dd8950036 ENVIJ0270NJ. HMSO, London.

Department of the Environment (1989). *Clean Technology.* Department of the Environment, D119NJ. HMSO, London.

Everest, D.A. (1990). The provision of expert advice to Government on environmental matters: the role of advisory committees. *Science and Public Affairs*, **4**, 17–40.

Haigh, N. & Baldock, D. (1989). *Environmental Policy and 1992.* Institute for European Environmental Policy, London.

House of Lords (1989). *Nitrate in Water.* Select Committee on the European Communities, Session 1988–89, 16th Report.

Royal Commission on Environmental Pollution (1983). *Ninth Report: Lead in the Environment.* Cmnd. 8852. HMSO, London.

Sheail, J. (1987). *Seventy-five Years in Ecology: the British Ecological Society.* Blackwell Scientific Publications, Oxford.

ten Hallers-Tjabbes, C. & Bijlsma, A. (Eds) **(1989).** Distress signals: signals from the environment in policy and decision-making. *Proceedings of the 3rd North Sea Seminar.* Werkgroep Nordzee, Amsterdam.

3. THE EVOLUTION OF CEREAL CROPS

G. FISCHBECK

Technische Universität München, Lehrstuhl für Pflanzenbau und Pflanzenzüchtung, 8050 Freising-Weihenstephan, München, Germany

INTRODUCTION

The term 'evolution' refers to heritable differences which have evolved from long-term genomic changes which have given rise to the distinction between different species of higher plants and other living organisms. The outline of the path of evolution from which the major cereal crop species in central Europe — wheat, barley, oats and rye — have originated is now clear. It is generally accepted that barley, together with einkorn and the tetraploid emmer wheats, belong to the primary crop species of the old world. They were domesticated within the Fertile Crescent around 8000 BC from their wild ancestors which have survived in this region until present times (Bell 1987). No wild progenitor is known for the hexaploid wheats, which in present times occupy the largest acreages in world agriculture. The genomic relationships indicate the amphiploid addition of the weedy *Aegilops squarrosa* genome (D) to the emmer genome (A + B) (Müller 1987). It is therefore concluded that the hexaploids originate from spontaneous crosses between the cultivated and the weedy species which must have occurred during the early phases of domestication. Among other weeds within the primary crop species weedy forms of rye and oats still exist and indicate the ancestry of these secondary crop species. They probably trace back to non-shattering mutations which may have had selective advantages as cereal culture extended beyond the early centre of domestication.

In view of the scope of this volume however interest should not focus upon long-term evolutionary trends which formed the basis of important genetic differences *between* the major cereal crops but concentrate upon the more recent trends of change which influence the genetic differences *within* the major cereal species. It therefore seems justified to confine this paper to the evolutionary impact of plant breeding efforts upon cereal species. Although such efforts do not trace back more than 150 years, the changes they have brought about in the genetic constitution of cereal species have consistantly accelerated. More than ever before, plant breeders' activities now form the major force of 'man-guided evolution' of cereal crops, a process which traces back to the most early settlements of

people who first tried to make their living from cultivation of plants about
10 000 years ago.

PLANT BREEDING AND
GENETIC DIFFERENTIATION OF
CEREAL CROPS IN EUROPE

Before an attempt is made to characterize the impact of plant breeding
upon cereals it is important to remember that arable land which now forms
a characteristic and often dominating constituent of the European
landscape was cleared and taken into cultivation over a period of 3–4000
years in context with the migratory movements and the increase of human
population which occurred throughout Europe. Cereals have been and
remain the major source of staple food throughout this time. The seeds
needed to extend the area of cereal cropping have been brought from the
old in order to sow the new cereal fields. The methods of cultivation
underwent only gradual changes during the long period in which arable
land was extended. Since seed supply did not exert very strong selection
pressure, uncounted numbers of generations of natural selection, based
mainly upon climatic and edaphic factors, acted to form numerous
landraces of each of the major cereal species. They differed mainly in
ecological adaptation while genetic differences for other characters have
remained largely untouched with possible exceptions of drift effects from
rather small seed lots.

Conscious efforts to select improved strains of cereals started during
the 19th century in most of the European countries as part of a general
trend to 'rationalize' agricultural production (Lupton 1987). They have
met with a wealth of genetic diversity stored within the numerous landraces. Since the landraces grown on countless numbers of cereal fields at
this time certainly offered much more genetic diversity than the early plant
breeders really needed in order to be successful, it is easy to understand the
purely empirical nature of the early steps of cereal improvement by plant
breeding which have been undertaken in almost all European countries by
practical farmers who had developed special interest in better seed quality
(BDP 1987).

It is for this reason that the gains obtained from the early steps in cereal
breeding were more dependent upon the methods of selection applied than
from the material available for it, even more so since improvement of
kernel yield — one of the genetically most difficult characters — was
already in the forefront of the objectives of most of the early cereal
breeders. As early as 1843 Louis de Vilmorin in France, founder of a

famous breeding company, derived the principle of selection between single plant progenies from the experience he gained in practical breeding work. This principle greatly helped to increase the efficiency of selection, especially for quantitative characters, and still forms an essential part of even most modern breeding schemes.

Another feature of the cereal crop species needs to be mentioned. The seed-set of wheat, barley and oats with only rare exceptions originates from self pollination within each fertile floret, while in rye self pollination is generally inhibited by certain genetic mechanisms and wind pollination from neighbouring plants is favoured by several features of the flowering process. These seemingly small deviations in the flowering habit are responsible for major differences in the genetic constitution of the landrace populations the early plant breeders started to work with. While self pollination in wheat, barley and oats inevitably leads to segregation and homozygosity, the regime of outcrossing in rye tends to maintain an equilibrium of gene frequency between different alleles of all non-homozygous loci.

As a result, selection within the landraces of wheat, barley and oats very rapidly exhausted the genetic variability stored in the original landraces, while successful selection within some of the superior populations which have been maintained from the outbreeding species rye still goes on even today (Hoffmann, Mudra & Plarre 1985). Since response to selection is always more effective in progenies of homozygous as compared to heterozygous plants it follows that rather rapid gains were made from selection within the landraces of the inbreeding cereal species almost everywhere in Europe. The relatively few superior genotypes secured from the original landraces soon occupied most of the areas sown to the individual cereal crops. The original wealth of genetic diversity decreased rapidly and many valuable genotypes may have been lost. The breeding process with inbreeding cereal species quickly depended upon the ability to restore genetic variability, the indispensible basis for genetic progress by means of selection. Although many of the early breeders accepted artifical crosses between selected genotypes only reluctantly (Fruwirth 1905), it was this method which provided new opportunities of restoring variability. Crossing, therefore, not only became the very basic part of breeding programmes with self-pollinating cereal species but also one of the major aspects of plant breeding research based upon the Mendelian principle of genetic recombination.

A look at the pedigrees of two important cultivars of winter wheat can be used to illustrate the major features of the breeding work with self-pollinating cereals from its very beginning until today (Figs 3.1 & 3.2).

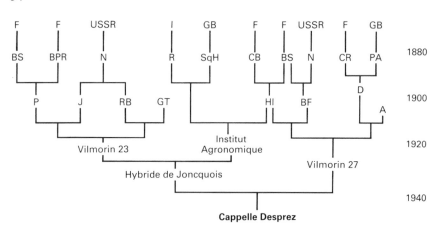

FIG. 3.1. Pedigree showing the landrace origin of winter wheat cultivar *Cappelle Desprez*. Capital letters in upper row refer to country of landrace origin. Abbreviations in the pedigree indicate initials of named varieties and breeding strains.

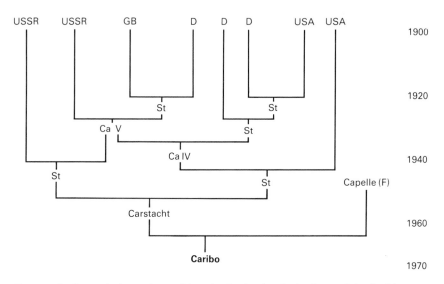

FIG. 3.2. Pedigree of winter wheat cultivar *Caribo* showing the landrace origin of cultivar *Carstacht*. Capital letters in the upper row refer to country of landrace origin. Abbreviations in the pedigree indicate initials of named varieties and breeding strains.

1   From selection and multiplication of single plant progenies out of the original landrace population superior genotypes were isolated and formed the early basis of plant breeders' contributions to cereal crop improvement.

2 A rapid loss of the genetic diversity which had been accumulated in the traditional landraces occurred from selective depletion as well as from replacement of landrace acreages by 'improved landraces' which originated from the early breeders selections.

3 New sources of genetic diversity became available from genetic recombination within the progenies of artificial crosses between well chosen parent lines.

4 In the early cycles of cross-breeding it was not very difficult to select improved genotypes from rather 'wide crosses' in which only one parent represents the original landrace complex of genetic traits in order to provide sufficient ecological adaptation.

5 In the later phases of the continued cross-breeding process recombinations between the more successful genotypes derived from the preceding generation within the same region of adaptation increased in importance. In this way continous recombination may have formed secondary complexes of traits for higher yield potential besides the primary gene complexes for ecological adaptation.

6 From the more recent experiences of the continuous cross-breeding process it is apparent that a rather limited number of parents have come to occupy a key position in the cross-breeding process. They often show an unrelated ancestry of genitors from landraces of very different geographic origin.

This process may explain the important position of the French variety Capelle in the wheat-breeding process in central and western Europe. The cultivar Caribo, which was released in Germany in 1967, descends from the cross Carsten VIII × Capelle. From sixty-nine cultivars of winter wheat on the 1989 official German list 35% are immediate progenies derived from Caribo crosses, and Caribo appears at least once in the pedigree of another 16%. Capelle again appears as a parent from which the Maris Ranger and Maris Widgeon family of British wheats originated and it also contributed, together with other French winter wheats, to the parentage of the high-yielding cultivar Maris Huntsman and the first semidwarf cultivar Hobbit (Lupton 1987).

In Fig. 3.3 a comparable example is given for the outbreeding species rye. The famous 'Petkus-rye' was released in 1895 (Hillmann 1910). It originated from the mixture of landrace populations obtained from three ecologically different regions in Germany. These populations obviously had excellent combining ability. Natural intercrossing within this synthetic landrace population provided sufficient diversity for more than eight decades of progress by selection from within this gene pool. Virtually all varieties of rye which have been released in later years in Germany and in

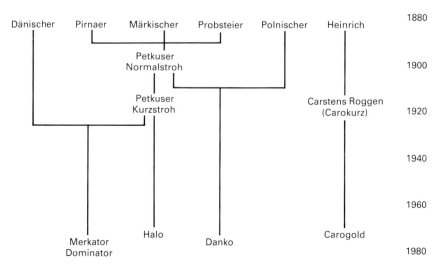

Fig. 3.3. Landrace origin of winter rye cultivars in Germany.

neighbouring countries carried Petkus-rye in their parentage, except for the cultivar Carstens. It therefore is very interesting to note that the new phase of breeding hybrid rye varieties is based upon the experience that successful hybrids have mainly been extracted from combinations between inbred lines derived from these two well-adapted but unrelated gene pools, each of which carries a comparably high yield potential (Geiger 1982).

In summary, breeding efforts certainly have reduced the genetic diversity which was originally present in European landraces, due to processes of selection. At the same time selection has maintained gene complexes related to the level of adaptation to the prevailing conditions of soil and climate and in addition may have created new gene complexes for higher yield potential derived from genetic recombination either by cross-breeding with the self-pollinating cereal species but also by outcrossing within natural or synthetic populations of the outbreeder rye.

EVOLUTIONARY TRENDS IN CEREAL BREEDING IN EUROPE

The early cereal breeders certainly did not expect to start or to accelerate evolutionary processes. Their major interest was simply to improve seed quality and provide a more effective method for better seed production (BDP 1987).

Therefore trends only become apparent if examined over a longer period of time. If landraces or the very early selections from landraces are compared with present day cereal cultivars, the general tendency to shorter stature becomes apparent (Donald & Hamblin 1983), although the degree of shortening varies considerably depending on the cereal species and to some extent on the geographical area.

Natural selection within landraces probably favoured long straw types as one possible manifestation of vigour. In addition long straw types also were better competitors against weeds and straw was of much higher value to the farmers than today. The shortening of plant stature which began only gradually but accelerated considerably during the last four decades in Europe was closely linked to the changes in cereal agronomy. With the application of yield-increasing inputs — mainly nitrogen fertilizer — there was insufficient resistance to premature lodging. For a substantial period of time the genetic effects of selection for shorter straw just compensated for the increasing height resulting from better plant nutrition (Table 3.1). With hindsight, it seems that until very recently most cereal-breeding activities in Europe were aimed at improved adaptation to the many changes in crop management which characterize the recent history of European cereal culture.

Changes in crop rotation, mechanization of field work from tilling to harvest, increase in use of fertilizers, the application of herbicides, growth regulators, insecticides and fungicides have all played a role and provided opportunities for new cultivars to outyield their predecessors because they are better adapted to the new set of growing conditions. For example, the German winter wheat cultivar Heine VII was released as early as 1943, but

TABLE 3.1. Performance of landraces and recent cultivars at different input levels in ancient cropping systems (5 year average). The cropping system was fallow, winter wheat and spring cereals with two input levels: (A) without fertilizer or manure, mechanical weeding; and (B) mineral fertilizer 70–80 N, 100–120 P_2O_5, 150–180 K_2O and application of herbicides. Data from Puch experimental farm, BLBP Freising-Weihenstephan (G. Bachthaler, personal communication)

Input level		Spring barley		Oats	
		(A)	(B)	(A)	(B)
Plant height	Landrace	76	100	93	132
(cm)	Cultivar	60	85	69	107
Kernel yield	Landrace	1·3	3·1	1·4	3·6
(t/ha)	Cultivar	1·3	4·0	1·5	5·0

only grown on very limited areas in its first years, which eventually even decreased because of its rather short stature that endangered yield stability throughout the meagre years from 1940 to 1950. The breeder was just about ready to withdraw this cultivar from the market in the early 1950s when more fertilizer became available and the era of combine harvesting of cereal fields arrived. In consequence more resistance to lodging, shattering and sprouting was required and by 1958 Heine VII, which excelled other cultivars in these characters, had higher seed-multiplication acreages than any other cultivar of winter wheat in West Germany, and kept this position until 1962.

Results from a long-term field trial at Dikopshof, the experimental farm of the University of Bonn provide another illustrative example of interactions between cereal breeding and cereal culture. A long-term crop rotation experiment with permanent differences in nutrient supply was introduced in 1904. While the five crop species included in the original rotation have been maintained throughout the trial period, the cultivars sown have been changed over the years, parallel to their rise and fall in practical agriculture. For two seasons the winter wheat part of this trial was sown to a series of five wheat cultivars which had been used during different phases of the experiment. The yield response to increased supply of nutrients accurately reflects the historic sequence in which the different cultivars were used (Fig. 3.4). The kernel yield increase with new as compared to the older wheat cultivars varied in relation to the level of nitrogen supply. At the lowest level no significant yield differences were found, but the yield increase obtained with higher doses of nitrogen with the old cultivars was only 39% of the yield response of the new wheat cultivars. This experiment therefore provides a very convincing example for the interaction between cereal breeding and cereal crop husbandry.

Adaptive changes in plant type in response to the changes in crop management explain most of the early contributions of plant breeding to the evolution of cereal crops. This is considerably less so later on, since the prospect of consistent yield increases has attracted more scientific interest. Much research work has been devoted to analyse the progress obtained in terms of growth analysis (Evans & Wardlaw 1976; Donald & Hamblin 1976). The results indicate that the breeders' share in the increase of kernel yield was largely based upon higher harvest indices (i.e. kernel yield as a percentage of biomass) which are largely determined by genotype, while the farmers share resulted from the increase of biomass production due to increased leaf area indices and leaf area duration, both obtained largely by improved crop management. The results of a series of trials with old and new cultivars of winter wheat carried out under close to optimum growing conditions and with measures to prevent lodging (which especially would

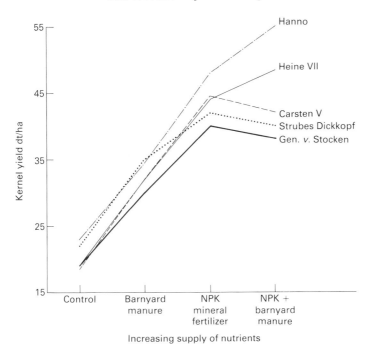

FIG. 3.4. Kernel yield obtained from old and new cultivars of winter wheat at different levels of nutrient supply. Cultivar *General von Stocken* is the oldest, then *Strubes Dickkopf, Carsten V, Heine VII* and the most recent is *Hanno.*

have endangered the older cultivars) revealed almost no changes in biomass production. However, the harvest indices increased from 0·36 to 0·51 (Table 3.2), giving an increase in kernel yield of 45% (see also Costigan & Biscoe, this volume, pp. 55–56). Since new cultivars generally do not produce more biomass than older ones, there is no indication as yet that cereal breeding has affected the efficiency of photosynthesis.

An attempt was made 20 years ago by Donald (1968) to deduce an 'ideotype' of a wheat plant, based upon the combined conclusions from the results obtained in yield physiology research (Fig. 3.5). None of the present day wheat cultivars closely resembles this ideotype. There is virtually no single culm variety, the percentage of awned types did not increase and narrow, erect leaves have not shown any immediate advantage. The only consistent trend is that of an increase of kernel weight per ear, largely resulting from more kernels per ear. Apparently it was possible to obtain such changes with genotypes which differed from all other features of Donald's (1968) ideotype.

Many genetic effects on straw length of wheat and other cereals have

TABLE 3.2. Genetic improvement of winter wheat as assessed by 3 years of experiments. Varieties susceptible to lodging were staked (Austin, Ford & Morgan 1989)

Variety	Year of introduction or first report	Grain yield (t/ha)	Height to base of ear (cm)	Harvest index (%)	Above-ground dry matter (t/ha)
Squareheads Master	1830	5·50	146	34	16·18
Browick	1844	5·86	138	31	18·90
Prince Albert	pre 1910	5·44	155	33	16·48
Partridge	1907	6·92	141	38	18·21
Little Joss	1908	6·61	134	36	18·36
Yeoman	1916	6·49	135	36	18·03
Capelle Desprez	1953	7·69	97	44	17·48
Maris Huntsman	1972	8·05	96	47	17·13
Norman	1981	9·48	76	52	18·23
Renard	1984	8·72	82	49	17·80
Brock	1985	9·57	73	52	18·40
Brimstone	1985	9·65	86	49	19·53
Slejpner	1986	9·89	71	53	18·66

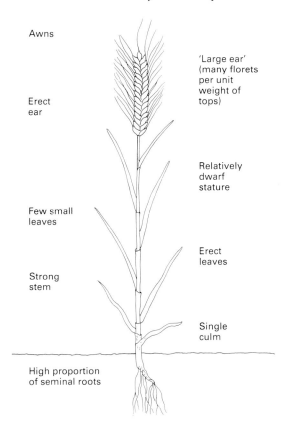

Awns

'Large ear'
(many florets
per unit
weight of
tops)

Erect
ear

Relatively
dwarf
stature

Few small
leaves

Erect
leaves

Strong
stem

Single
culm

High proportion
of seminal roots

FIG. 3.5. A design for a wheat ideotype. After Donald (1968).

been described. It appears that the differences in plant height observed between two cultivars are based upon a number of height-promoting as well as height-reducing effects exerted by loci on different chromosomes. Several of the major genes responsible for reduced height (Rht) in the wheat plant have been located on specific chromosomes (Worland, Gale & Law 1987). Major dwarfing genes are located on chromosome 4 D (Rht 2) and chromosome 4 A (Rht 1) and have been introduced from the Japanese wheat cultivar Norin 10 into many wheat-breeding programmes around the world. They form the genetic basis of the widely used and successful 'semidwarf' wheat varieties. In contrast to earlier studies which show that short straw correlates with low yield, it has been established beyond any doubt that the Rht 1 and Rht 2 genes reduce plant height and increase the number of kernels per ear (Fig. 3.6). Many of the more recent

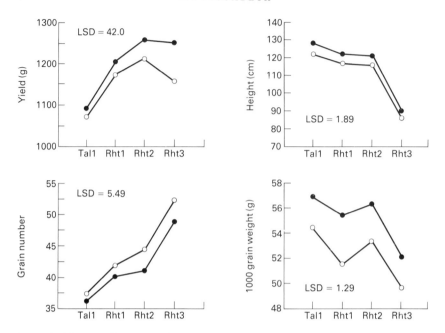

FIG. 3.6. Rht gene effects on yield, plant height and yield components of wheat. Open circles, intravarietal hybrids; closed circles, intervarietal hybrids. From Law (1986).

winter wheat cultivars bred in the UK carry the Rht 2 gene, but other high-yielding winter-wheat cultivars from breeding programmes in central Europe are known *not* to carry Rht 2 or Rht 1. They often are not as short in stature than most of the semidwarfs yet they produce very high numbers of kernels per ear due to other factors which influence the partitioning of photosynthetic products within the cereal plant (Snyder & Carlson 1984).

Other genes have been identified which affect the low temperature requirements (Vrn) and day-length response (Ppd) of the wheat plants (Worland, Gale & Law 1987). Their effects upon ecological adaptation are certainly confounded with favourable effects upon kernel yield within and negative-yield effects beyond their region of adaptation. This illustrates that present knowledge about the physiological basis of kernel yield of cereals is not sufficient to define conclusively the ideotype for a given environment and we still lack much of the detailed knowledge needed for a comprehensive assessment of the genetic basis for higher kernel yield potential.

As discussed earlier, the selection by plant breeders has inevitably caused a reduction of the genetic diversity between and within modern cultivars. This reduction has affected the interrelationship between the recent evolution of cereals and the epidemiology of plant pathogens. Plant breeders are inclined to look at plant pathogens only in terms of susceptibility or resistance of the host plants they are working with. Although improvement of resistance against the major diseases has been given high priority among breeding objectives, especially in wheat and barley, it often gave rise to a long chain of repeated successes and failures. The early breeders looked at disease resistance more or less as a quantitative character which offered chances for gradual improvement by selection similar to the improvement of grain yield. In the first decade of the 20th century Biffen (1906), a British scientist, was able to demonstrate a one-factor Mendelian inheritance for yellow rust resistance in wheat. Hypersensitive forms of 'resistance' against obligatory pathogens were regarded as hereditary stable characters of the host plant. The conclusion seemed justified that the rapid increase and dissemination of such types of resistance might offer a rapid and final solution for preventing the losses caused by such pathogens. When the first failures became apparent, the search for 'new' and possibly more reliable sources of resistance began. Many outstanding scientists have worked on this problem and essential components of cereal-breeding programmes have been and still are occupied with efforts to incorporate new sources of resistance into future breeding materials.

Mildew resistance of barley serves as a well-documented example. By means of a differential set of mildew cultures it is possible to determine the presence of thirteen different sources of mildew resistance in spring or winter barley cultivars (Table 3.3). For example, each of the forty-seven cultivars of spring barley and thirty-one out of fifty-four cultivars of winter barley mentioned in the *Beschreibende Sortenliste 1989* carry at least one, sometimes two, but rarely three, of the major genes for mildew resistance listed in Table 3.3. But if mildew reaction in the field is taken into account, only seventeen cultivars of spring barley and seven cultivars of winter barley are found to suffer little or no mildew attack. It is clear from Flor's (1956) gene-to-gene hypothesis that not only the presence of a certain gene for resistance in the host plant is needed to protect the host plant from disease attack, but also the absence of the corresponding gene for virulence in the pathogen population.

In a co-operative research project with the University of Tel Aviv, the reaction of representative samples of *Hordeum spontaneum* accessions from 77 locations in Israel to the infection from a countrywide mixture of mildew inoculum has been observed (Fig. 3.7). Very rarely hypersensitive defense reactions appeared, which indicates the presence of matching genes for virulence in the endemic pathogen population. At the same time

TABLE 3.3. Genes (sources) for mildew resistance in European barley cultivars. From Torp, Jensen & Jørgensen (1978); Schwarzbach & Fischbeck (1981)

Gene/source	Located on chromosome	First release	Country	Reaction type	Spring/winter barley cultivars
Ml-g/Ml (CP)	IV	1940	Germany	R/r	S
Ml-ra	V	1950	Germany	s	W
Ml-a6	V	1963	Germany	R	S + W
Ml (La)	?	1963	Netherlands	r	S
Ml-a7 + Ml-k	V	1968	Germany	r	S
Ml-a12	V	1968	Netherlands	r	S (W)
Ml-(WO)	?	1974	Germany	s	W
Ml-a	V	1974	Denmark	R	S
Ml-a9	V	1976	Sweden	R	S
Ml-a13	V	1980	Czechoslovakia	R	S (W)
ml-o	IV	1980	Netherlands	0/4	S
Ml-a3	V	1985	Denmark	R	S
Ml-a (HT)	V	1986	Germany	R	S

Reaction type:

$R = 0$, no visible symptoms

$r = I-II$, necrotic spots, chlorotic spots showing some mycelial growth

$s = II-III$, chlorotic spots showing mycelial growth and some sporulation

$0/4 = $ no visible symptoms, except sometimes small pustules

Distributed in:

S = spring barley cultivars

W = winter barley cultivars

a characteristic regional pattern of quantitative differentiation in disease incidence was observed (Fischbeck 1981). Apparently, natural selection for more protection against mildew attack has been operative at a secondary level, independent from the virulence pattern. On the other hand when random samples of the same sets of *H. spontaneum* lines were infected with mildew of European origin, hypersensitive reactions predominated sometimes at frequencies of more than 50% (Fischbeck *et al.* 1976). This indicates the lack of corresponding genes for virulence. It may be for similar reasons that the majority of the known genes for mildew resistance present in the European barley cultivars have been derived from barley plants originating from rather remote areas with different virulence spectras. Large-scale spore samplings during the vegetative period together with suitable virulence tests has allowed us to draw a representative picture of the geographic distribution of virulence genes in the mildew population present in different European regions. Distinctive differentiation in virulence frequency is seen (Fig. 3.8) and substantial changes occur in a given region sometimes within the course of only 3–4

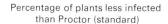

Percentage of plants less infected
than Proctor (standard)

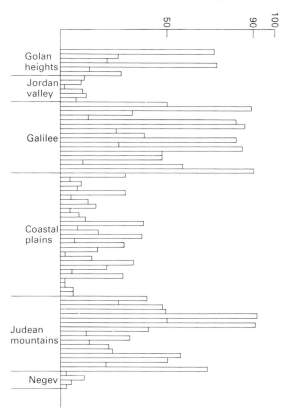

FIG. 3.7. Mildew resistance of seventy-seven subpopulations of *Hordeum spontaneum* from different areas of Israel. From Fischbeck (1981). The degree of mildew infection was assessed in infection trials including 100 unselected single head progenies from each subpopulation. Frequent plantings of spring barley cultivar *Proctor* were included in the trials for comparison.

years. Two major forces explain most of the differentiation and change in virulence frequency: the concentration of a given (effective) major gene for resistance within a region and the main wind direction.

Since progress in cereal breeding inevitably includes the spread of new varieties over large areas it clearly correlates with an evolutionary trend for an increased 'genetic vulnerability'. If the many efforts to introduce favourable genes from unrelated sources into the cereal cultivars are set against the selective response in the virulence pattern from many plant pathogens, it becomes apparent that genetic vulnerability is not based

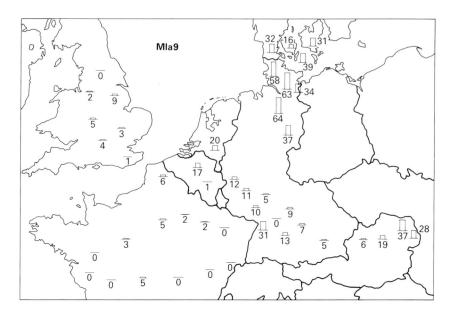

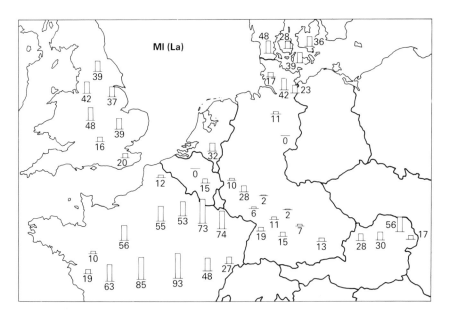

FIG. 3.8. Virulence frequency shown as percentages and bars against two sources of mildew resistance (Mla9 and Ml (La)) in barley in 1985. From Limpert & Fischbeck (1987).

upon the loss of genes for resistance from the original landrace populations, as is sometimes assumed. Rather, it originates from the insufficient level of genetic diversity which has remained with the preferential use of only a few cultivars in the cultivation of cereals at a given time in a certain region (Table 3.4). Detailed studies of mildew resistance in a set of *H. spontaneum* derived barley lines have shown that thirty-nine out of forty-one lines tested showed different patterns of mildew reaction as compared to twenty-eight tester lines carrying most of the known genes for mildew resistance (Jahoor & Fischbeck 1987). It appears therefore that a substantial increase in genetic diversity for mildew resistance in barley can be obtained which would overcome the present insufficiency. Furthermore, from results of virulence analysis in different parts of Europe, substantial decreases of virulence frequency in the pathogen population have been observed to follow the withdrawal of corresponding genes for mildew resistance within the replacement of older generation cultivars (Wolfe 1987). It therefore should be possible to introduce a recycling process into cereal breeding which allows for the repeated use of resistance genes if once a sufficient level of genetic diversity has been obtained and a monitoring survey of changes in virulence frequencies becomes available (Wolfe & Limpert 1987).

FUTURE TRENDS IN THE EVOLUTION OF CEREAL CROPS

To some extent future trends will follow the major lines of development in the recent past. This general observation certainly constitutes a dominating component in the changes to be expected in cereal breeding in the next years to come. But recent progress in cell biology and molecular genetics may create a similar situation as in the early decades of this century when

TABLE 3.4. Crop area sown to the three most popular cultivars in the Federal Republic of Germany and in two states 1986–1988 (percentage of total crop acreage). BMELF (1987, 1988, 1989)

	1986	1987	1988
Winter wheat			
FRG	45·8	44·4	47·8
Schleswig Holstein	81·8	84·5	91·7
Spring barley			
FRG	48·8	49·3	52·7
Bayern	79·7	83·9	76·9

genetic recombination by cross-breeding gained its position as being much more powerful than selection within landraces as was practised before. Biotechnology has already been applied in certain areas of cereal breeding and will rapidly occupy other important parts in breeding methodology because of its potential to:

1   accelerate breeding procedures;
2   increase the efficiency in selection; and
3   enlarge genetic variability for breeding purposes.

Breeding procedures

At present it generally takes 10–12 years to complete a breeding cycle from crossing to the release of a new cereal cultivar. With new techniques (e.g. anther culture) it is possible to extract homozygous individuals from the anthers which develop in the immediate (F_1) progenies of a cross. Additional cultivation methods can be employed which speed up the life cycle of a cereal plant in order to raise more than one generation per year. It therefore will become possible to shorten the conventional breeding cycle by 2–4 years (Table 3.5) (Wenzel & Faroughi-Wehr 1984). The anther culture method is now applied by many barley breeders. It seems that wheat will follow within the next few years, but rye has so far proved unamenable to anther culture and not very many attempts have been made with oats. Anther culture methods will generally enable cereal breeders to respond more quickly to future changes in cereal agronomy. It also may favour a trend to increase the number of varieties offered to farmers, which may provide an opportunity to reintroduce higher levels of genetic diversity into the cereal acreages grown in future.

TABLE 3.5. Breeding scheme using double haploid progenies from crosses of spring barley

Year	Time scale (months)	Working procedures
1	0–5	Production of 15–20 F1 seeds by artificial crossing
	9	Cultivation of F1 plants until anther initiation
	10	Callus initiation induced on 2–4000 cultivated anthers
	12	Regeneration of c. 400 double haploid, green seedlings
2	16	Harvesting seeds from c. 300 double haploid plants (A1)
	21	Seed multiplication (A2) from A1 plants sown in green-(or net-) houses
3		A3 seeds from c. 300 homozygous segregants out of the original cross are available for comparative field tests
4		A4 seeds available from selected A3 lines for replicated yield trials

Efficiency of selection

Selection in present-day cereal breeding is mainly based upon identifying single plants or single plant progenies during the segregating generations which have the desired phenotype. This requires a great deal of space in greenhouses and in the fields and leaves much room for inaccuracy through environmental effects. The efficiency of selection may be increased using several different methods.

With sufficient progress in cell culture methods it will be possible to apply *in vitro* selection against toxins produced by plant pathogens, against new herbicides or other stress factors for virtually unlimited numbers of single cells or cell clusters provided that normal plants can be regenerated from the survivors.

Another way to improve the efficiency of selection will be based upon the development and use of reliable gene markers which are being developed at present on the basis of polymorphisms in the length of DNA fragments obtained from digestion with certain restriction enzymes termed Restriction Fragment Length Polymorphisms (RFLPs). This method will offer new opportunities for identifying very accurately linkage relationships between closely associated genes of economic importance as soon as suitable DNA fragments become available. An example of polymorphism obtained from 18 RFLP markers located on chromosome 7 D of wheat as determined by Chao *et al.* (1989) is shown in Table 3.6. RFLP chromosome mapping of maize has already reached a very advanced stage (Helentjaris 1987) and barley as well as wheat will follow in the next few years. Linkage with RFLP markers will enable the breeder to save much space and labour as compared to present methods. But biotechnology certainly will not replace the need for final evaluation of new breeding materials in field experiments and will require more capital investment in laboratory space and equipment than was needed before. A major effect that goes along with the development of RFLP techniques will be a continous increase of knowledge about the genetic differences between the cultivars of our cereal crops. It therefore will open the way for a more systematic increase of genetic diversity for all traits of economic and/or ecologic importance.

Genetic variability

Additional techniques have become available which allow the extension of genetic variability of higher plants beyond the conventional limits of crossability within and between different species and more closely related genera. Somatic fusion of protoplasts can be used to increase genetic variation within a given species. Useful results are expected if applied to

TABLE 3.6. Polymorphism detected with thirteen enzymes and eighteen genomic DNA clones on chromosome 7D of wheat. (1) Pairwise comparison between six cultivars and breeding strains of *Triticum aestivum*. (2) Comparison of the short arm of chromosome 7D of wheat cultivar Hobbit and strain VPM1 which carries among other DNA introgressions from *Aegilops ventricosa* resistance to *Cercosporella herpotrichoides*. ND; no data. Data from Chao *et al.* (1989)

	Percentage polymorphism	
Enzyme	1	2
ApaI	0	7·1
BamHI	5·2	21·4
BgIII	6·0	23·5
DraI	9·2	35·3
EcoRI	5·2	35·3
EcoRV	5·6	29·4
HindIII	9·6	23·5
KpnI	0	14·3
PvuII	0	21·4
SspI	14·3	ND
SstI	1·6	21·4
XbaI	2·8	ND
XhoI	1·4	ND
Average	4·7	23·3

overcome the crossing barriers which have developed during the more recent evolutionary differentiation of otherwise related plant species and genera. A second method, the transfer of isolated DNA sequences does not meet with any restriction from the donor species. It therefore allows the incorporation of genes which have been isolated from any living organism as soon as suitable vectors for a gene transfer become available and the regeneration of plants from transformed cells succeeds. Although this has proved more difficult for monocotyledonous plant species than dicots (Table 3.7), continous improvements in methodology seem likely to yield practical results (Austin *et al.* 1986). It therefore is generally expected that within the next decade gene transfer will be an available tool for breeding work with the major cereal species. It is not possible to predict with confidence the range of new opportunities this will provide, but it seems justified to expect useful results in the near future for simply inherited traits affecting quality characteristics as well as resistance to viruses, insect pests and new types of herbicides. The future development in this area will also depend very much on the progress made in the isolation of useful genes for breeding purposes and progress in knowledge about gene regulation (Gasser & Fraley 1989).

TABLE 3.7. Survey of the number of plant species and vectors used for the production of transgenic plants. Data from Gasser & Fratley (1989)

		Vectors					
	Species	At	Ar	Fp	Pg	Mi	IR
Herbaceous dicots	21	16	5	2	2	1	—
Woody dicots	3	—	3	2	2	—	—
Monocots*	5	1	—	3	—	—	1

* Asparagus, rice, corn, orchardgrass, rye.
Vectors used: At, *Agrobacterium tumefaciens*; Ar, *Agrobacterium rhizogenes*; Fp, free DNA introduction into protoplasts; Pg, particle gun; Mi, microinjection; IR, injection of reproductive organs.

BREEDING GOALS

The possibility of increasing kernel yield of cereals by selection within landrace populations was the leading objective of the early cereal breeders. Higher kernel yield has remained a dominant goal of cereal breeding in the later phases of its history, supported by the consistent improvement in methods of cereal cultivation (e.g. lodging resistance). A high-price system within the European community has favoured the use of additional inputs which are helpful in making the maximum use of the high yield potential of modern cereal varieties by applications of more fertilizer, herbicides, fungicides, insecticides and growth regulators which eventually resulted in the present situation of surplus production and often inadequate attention to the needs of environmental protection (Costigan & Biscoe, pp. 64–66; Murphy, pp. 89–91; Potts, pp. 3–21, all this volume).

Important changes in cereal crop management are now being considered and will require changes in the future goals of cereal breeding (Fischbeck 1988) if they are to become general practice. But the use of new methodologies together with the increasing body of genetic information about the breeding materials will allow much better opportunities for a more active role by plant breeders in future changes of the man-guided evolution of cereals.

Although it is too early to draw definite conclusions, some possible trends in future cereal breeding may be mentioned. Because of the present surpluses, an increase of importance of quality traits as compared to yield potential should be expected; and in part this is already underway. This might include the development of distinct cultivars for specific uses for food, feed and in industry.

More reliable protection against losses from the major cereal pathogens and parasites with less support from pesticides will be needed. From the increase of knowledge about the genetic basis of the different levels of disease reactions of the cereal plants as well as virulence frequencies of the pathogens it should be possible to establish rationalized levels of genetic diversity in heritable traits for disease resistance against the major cereal pathogens and parasites. This might include the preferential use of variety mixtures differing in genes for disease resistance.

Ecological and social objections against the use of herbicides may require the development of plants which are more competitive against weeds preferably in combination with greater tolerance to edaphic and climatic stresses. If new herbicides should become available which meet all requirements of ecological safety, resistance against such herbicides will be an important trait of future cereal cultivars.

Finally, more knowledge about the molecular basis of genetic variability in the major cereal species will provide more powerful means of safeguarding and increasing the resources for unknown future needs in the evolution of cereal crops.

REFERENCES

Austin, R.B., Flavell, R.B., Henson, J.E. & Lowe, H.J.B. (Eds.) (1986). *Molecular Biology and Crop Improvement*. Cambridge University Press, Cambridge.

Austin, R.B., Ford, M.A. & Morgan, C.L. (1989). Genetic improvement in the yield of winter wheat: a further evaluation. *Journal of Agricultural Science, Cambridge*, **112**, 295–301.

BDP, Bundesverband Deutscher Pflanzenzüchter (Hgb) (1987). *Landwirtschaftliche Pflanzenzüchtung in Deutschland*. Mann, Gelsenkirchen-Buer.

Bell, G.C.H. (1987). The history of wheat cultivation. *Wheat breeding* (Ed. by F.G.H. Lupton), pp. 31–49. Chapman & Hall, London.

Biffen, R.H. (1906). Mendel's law of inheritance and wheat breeding. *Journal of Agricultural Research*, **1**, 4–48.

BMELF (1987, 1988, 1989). *BML Daten-Analysen, Besondere Ernteermittlung bei Getreide und Kartoffeln, 1987, 1988, 1989*. Bundesministerium für Ernährung, Landwirtschaft und Forsten, Bonn.

Chao, S., Sharp, P.J., Worland, A.J., Warham, E.J., Koebner, R.M.D. & Gale, M.D. (1989). RFLP based genetic maps of wheat homologous group 7 chromosomes. *Theoretical and Applied Genetics*, **78**, 495–504.

Donald, C.M. (1968). The design of a wheat ideotype. *Proceedings of the 3rd International Wheat Genetic Symposium* (Ed. by F.W. Finlay & F.W. Shepherd), pp. 377–387. Butterworth, Sydney.

Donald, C.M. & Hamblin, J. (1976). The biological yield and harvest index of cereals as agronomic and plant breeding criteria. *Advances in Agronomy*, **28**, 361–404.

Donald, C.M. & Hamblin, J. (1983). Convergent evolution of annual seed crops in agriculture. *Advances in Agronomy*, **36**, 97–143.

Evans, L.T. & Wardlaw, I.F. (1976). Aspects of comparative physiology of grain yield in cereals. *Advances in Agronomy*, **28**, 301–359.

Fischbeck, G. (1981). Ergebnisse von Untersuchungen über Krankheitsresistenz in israelischen Populationen von *Hordeum spontaneum* im Hinblick auf die Gegenwartsprobleme der Resistenzzüchtung. Bericht über die 32. *Arbeitstagung der Arbeitsgemeinschaft Saatzuchtleiter, Gumpenstein*, 3–12.

Fischbeck, G. (1988). Cereal breeding and input reductions in cultivation of cereals. *Cereal Breeding Related to Integrated Cereal Production* (Ed. by M.L. Jona & L.A.J. Slootmaker), pp. 9–27. Pudoc, Wageningen.

Fischbeck, G., Schwarzbach, E., Sobel, Z. & Wahl, I. (1976). Mehltauresistenz aus israelischen Populationen der zweizeiligen Wildgerste (*Hordeum spontaneum*). *Zeitschrift für Pflanzenzüchtung*, **76**, 163–166.

Flor, H.H. (1956). The complementary genetic systems in flax and flax rust. *Advances in Genetics*, **8**, 29–54.

Fruwirth, C. (1905). *Allgemeine Züchtungslehre der landwirtschaftlichen Kulturpflanzen*, Vol. 2. Auflage, Parey, Berlin.

Gasser, C.S. & Fratley, R.T. (1989). Genetically engineering plants for crop improvement. *Science*, **244**, 1293–1299.

Geiger, H.H. (1982). Zuchtmethoden bei diploidem Roggen (*Secale cereale*). *Eucarpia Tagung, Berichte der Akademie der Landwirtschaftswissenschaften der DDR, Berlin*, **198**, 305–332.

Helentjaris, T. (1987). A genetic linkage map for maize based on RFLP's. *Trends in Genetics*, **3**, 217–221.

Hillmann, P. (1910). *Die deutsche landwirtschaftliche Pflanzenzüucht. DLG, Berlin.*

Hoffmann, W., Mudra, A. & Plarre, W. (1985). *Lehrbuch der Züchtung landwirtschaftlicher Kulturpflanzen*, Vol. 2, pp. 15–38. Auflage, Parey, Berlin.

Jahoor, A. & Fischbeck, G. (1987). Sources of resistance to powdery mildew in barley lines derived from *Hordeum spontaneum* collected in Israel. *Plant Breeding*, **99**, 265–273.

Law, C.N. (1986). The effect of dwarfing genes on the expression of heterous for grain yield in F. hybrids. *Annual Report 1986*, 53–55. Plant Breeding Institute, Cambridge.

Limpert, E. & Fischbeck, G. (1987). Distribution of virulence and fungicide resistance in the European barley mildew population. *Integrated Control of Cereal Mildews: Monitoring the Pathogen* (Ed. by M.S. Wolfe & E. Limpert), pp. 9–30. Martinus Nijhoff Publisher, Dordrecht.

Lupton, F.G.H. (1987). History of wheat breading. *Wheat Breeding* (Ed. by F.G.H. Lupton), pp. 51–70. Chapman & Hall, London.

Müller, T.E. (1987). Systematics and evolution. *Wheat Breeding* (Ed. by F.G.H. Lupton), pp. 1–30. Chapman & Hall, London.

Schwarzbach, E. & Fischbeck, G. (1981). Die Mehltauresistenzfaktoren von Sommer- und Wintergerstensorten in der Bundesrepublik Deutschland. *Zeitschrift für Pflanzenzüchtung*, **87**, 309–318.

Snyder, F.W. & Carlson, G.E. (1984). Selecting for partitioning of photosynthetic products in crops. *Advances in Agronomy*, **37**, 47–72.

Torp, J., Jensen, H.P. & Jørgensen, J.H. (1978). Powdery mildew resistance genes in 106 northwest European spring barley varieties. *Royal Veterinary and Agricultural University Yearbook*, 75–102.

Wenzel, G. & Faroughi-Wehr, B. (1984). Anther culture of cereals and grasses. *Cell Culture and Somatic Cell Genetics of Plants* (Ed. by I.K. Vasil), Vol. 1, pp. 311–327. Academic Press, New York.

Wolfe, M.S. (1987). Trying to understand and control powdery mildew. *Populations of Plant

Pathogens. Their Dynamics and Genetics (Ed. by M.S. Wolfe & C.E. Caten), pp. 253–273. Blackwell Scientific Publications, Oxford.

Wolfe, M.S. & Limpert, E. (Eds) **(1987).** *Integrated Control of Cereal Mildews: Monitoring the Pathogen.* Martinus Nijhoff, Dordrecht.

Worland, A.J., Gale, M.D. & Law, C.N. (1987). Wheat genetics. *Wheat Breeding* (Ed. by F.G.H. Lupton), pp. 129–171. Chapman & Hall, London.

4. WEATHER, INPUTS AND CEREAL PRODUCTIVITY

P.A. COSTIGAN*[1] AND P.V.BISCOE†[2]

*ICI Fertilizers, Jealott's Hill Research Station, Bracknell, Berkshire RG12 6EY, UK; †ICI Fertilizers, PO Box 1, Billingham, Cleveland TS23 1LB, UK

INTRODUCTION

The purpose of this paper is to examine the effects of the weather and of farmers' inputs on cereal productivity. The extent of variation in cereal yields is amply demonstrated by the changes in wheat yields that have been recorded over the past 50 years (Fig. 4.1a). There are two components to the changes in yield from year to year. Firstly, there is an average trend that is shown in the 7-year moving average (Fig. 4.1b), and secondly, the deviation from the moving average in each year (Fig. 4.1c). The general increasing trend (Fig. 4.1b) has largely been enabled by improvements in technology. About half of the improvement can be attributed to the use of

TABLE 4.1. Average performance of 13 varieties of wheat in 3 years of experiments in which varieties susceptible to lodging were staked. Adapted from Austin, Ford & Morgan (1989)

Variety	Year of introduction or first report	Grain yield (t/ha)	Harvest index (%)	Above-ground dry matter (t/ha)
Squareheads Master	1830	5·50	34	16·18
Browick	1844	5·86	31	18·90
Prince Albert	pre 1910	5·44	33	16·48
Partridge	1907	6·92	38	18·21
Little Joss	1908	6·61	36	18·36
Yeoman	1916	6·49	36	18·03
Capelle Desprez	1953	7·69	44	17·48
Maris Huntsman	1972	8·05	47	17·13
Norman	1981	9·48	52	18·23
Renard	1984	8·72	49	17·80
Brock	1985	9·57	52	18·40
Brimstone	1985	9·65	49	19·53
Slejpner	1986	9·89	53	18·66

Present address: [1] Ministry of Agriculture, Fisheries and Food, Nobel House, London SW1P 3JR
[2] Silsoe Research Institute, Wrest Park, Silsoe, Bedford MK45 4HS

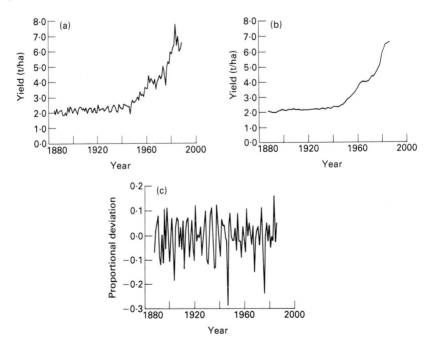

Fig. 4.1. Wheat yields from 1885 to 1989 in England and Wales. (a) Average wheat yield for each year, (b) 7-year moving average of wheat yield, (c) deviation of individual years from the 7-year moving average ((b − c)/b).

better varieties. In a comparison of the performance of varieties introduced between 1830 and 1986, Austin, Ford & Morgan (1989) showed that grain yields had improved by about 80% over the period (Table 4.1). This improvement was virtually all attributable to a change in the harvest indices of the crops, so that the percentage of the top weight found in the grain rose from 32 to 50%. There was virtually no increase in the above-ground biomass production (see also Fischbeck, this volume, pp. 38–42).

The year to year deviation in yield from the moving average (Fig. 4.1c) is largely a reflection of the effects of seasonal weather conditions on wheat yields. It is noticeable that, despite the large increases in yield that have occurred, the deviation in yield from year to year has stayed remarkably consistent, and in only about 1 in 15 years is the deviation greater than 10% of the moving average. Two years produced yields that were markedly (>20%) lower than the moving average, 1947 and 1976. Both of these years were characterized by hot conditions resulting in a severe drought. In contrast, 1984 was an exceedingly favourable year with a yield about 15% greater than the moving average.

EFFECTS OF WEATHER ON YIELD

The effects of weather on yield need to be divided into two categories. There are systematic effects (temperature, day length, radiation receipt, etc.) about which a great deal is known and which can explain most of the patterns of growth of cereal crops. There are also catastrophic effects of weather, such as storm, flood, frost, etc., which often occur as single events and are very difficult to predict and model but can have marked effects on crop yields. This paper will concentrate on the systematic effects.

Dry matter accumulation is largely a function of the amount of solar radiation that a crop can intercept and utilize (Fig. 4.2). The efficiency of conversion of solar radiation into dry matter is determined by the type of photosynthetic pathway. Temperate crops tend to use the C3 pathway, so-called because a three-carbon compound is the initial product of CO_2 fixation, whereas some tropical crops use the C4 pathway which initially produces a four-carbon compound (Milthorpe & Moorby 1979). The fact that temperate crops use the same photosynthetic pathway explains why the efficiency of photosynthetic conversion is similar from crop to crop.

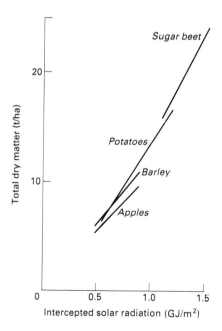

FIG. 4.2. Relationship between total dry matter production of four crops sown in England and intercepted solar radiation. From Monteith (1977).

Many studies have shown that such crops accumulate dry matter in proportion to their absorption of radiant energy (Monteith 1977; Gallagher & Biscoe 1978). There are a number of models that predict canopy development, interception of radiation, and dry matter accumulation for cereal crops (e.g. Ritchie & Otter 1984; Weir *et al.* 1984). An important feature of these models is their ability to partition growth between the vegetative and reproductive organs of the cereal. This is only possible because of our relatively good understanding of the processes involved.

There are a number of distinct developmental stages between sowing the seed and harvesting the mature grain (Fig. 4.3). As long as water and nutrients are not limiting, the duration of each developmental stage is largely determined by the accumulation of thermal time or day degrees. The actual number of day degrees between each developmental stage will be different for different varieties. The time between emergence and flowering can also vary within a variety depending upon the prevailing conditions of weather and day length. Some varieties will not flower unless they have experienced a 'vernalization' period with cold conditions (Flood & Halloran 1986). In general, the development rate between emergence and flowering is increased by cold conditions in the period after sowing. The pattern and extent of this response differs between varieties.

The various developmental stages are best dealt with separately.

Seedling growth and tillering

This is the phase that establishes the framework of the plant. Leaves are produced in a regular way. Leaf production throughout growth is invariably well correlated with accumulated thermal time (Fig. 4.4). Although

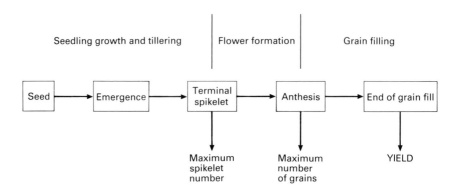

FIG. 4.3. Some of the key stages in the development of winter wheat.

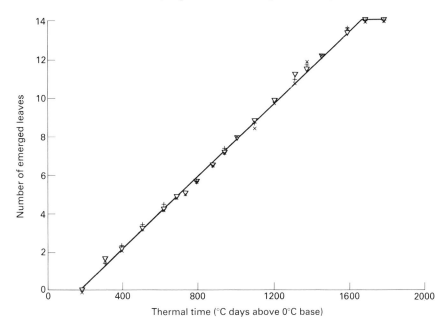

FIG. 4.4. Rate of leaf emergence on the mainstem related to accumulated mean air temperature above 0 °C for three crops of winter wheat (cv. Avalon) receiving either 90 (+), 180 (×) or 330 (▽) kg N/ha. From Biscoe & Willington (1985).

there are marked varietal differences in the rate of leaf emergence, soil fertility seems to have relatively little effect. The benefits of nitrogen fertilizers in increasing yields seem to come principally from delaying senescence of green tissue (Fig. 4.5), rather than by any direct effects on development rate (Biscoe & Willington 1985).

Tillers are also produced in a regular way, but their emergence is much more dependent upon dry matter production than is leaf emergence (Porter 1985). As the crop changes from a vegetative to a reproductive mode there is a period of rapid stem extension with consequent severe competition between the tillers for light, water and nutrients. This precipitates a period of rapid tiller death (Fig. 4.6) leaving the crop with a sustainable number of ear-bearing tillers, (which for optimal UK cereal production needs to be about 500–600/m^2). This overproduction of tillers and subsequent tiller death is a major reason for the ability of cereal crops to compensate for adverse winter conditions. This remarkable ability of cereals for compensatory growth has been well demonstrated by Puckeridge & Donald (1967) who found that a five-fold difference in wheat seedling density had no significant effect on final grain yield, and by Prew *et al.*

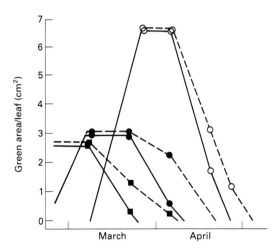

FIG. 4.5. Green areas of individual leaves (leaves 7 (■), 8 (●) and 9 (○)) on the main stem of winter wheat after the application of 90 kg N/ha (solid lines) or 330 kg N/ha (dashed lines). From Biscoe & Willington (1985).

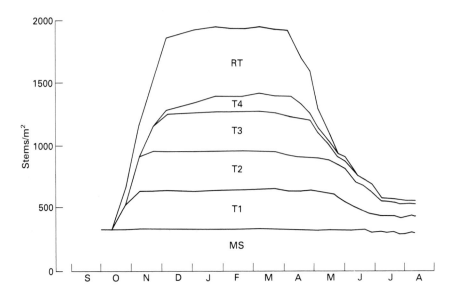

FIG. 4.6. Pattern of tillering of a September sown crop of winter wheat (cv. Avalon) showing the numbers of mainstems (MS), tiller 1 (T1), tiller 2 (T2), tiller 3 (T3), tiller 4 (T4) and remaining tillers (RT). From Biscoe & Willington (1985).

(1985) who found that large differences in early growth from season to season produced only small differences in grain yield. However, such effects cannot be guaranteed and it is likely that a well-established crop of adequate density is better able to withstand periods of adverse conditions without loss of yield than is a poorly established or 'thin' crop.

Terminal spikelet stage

Just prior to its phase of rapid stem extension the plant reaches the 'terminal spikelet' stage. This is when the last spikelet has been formed on the embryo seed head, and it marks the transition between the crop establishment and the ear development phase. The time after terminal spikelet production can be divided into two important stages, flower formation and grain filling. Each of these stages is crucial for yield production and each is controlled by two principal environmental factors. The *duration* of each stage is controlled by accumulation of thermal time so that a period of warm temperatures will shorten the period, whilst the *amount of activity* (whether in producing flowers or filling grains) is determined by the amount of radiation intercepted during the period. Optimum conditions are therefore relatively cool temperatures (to extend the length of each stage) combined with high radiation intensity (for maximum growth).

Flower formation

The number of spikelets that are formed before the 'terminal spikelet' places a limit on the potential yield of the crop. Each spikelet will produce between zero and six potentially fertile flowers, and each flower can produce one grain of wheat. The number of flowers produced on each spikelet depends upon radiation interception during flower formation. In experiments under controlled temperature conditions the number of florets produced from a spikelet in the centre of the spike varied from two to four, depending upon the light intensity that the plants received (Stockman, Fischer & Brittain 1983). The length of the flower formation stage is largely governed by accumulated temperature, but long days also shorten its duration (Porter *et al.* 1987). No matter how good the growing conditions after this point, if insufficient flowers have been produced yields will be low. The successful cereal grower will have controlled many factors, such as crop rotation, seed-bed preparation, seed quality, plant population density, and the fertilizer and pesticide regime, in order to ensure that his crop has grown well up to this point. A large number of flowers should then be produced which presents the potential for a high yield. Once the flowers have formed anthesis quickly occurs and grain filling commences.

Grain filling

The period between anthesis and grain maturity is crucial for yield production. Over 70% of the grain weight comes from photosynthesis occurring during this grain-filling period (Austin *et al.* 1977). The duration of the period is determined by factors that hasten senescence and ripening. The principal factor is the average temperature during the period; the higher the temperature, the shorter the period of grain filling. Therefore, at anthesis a race starts between grain filling which is hastened by radiation interception and grain ripening which is hastened by high temperatures. Consequently, seasons which have a cool grain-filling period and a high level of solar radiation will tend to produce higher yields. Mean temperature and sunshine hours for the grain-filling period display marked differences from year to year (Fig. 4.7), and this probably accounts for a considerable amount of the year to year variation in yields. National wheat yields were particularly high in 1984, which coincided with a grain-filling period that was particularly sunny but not particularly warm. Other factors which hasten senescence, such as nutrient starvation or drought can also limit the length of the grain-filling period and hence yields. In pot

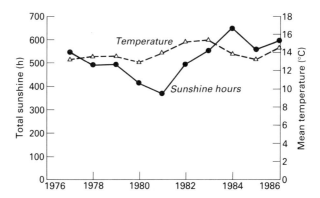

FIG. 4.7. Pattern of occurrence of sunshine hours and air temperature for the flowering and grain-filling period (May–July) over a 10 year period for a single 40 × 40 km square (from the Meteorological Offices's MORECS system) covering parts of Oxfordshire, Buckinghamshire and Berkshire.

experiments, wheat and barley subjected to water stress after anthesis showed no differences in rate of grain growth but a reduction in final grain dry matter owing to a premature cessation of grain filling (Brooks, Jenner & Aspinall 1982). Examination of data from the ICI Fertilizer accounting service 'Cropcheck' shows that yield variation from year to year between 1983 and 1988 was almost totally explained by differences in mean grain weight (Fig. 4.8). This implies that good farmers can achieve the correct density¨f crop by the time of anthesis, but they then depend upon the weather between anthesis and maturity to determine yield.

A largely unknown factor for the future is what will happen to UK weather patterns in an era of possible global warming. It has been suggested that UK wheat yields will be reduced by over 20% if the average temperature rises by 4 °C (Treharne 1989) largely because of a shorter period of grain filling. However, there may be subtle climatic effects which might have a more marked impact on cereal productivity. Morrison & Butterfield (1990) showed that barley crops growing through a warm winter developed more rapidly than usual and were consequently exposed to a greater risk of frost damage in the succeeding spring. If the climate does change, even by just a few degrees, it is likely that we will need a ready supply of new cereal varieties adapted to cope with the new conditions or that can easily adapt to a wide range of weather conditions.

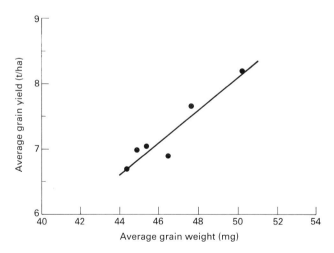

FIG. 4.8. Relationship between average grain weight and average grain yield reported by farmers to the ICI 'Cropcheck' service 1983–88.

EFFECTS OF INPUTS ON YIELDS

In order to provide an economic context it is useful to examine the costs of inputs and the value of cereal production from the farmer's point of view. The ICI 'Cropcheck' system enables average input costs of participating farmers to be determined (Table 4.2). Average variable costs for cereal production amount to about £200/ha, very much in line with those given by Nix (1989). The increase in yield that must be achieved by each of these inputs for them to be economically viable is also given in Table 4.2, and can be seen to be fairly modest. However, the farmer must pay his fixed costs in addition to his variable costs. The average fixed cost for a medium-sized arable farm is £520/ha (Nix 1989) which means that with average input costs a farmer needs to achieve a yield of at least 6·6 t/ha to be reasonably profitable. Of course, if he cuts back on his inputs he will save on his costs but this needs to be finely balanced against the overall profitability of his crop, as seen by examining the effect of varying the nitrogen fertilizer rate (Fig. 4.9). The farmer only makes a profit from growing his crop if the value of the yield exceeds about £640/ha. There is a fine margin between profit and loss and a difference in yield of just a fraction of a t/ha can make a large difference to the profitability or otherwise of an enterprise.

The very high fixed costs involved in cereal farming force the farmer to maximize his production efficiency. Over recent years input costs have tended to rise with inflation whilst crop prices have stayed fairly stable. Farmers are forced to either increase their productivity or to somehow

TABLE 4.2. Average costs (and required returns calculated at a wheat price of £109/t) for variable inputs recorded by farmers using the ICI 'Cropcheck' service in 1988

Input	Cost (£/ha)	Required return for 'break-even' (kg grain/ha)
Seed	45·00	409
N fertilizer	56·68	515
P fertilizer	16·23	147
K fertilizer	4·85	44
Herbicide	33·26	302
Fungicide	34·47	313
Insecticide	6·13	56
Growth regulator	14·46	131
Other sprays (e.g. trace elements)	6·67	61

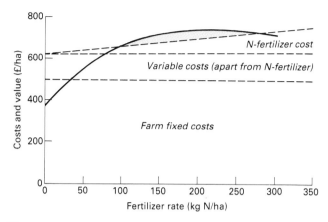

FIG. 4.9. The balance between profit and loss in UK wheat farming as nitrogen fertilizer rate is adjusted. The dashed lines show the fixed and variable costs, the solid line shows the value or grain produced (at a grain price of £109/t) and the shaded area denotes profit. After Jollans (1985).

reduce their costs. This choice has contributed to the marked and continuing change in the human ecology of farming in Great Britain. In 1920 there were about 1 million farm workers employed, but by 1989 the number had fallen to 300 000 (MAFF 1990).

It is apparent from Table 4.2 that most inputs only have to achieve a modest improvement in yield to be economically worthwhile. The actual yield improvements achieved will vary from crop to crop. In a multifactorial trial at Rothamsted an aphicide treatment increased yields by 1·3 t/ha, and a fungicide treatment by 1·0 t/ha (Prew *et al.* 1983). Variety screening trials typically show a 10–30% increase in cereal yields attributable to fungicide use (e.g. NIAB 1990). Wilson & Cussans (1978) showed yield increases from herbicide use for winter wheat of 0·5–2 t/ha. Growth regulators can produce benefits of 0·4–0·8 t/ha in winter barley (Matthews & Thomson 1984). The response of cereals to nitrogen fertilizer will depend upon soil fertility but is typically between 2 and 5 t/ha (Sylvester-Bradley, Dampney & Murray 1984). The inputs of phosphorus and potassium fertilizers are usually set at a level that will replace the removal of these elements by the crop, so apart from any direct benefit in their year of application they need to be considered in the context of maintaining soil fertility (Cooke 1985). It is more difficult to predict what the actual benefit to yields will be in advance of applying an input, but it is clear that these typical levels of benefit more than pay for the inputs listed in Table 4.2. However, the farmer cannot afford to be profligate with these variable inputs, and needs to ensure that he only applies the minimum necessary for the optimum return in each crop situation.

CONCLUSIONS

Weather has a marked effect on the productivity of cereals both in terms of dry matter production and in how that dry matter is partitioned. Farmers' inputs also have a marked effect on productivity. As farmers face continuing economic and environmental pressures to use their inputs more efficiently, there is an increasing need for information as to how the weather, crop growth, and farming inputs interact to produce the final yield. Systems need to be developed to enable farmers to match their inputs for each particular field to the crop requirements on that field. One example of such a system is the ICI Fertilizers 'N-Sure' service which, using a crop development model, predicts the terminal spikelet and second node development stages so that nitrogen fertilizer timing can be closely matched to crop uptake requirements on each field of winter wheat (Biscoe 1988). Other examples are under development by other organizations to improve the forecasting of disease and aphid attacks so that insecticides or fungicides are only applied when they are most needed by the crop. The main disadvantage of such services is that they cost money to run and ultimately that money must come from the farmer, yet at the same time farming income is being seriously eroded. Only if such services can demonstrate an economic value to the farmer will they be successful. High agricultural productivity does not have to be damaging to the environment, as long as that productivity is achieved in an efficient and environmentally responsible way. It is ironic that just at the time when environmental concerns are pressing hard upon farming and there is a need for more technical information for farmers, farming finds itself with reducing real incomes and consequently little money available for taking a more complex approach to the management of inputs. The more that farm profitability is squeezed, the more pressure there is on the farmer to intensify his farming activities.

REFERENCES

Austin, R.B., Edrich, J.A., Ford, M.A. & Blackwell, R.D. (1977). The fate of the dry matter, carbohydrates and ^{14}C lost from the leaves and stems of wheat during grain filling. *Annals of Botany*, **41**, 1309–1321.

Austin, R.B., Ford, M.A. & Morgan, C.L. (1989). Genetic improvement in the yield of winter wheat: a further evaluation. *Journal of Agricultural Science, Cambridge*, **112**, 295–301.

Biscoe, P.V. (1988). 'N-Sure' — the selective N-fertilization for wheat. *Proceedings of the International DLG Congress for Computer Technology Knowledge Based Systems in*

Agriculture, pp. 401–413. Deutsche Landwirtschafts-Gesellschaft, Frankfurt.

Biscoe, P.V. & Willington, V.B.A. (1985). Crop physiological studies in relation to mathematical models. *Wheat Growth and Modelling* (Ed. by W. Day & R.K. Atkin), pp. 257–269. Plenum Press, New York.

Brooks, A., Jenner, C.F. & Aspinall, D. (1982). Effects of water deficit on endosperm starch granules and on grain physiology of wheat and barley. *Australian Journal of Plant Physiology*, **9**, 423–436.

Cooke, G.W. (1985). The present use and efficiency of fertilizers and their future potential in agricultural production systems. *Environment and Chemicals in Agriculture* (Ed. by F.P.W. Winteringham), pp. 163–206. Elsevier, London.

Flood, R.G. & Halloran, G.M. (1986). Genetics and physiology of vernalization response in wheat. *Advances in Agronomy*, **39**, 87–125.

Gallagher, J.N. & Biscoe, P.V. (1978). Radiation absorption, growth and yield of cereals. *Journal of Agricultural Science, Cambridge*, **91**, 47–60.

Jollans, J.L. (1985). *Fertilizers in UK Farming*, CAS Report No. 9. Centre for Agricultural Strategy, Reading.

MAFF (1990). *Agriculture in the United Kingdom: 1989.* HMSO, London.

Matthews, B.J. & Thomson, W.J. (1984). Growth regulation: control of growth and development. *Cereal Production* (Ed. by E.J. Gallagher), pp. 259–266. Butterworths, London.

Milthorpe, F.L. & Moorby, J. (1979). *An Introduction to Crop Physiology*. Cambridge University Press, Cambridge.

Monteith, J.L. (1977). Climate and the efficiency of crop production in Britain. *Philosophical Transactions of the Royal Society, B*, **281**, 277–294.

Morrison, J.I.L. & Butterfield, R.E. (1990). Cereal damage by frosts, Spring 1990. *Weather*, **45**, 308–313.

NIAB (1990). *Recommended Varieties of Cereals*. Farmers' leaflet No. 8. National Institute of Agricultural Botany, Cambridge.

Nix, J. (1989). *Farm Management Pocketbook*, 20th edn, 1990. Wye College, University of London.

Porter, J.R. (1985). Approaches to modelling canopy development in wheat. *Wheat Growth and Modelling* (Ed. by W. Day & R.K. Atkin), pp. 69–81. Plenum Press, New York.

Porter, J.R., Kirby, E.J.M., Day, W., Adam, J.S., Appleyard, M., Ayling, S., Baker, C.K., Beale, P., Belford, R.K., Biscoe, P.V., Chapman, A., Fuller, M.P., Hampson, J., Hay, R.K.M., Hough, M.N., Matthews, S., Thompsom, W.J., Weir, A.H., Willington, V.B.A. & Wood, D.W. (1987). An analysis of morphological development stages in Avalon winter wheat crops with different sowing dates and at ten sites in England and Scotland. *Journal of Agricultural Science, Cambridge*, **109**, 107–121.

Prew, R.D., Church, B.M., Dewar, A.M., Lacey, J., Penny, A., Plumb, R.T., Thorne, G.N., Todd, A.D. & Williams, T.D. (1983). Effects of eight factors on the growth and nutrient uptake of winter wheat and on the incidence of pests and diseases. *Journal of Agricultural Science, Cambridge*, **100**, 363–382.

Prew, R.D., Church, B.M., Dewar, A.M., Lacey, J., Morgan, N., Penny, A., Plumb, R.T., Thorne, G.N., Todd, A.D. & Williams, T.D. (1985). Some factors limiting the growth and yield of winter wheat and their variation in two seasons. *Journal of Agricultural Science, Cambridge*, **104**, 135–162.

Puckeridge, D.W. & Donald, C.M. (1967). Competition among wheat plants sown at a wide range of densities. *Australian Journal of Agricultural Research*, **18**, 193–211.

Ritchie, J.T. & Otter, S. (1984). *CERES-Wheat: A User-Oriented Wheat Yield Model*. Preliminary documentation, Agristars Publication No. YM-U3-04442-JSC-18892, USDA-ARS, Temple, Texas.

Stockman, Y.M., Fischer, R.A. & Brittain, E.G. (1983). Assimilate supply and floret development within the spike of wheat (*Triticum aestivum* L.). *Australian Journal of Plant Physiology*, **10**, 585–594.

Sylvester-Bradley, R., Dampney, P. & Murray, A.W.A. (1984). The response of winter wheat to nitrogen. *The Nitrogen Requirement of Cereals*, pp. 151–174. MAFF Reference Book 385. HMSO, London.

Treharne, K. (1989). The implications of the 'greenhouse effect' for fertilizers and agrochemicals. *The 'Greenhouse Effect' and UK Agriculture* (Ed. by R.M. Bennett), pp. 67–78. Centre for Agricultural Strategy, Reading.

Weir, A.H., Bragg, P.L., Porter, J.R. & Rayner, J.H. (1984). A winter wheat crop simulation model without water or nutrient limitations. *Journal of Agricultural Science, Cambridge*, **102**, 371–382.

Wilson, B.J. & Cussans, G.W. (1978). The effects of herbicides applied alone and in sequence, on the control of wild oats (*Avena fatua*) and broadleaved weeds, and on yields of winter wheat. *Annals of Applied Biology*, **89**, 459–466.

5. AN ECONOMIC PERSPECTIVE ON CEREAL FARMING, FOOD CONSUMPTION AND THE ENVIRONMENT

M. MURPHY

*Department of Land Economy, University of Cambridge,
Cambridge CB3 9EP, UK*

THE QUEST FOR SELF-SUFFICIENCY IN CEREAL PRODUCTION

In a farming context the drive for self-sufficiency in indigenous food production in the UK/EC plainly required more inputs of agrochemicals and capital, because the greater part of extra output has resulted from more intensive farming. In the UK the area of cereals harvested in the last two decades increased by a mere 6–8%, but production has more than doubled. Without agrochemicals, efficient field drainage, improved mechanization and farm structure this triumph would not have been possible. But now excess capacity in farming is a growing problem in many industrialized countries. Overproduction and declining levels of consumption, exacerbated by more sedentary, ageing, diet-conscious and declining populations, are growing problems for policy makers. These problems are further exacerbated by the burgeoning cost of public support for farming throughout the world, but especially in those industrially advanced countries who are members of the Organization for Economic Co-operation and Development (OECD) such as the USA, the EC, Canada, Australia and Japan, much of which is incurred to pay for the storage and subvention of surpluses onto fiercely competitive export markets. This situation prompts a removal of resources from farming as a possible solution towards curbing public expenditure. There are several options open to policy makers: swingeing price reductions at the farm gate through the reduction or elimination of public price support (effectively the objective of the General Agreement on Tariffs and Trade (GATT) proposals); lowering or abolition of production incentives, such as fertilizer subsidies (still operating in some EC countries); and reduction of tax allowances on capital expenditure. Introduction of controls on the use of agrochemicals through a system of permits and licenses, or the imposition of taxes on nitrogen, thereby making it more expensive, are hotly debated policy options. A levy in excess of 80% has been suggested for farms in the

Netherlands (Dietz & Hoogervorst 1990), and in the UK some researchers suggest the levels would have to be well in excess of 100% just to peg nitrogen use at present levels. Incentives to farm more extensively, to pursue alternative farm enterprises, and the development of organic farming are also under consideration (Becker 1990; Laan & Thiyssen 1990).

An emerging consensus on farming in the industrialized economies is that protectionist national farm policies funded by taxpayers, buoyed up by transfers from consumers to producers, have failed. The abandonment of laws of competition is now alleged to be the principle cause of the growing tension in international trade in agriculture. Excess capacity in farming abounds in the industrialized world manifested by overproduction, though there are incorrigible Malthusian cynics who will disagree. Declining and unstable prices, threats of trade wars, neomerchantilist tactics, burgeoning costs of government support worldwide are all heaped at the door of protectionism. The charge is that protectionism has gone too far. By encouraging farmers to boost output, production systems intensified and may have had a damaging impact on the environment. For example, the quality of water supplies in a growing number of agriculturally intensive regions in the EC, and possibly in eastern European countries, may be unsatisfactory for human consumption. There is also concern about food quality. The question now being raised is why not couple measures to control farm output with measures to safeguard the rural environment? There is increasing doubt that farmers can be persuaded to reduce output quickly enough by just using farm gate price reductions. These would have to be severe to achieve a withdrawal of agrochemicals — the most potent and controversial inputs in modern farming. For a number of years what has been dubbed a 'cross compliance' programme has been discussed in the US whereby receipt of 'subsidized prices' is conditional upon the adoption of environmentally friendly farming methods. If additional conservation objectives are sought 'extra subsidies' could be made available (Russell 1990). Understandably some commentators warn that the world is still short of food and that while pockets of famine and starvation persist, land should not be removed from cereal production in the UK or EC. They further charge that the 'food mountains' have been all but eliminated and that it is now back to business as usual. If reductions are to be made in production levels, the cry is 'not in my backyard!' UK/EC producers are 'efficient' and more should be done to promote an expansion of grain exports. Taxpayers, who outnumber grain farmers, may ask however if it makes economic sense to continue to use resources to produce grain for which markets are difficult to find, and at best can only be sold on the world market at knock down prices and generous subsidies. If in doing so farmers, not in any way intentionally, are using agrochemicals to the detriment of the environment and the quality of food raw materials, then perhaps there is a more intelligent and socially acceptable way of maintaining farmers in a greener and gentler countryside. This in effect is a proposal to switch part of the present funding for conventional farming to conservation-linked but lower levels of production (see Potts, this volume, pp. 15–18). Many may say, why pay farmers for merely looking after the landscape (doing nothing), but it could be more realistic to 'mothball' farmland rather than letting it revert to scrubland, and perhaps something could be done to slake the reputed thirst of the urban dweller for access to more parkland.

PROSPECTS FOR WORLD TRADE IN GRAIN TO THE YEAR 2000

Over the past three decades (1962–1990) world wheat production more than doubled to 594 million tonnes in 1990. Rising production was matched by rising levels of demand and in most years production was greater than consumption (Table 5.1). Consequently year ending world stockpiles of wheat expressed as a percentage of annual consumption were on average over 28% for the years 1962–90. The lowest levels recorded were closer to 20% in 1972/73 and 1989/90. Most of the increase in production resulted from increasing yields per hectare, the area harvested worldwide expanded by a mere 8–9% in 28 years. As production and consumption expanded so did world trade especially in the 1960s and 1970s, doubling between 1962/63 and 1981/82. Annual world trade in wheat began to stagnate from the mid 1980s onwards and is expected to be no more than 94 million tonnes in 1990/91 (Fig. 5.1). Uppermost in many grain producers and policy makers minds is the question of a much needed revival in world wheat trade in the 1990s. Present dietary trends in OECD countries

TABLE 5.1. World wheat production and trade 1962–2000

	Area (million ha)	Yield (t/ha)	Production	Trade (million tonnes)	Consumption	Stocks
1962	206	1·22	251	44	245	76
1972–74	216	1·66	358	65	360	79
1982–84	232	2·08	483	100	467	137
1989–90	228	2·49	568	97	550	122
1992–95	230	2·59	596	110	560	140
1998–2000	240	2·60	624	120	614	150

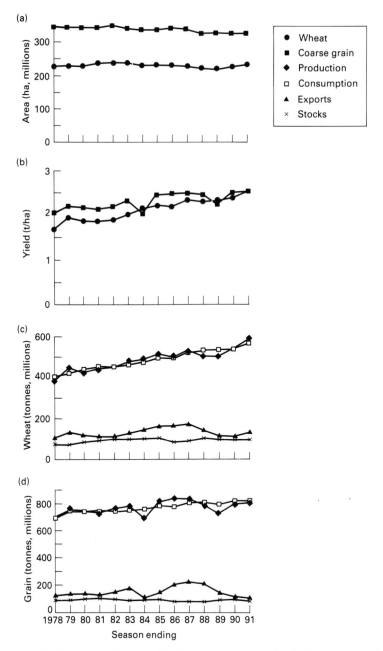

FIG. 5.1. World cereal production and trade from 1978 to 1991 (preliminary estimates for 1990 and projected for 1991 — August 1990 data). The areas (a) and mean yield per hectare (b) sown to wheat and coarse grain (rye, barley, oats, maize, sorghum, millet, mixed grains but not rice). Global production, consumption, stock levels and amounts exported around the world are shown for wheat (c) and coarse grains (d) From US Department of Agriculture.

suggest that consumption will at best remain static while output will rise. Plainly unless world exports can be expanded costly stockpiles will grow to unreasonable levels. Here opinion is divided; there are those who say that stocks should be kept at high levels to prevent famine in the next decade during which the world population will increase further. A more reasonable expectation is that production will expand in the major importing countries accompanied by a further modest expansion in world exports. Consequently annual world trade in wheat is not expected to grow by more than 20 million tonnes by the year 2000.

It was between 1970 and 1980 that the great expansion in world wheat trade occurred, by over 45 million tonnes to close on 100 million tonnes in 1980–82. However it has been stagnating ever since. If the distribution of world exports by supplier remained roughly as it is was in the mid 1980s, the EC could expect to contribute annually about an extra 4 million tonnes to this expected further expansion in world trade. This would be achieved with the present level of resources committed to the EC cereal sector. Furthermore it is widely expected that EC wheat production by the late 1990s will be about 30 million tonnes in excess of consumption. At present, EC human consumption amounts to no more than 31–32 million tonnes (1989/90) out of a total crop of about 78–80 million tonnes a year. Plainly the EC, even with the present level of resources committed to wheat production, could quite easily satisfy any predicted expansion in world trade in wheat in the coming decade.

The 1990 world wheat harvest was approximately 594 million tonnes — more than enough to satisfy the consumer demand of 566 million tonnes. Despite soil moisture deficits in major wheat producing countries, the level of stocks were expected to recover to about 22% of annual world consumption to 124 million tonnes: wheat production is once again approaching its normal state of oversupply (Fig. 5.1) (International Wheat Council 1990). For the third year in succession cereal production in the EC approached or exceeded the maximum guaranteed quantity of 160 million tonnes (Fig. 5.2) and stagnating world markets could result in a rapid build up of intervention (mostly wheat) stocks. Export refund levels are now approaching inflammatory levels; looking forward to 1991 these will reach 60% of the intervention price. The USA is likely to respond with special export enhancement payments which would allow US grain traders to sell US grain at $74 per tonne. These are the opening shots of a cereal war which could be the costliest yet. Plainly excess capacity in the EC/US cereals sectors must be tackled with determination.

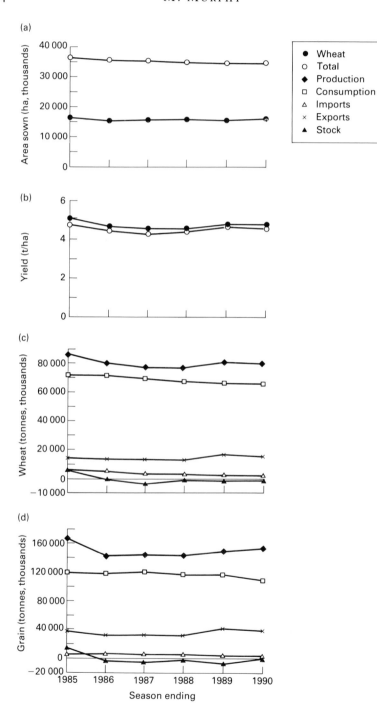

World production and trade in
coarse grains 1990–2000

World production of coarse grains (principally maize, barley, oats and rye) increased by over 370 million tonnes since the early 1960s to about 830 million tonnes in 1990/91 (Fig. 5.1). From 320 million hectares in 1962, the area harvested increased to 350 million hectares in 1981/82, but later declined to about 325 million hectares in 1990/91. The US set-aside programme was partly responsible for this downturn following the glut in supplies in the mid 1980s. In 1986/87 year ending stocks as a percentage of annual consumption reached 29%. Indeed stocks are still tight, in the region of 14% in 1990/91, about 4 percentage points below the 28 year average of 18·5% (USDA 1990). As production increased so did consumption and world trade in exports. From a mere 31 million tonnes in 1962 world exports reached 108 million tonnes in 1980/81. Thereafter the trend has been erratic; between 1985 and 1988 export trade declined by almost a fifth, recovering to just under 100 million tonnes in 1989–91. It is unlikely that the rapid growth seen in the 1970s will be repeated in the 1990s; at best world trade could expand by an additional 2–5 million tonnes per annum. Most of this increase in demand will be for maize. However the USSR was responsible for importing one fifth of the world trade, or 24 million tonnes, in 1989/90. Japan, Mexico, Taiwan and Korea are the next most important importers at close on 40 million tonnes. Much will depend on what developments take place in the USSR, and the ability of some developing countries in Africa and Asia to finance imports.

In summary, even if world trade in wheat and coarse grains were to recover and possibly expand to about 230–240 million tonnes by 2000, this is well within the present productive ability of the world's major export suppliers, the USA, Canada, the EC, Australia and Argentina. Furthermore, since 1985, the US has removed 4 million hectares from wheat production thereby reducing the level of world output by 10 million tonnes. It may not be appreciated that the accumulated impact (up to 1989/90) of the USA set-aside programme was to remove 20.48 million hectares of cereal land from production (OECD 1989). The greater proportion of this was land taken out of wheat and maize production which had the effect of

FIG. 5.2. EC-12 cereal production and trade from 1985 to 1990 (preliminary estimates for 1990 and projected for 1991 — August 1990 data). The areas (a) and mean yield per hectare (b) sown to wheat and total grain production (wheat, rye, barley, oats, maize). EC-12 production, consumption, stock changes, imports and exports are shown for wheat (c) and total cereals (d). From Eurostat and Commission Bilan.

keeping almost 40 million tonnes of grain off the world market. The OECD estimated that had the USA not done so, world prices would have been at least 14% lower. In summary, the USA set-aside programme, the absolute level of grain produced in the United States, and the volume of trade with the USSR and possibly China, are major determinants in the price formation of world grain trade. There is the notion that the USA set-aside programme fuels further intensification of cereal production in the UK/EC (Potts, this volume, p. 15). This proposition presupposes that UK/EC producers would seek to compensate for the ensuing shortfall in USA production, but to what end? Despite set-aside the USA has always maintained more than adequate stocks to meet world export needs. Additionally but more importantly EC cereal production is sealed off from the mainstream of world production by a system of import levies, and producer prices set at levels well above world prices. It is the EC cereal price which encourages high output and intensification within the EC grain sector; inputs on profit maximizing cereal farms are mainly determined by product price levels.

Medium term outlook for
world grain production

The outlook for grain production over the next decade in OECD countries will be influenced by the relative levels of protection and comparative costs of production. Production will continue to be driven by improving technology but hopefully by more market-determined rather than administered prices. OECD wheat production is expected to reach 230 million tonnes by 1992 and coarse grain 420 million tonnes. At the same time if non-OECD cereal production were to continue to grow at around 2% per year, world production would reach about 1900 million tonnes by 1992. More importantly, however, the surplus over OECD consumption of about 443 million tonnes would be available for export: 207 million tonnes. Non-OECD exports, principally from Argentina, could reach 14 million tonnes, making a total of 220 million tonnes available for export, at least 20 million more than exported in the late 1980s.

The OECD countries are the world's largest cereal exporters. Wheat exports over the past 3 years (1986–89) approached 87 million tonnes, while those for coarse grain were 70 million tonnes, a total of 157 million tonnes, the bulk of world exports. It is not unreasonable to suggest that if the USA were to bring back all the land put into 'set-aside' and 'conservation programmes' over the past decade or so, not only would they dominate world cereal trading, they would emerge as the world's largest producer. Developing countries by contrast exported about 11 million tonnes of wheat and 18 million tonnes of coarse grain, a total of 29 million tonnes. The combined level of exports from developing and developed countries averaged about 186 million tonnes over the last 3–5 years. However it is the developing countries which are the world's largest importers, accounting for 80% of world trade in 1987–90. If Soviet imports are excluded then they represent about 60% of world trade. If it is considered that by 1992 that OECD countries are expected to have a surplus over 200 million tonnes of grain, and assuming that world production remains more or less neutral to world climate changes, then even if developing countries could afford to increase their present share of grain imports by a further 30 million tonnes, the world's major exporters could confidently meet this level of demand without vast increase in the present resource used in cereals production.

Climatic change has the potential to greatly affect cereal production. However, much of the discussion on the likely impact of the greenhouse effect on agricultural production is statistically imprecise. US reports (USDA 1990) suggest that

> Air concentrations of carbon dioxide have dramatically increased during the past 20 years, mainly because carbon dioxide is a byproduct of oil, gas, and coal combustion . . . Today's air contains about 350 parts per million of carbon dioxide — higher compared with a level of 265 parts in pre-industrial times and 314 as recently as 1958. There's every reason to believe that these levels will continue to increase.

Increasing livestock numbers upon which organic farming methods are to some extent dependent also exacerbate the greenhouse gas syndrome.

Such heightened levels of carbon dioxide are no direct threat to humans. Air contains 78% nitrogen, and 21% oxygen; the remaining 1% is a mix of carbon dioxide, argon, neon and helium.

> Almost everyone agrees that the amount of carbon dioxide in the atmosphere will double sometime in the mid or late 21st century, but few people agree on how to interpret this — how much warmer the Earth will get and when this will occur.

Our climate, like our rapidly changing daily weather, is dynamic rather than static in nature. The climate of the world has continually changed over the aeons of time. However, it has only been systematically documented by instrument observations for less than 150 years. Climate models suggest rainfall will increase about 10% worldwide when carbon dioxide doubles. This would probably not be felt equally everywhere — perhaps some areas might experience even less than now. No one has great confidence in predicting what will happen in any specific region. Most informed opinion suggests that the effects, if any, of the so-called greenhouse effect would not be manifested until sometime in the 21st century. Consequently it would not be wise for individual cereal farmers and their political representatives to rely on this phenomenon to bail out the hard-pressed world cereal sector in the 1990s.

Implications for UK/EC-12 cereal producers in a
more liberalized trading system for world wheat
and coarse grain in the 1990s

Over the last three decades annual world trade in grain increased by 100 million tonnes, if erratically so. How realistic then is it to expect a further expansion of say 3–5 million tonnes per year to a total of 230 million tonnes per year in the next decade? The answer plainly depends very much on whether those who wish to import can afford to do so and whether more liberal trading arrangements, talked about for so long, can help to boost world grain trade and by implication, consumption. Could the boom in world grain trade in the seventies, when exports expanded from a mere 100 million tonnes to over 180 million tonnes a year (but then followed by the relative stagnation the 1980s, when annual world exports of wheat and coarse grain, altered little from around 190 million tonnes per year), be repeated? If world exports could be expanded by 35 million tonnes, from 195 to about 230 million tonnes per year, what are the implications for hard-pressed UK and EC grain growers in the early nineties? It would be naïve to assume that exports would expand instantaneously by 35 million tonnes; more likely the build up would be gradual throughout the next decade. If so then the total amount of extra grain marketed might be in the region of 200 million tonnes over a period of 10 years. On the assumption that the EC maintained its present share of world trade in grain, or notched it up by a further 4 to 20%, UK/EC grain growers would eventually have the opportunity to export an additional 4 million tonnes annually. A glance at the present trends in consumption and production for EC grain (Fig. 5.2) and even a superficial examination of where the underlying trend might take EC grain production by the mid 1990s, inescapably reveals that an additional 4 million tonnes of exports per year would have but a trivial impact on farm gate sales, real prices and real incomes. Even if world exports were to expand further to, say, 250–270 million tonnes per year over the next decade, and EC cereal exports were to reach 40 million tonnes per year, this increase would make but a small dent on cereals farmers' fortunes, given the pent-up levels of potential production which exist on a great number of modernized cereal farms in the UK, France and

Germany, the three major cereal suppliers in the EC. Indeed that level of expansion in world grain trade would be unlikely to take place without a reduction in real world cereal prices or without very heavy export subsidies or bonuses, plainly unrealistic in the 1990s. Consequently it is not surprising that there is a growing consensus amongst many economists that the rat race in burgeoning world cereal production, especially in OECD countries, is leading many farmers closer to insolvency. A further view is that unless the GATT proposals for a lowering of protection levels are adopted by OECD countries, however painful this would be, the surplus problem will become graver. World market prices would continue to fall and the real net margin on every tonne of grain produced on UK/EC farms would relentlessly decline to paultry levels. Surely there must be a more economically rational system of world grain production. More reflective farmers will not be surprised to learn that if all protection now known as the 'aggregate measure of support' (AMS) is removed too abruptly, UK/EC farmers would be grievously injured. EC output would fall by at least 20%. The knock-on effect on the food processing sector would be even greater; output would decline by 25%; and jobs would be lost. Land values would fall by at least 50%, and UK/EC farm incomes would plummet by more than 20%, even after allowing for adjustment in the cost of some inputs. The gainers would be those countries with low levels of protection; Australia, New Zealand, and some Latin-American countries. The USA, which is now expected to reveal that it has made substantial reductions in direct support for farmers since 1986 of 68%, would be less affected. UK/EC farming would be emasculated and plunged into distress unless there were swingeing reductions in input costs. There is a ray of hope in this scenario, as the world output of cereals would fall, and as a consequence world prices would rise — anything from 7 to over 25%. An estimate of 12% now seems most likely.

Trade liberalization in world grain markets

Trade barriers are common in the world market for grain. They are linked in complex webbing to domestic farm policies and help alter production and consumption signals. Domestic policies include price and income supports and production subsidies. Trade policy constraints include export subsidies, import quotas, variable levies and export licences. In the EC, relatively high cereals support prices are an integral part of the common agricultural policy. Variable levies, a type of import tax, are necessary to prevent foreign cereals from entering the EC at prices that would undercut those received by EC farmers. Without a trade barrier such as the variable levy, the commission in Brussels would have to alter the common agricultural policy radically. Domestic farm policies can also affect the development of trade policies. Relatively high USA target prices and loan rates in the early to mid 1980s led to greater USA and non-USA grain production, USA export losses and high stockpiling. More importantly perhaps, the export enhancement programme built into the 1985 Security Act provides a means whereby the USA can match other countries' export subsidies and helps reduce surpluses, albeit at great cost to taxpayers. Domestic and trade policies can reallocate production, alter consumption, and shift output from efficient producers who can compete with little protection to less efficient producers who often may need protection to survive. Producers who are guaranteed high and stable prices despite surpluses and shortages hold an artificial advantage over other producers.

The most extreme example of economic protection can be traced to Switzerland (where Government subsidies accounted for 75% of net value added of Swiss agriculture in 1990), Finland and Japan and to a much lesser extent to the EC. These policies can also destabilize world markets. When the link between world and domestic cereal prices is severed, wider variations in world prices are necessary to adjust supply and demand to stocks. On the other hand developing countries have been the major beneficiaries of export subsidies and other related policies. Paradox abounds in discussions of this nature. The USA loan rate target price system has frequently in the past contributed to excess stockpiling, which in return prompts supply control measures such as reduction programmes. It could, however, also be argued that USA stocks have helped to cushion shocks to the world grain trading systems such as the Soviet entry into the world market as a major importer of grain in the mid 1970s, and have possibly reduced the escalation in prices that would have occurred under a free market system.

Levels of support vary widely among countries. Most of the major grain exporters are involved in policies to assist or protect their farmer producers. The role of subsidies or their equivalents can be neatly divided into producer subsidy equivalents (PSEs) and consumer subsidy equivalents (CSEs). The former provide a measure of government support and vary widely between major grain exporters. A PSE is defined as the level of subsidy that would be necessary to compensate producers for the removal of government programmes for a particular cereal commodity. It is often measured as the total value of policy transfers to farmers (grain producers) as a percentage of total producer agricultural returns, cash receipts plus government payments. A CSE measures the level of subsidy that would have to be paid to consumers to compensate them for the removal of agricultural policies. It is measured as the ratio between the value of government policy transfers to consumers and total consumer expenditure for a particular commodity (cereals). A positive CSE represents consumer support while negative values refer to consumer taxation. As a study of Table 5.2 will show, the level of PSE and CSE subsidy equivalents vary widely across the globe. Between 1982 and 1986 the EC, USA and Canada offered the highest levels of protection to their grain producers; Australia had a much lower level of PSE, followed by Argentina, although circumstances there were not comparable with elsewhere. Consumer subsidies are often inversely related to PSEs, particularly in countries where border measures are important to sustain production and farm income level. In Japan and the EC where consumers are implicity taxed through high border levies, producer support is relatively high. In contrast the high levels of consumer support found in India and Nigeria are paired with relatively high producer taxation. Support for farmers is widespread throughout the industrialized world (e.g. Table 5.3). The policies used for grain-grower assistance also differ widely among exporting countries. A study of Table 5.4 shows price and income instruments as the main source of support for farmers in the Community, whereas Canada uses a combination of price and income support and marketing subsidies as the major source of assistance and encouragement for cereals producers.

TABLE 5.2. Producer and consumer subsidy equivalents (PSEs and CSEs) for wheat; 1982–86 averages for selected countries. From USDA (1990)

	PSE	CSE
Exporters		
EC Durum wheat	38·4	−28·2
Soft	47·1	−14·8
USA	36·5	−2·0
Canada	30·4	−2·7
Australia	6·8	NA
Argentina	4·8	9·1
Importers		
Japan	97·8	−31·5
Taiwan	64·8	−9·2
Brazil	63·4	NA
South Korea	59·9	17·2
Mexico	18·8	NA
South Africa	18·3	21·9
India	−35·5	20·8
Nigeria	−18·7	217·3

NA, not available.

Table 5.3. Total financial transfers associated with agricultural policies in the USA and EC ($billion). From Dunne (1990)

	EC-12	USA
Transfers from taxpayers (1)		
1986	31·7	59·4
1987	38·2	50·3
1988	45·8	49·1
1989	44·1	46·3
Transfers from consumers (2)		
1986	71·9	29·6
1987	78·3	30·4
1988	63·7	26·0
1989	54·1	21·6
Total transfers (1 + 2 — budget revenue)		
1986	102·9	88·1
1987	115·9	80·0
1988	108·8	74·3
1989	97·5	67·2

Table 5.4. Percentages of policies used for wheat producer assistance, 1982–86 average. From USDA (1990)

Policy	Australia	Canada	EC	Japan	USA
Price/income support	65	43	99	87	73
Input subsidies	13	3	0	13	16
Marketing subsidies	3	41	0	0	2
Long-term research	19	6	1	0	4
Other	0	7	0	0	5

Effects of policy reform on
world cereals markets

Economic theory suggests that the elimination of government programmes would produce benefits to society through increased efficiency and improvement in the economic use of resources. Production would shift to those areas that can produce and deliver to consumers at the lowest cost. In any specific country the most efficient farmers would fare best or would be least affected by cuts in government programmes. Consumption patterns would probably remain more or less the same. They are now somewhat unresponsive to changes in price. The key issue in this debate however is that economic theory gives no clear direction as to whether world grain trade would rise or fall after substantial trade reforms. The result would depend on whether importers or exporters protect their producers more. The domestic prices would fall, production would decline and imports would increase within importing countries with domestic prices initially above world prices as they removed protection. The result would be to increase world prices. Simultaneously, despite higher world prices, some exporting countries could experience a decline in production as subsidies are removed and domestic prices fall towards world prices. This would reduce the quantities available for export. If the cut-back in production among the major world cereal exporting countries were to be greater compared with that in importing countries, world trade would contract rather than expand. Since most of the world grain, more importantly grain for export, is produced in OECD countries, it is likely that average grain prices, in particular for wheat for export, would increase as these major producers reduced their production. Importers among the developing countries would look more to the world market. The volume of world trade would decline but exports from all the present OECD exporting countries may not necessarily fall. For example if EC production declines, then that in the USA, Australia and Argentina could expand. It is likely that prudence will dictate that dismantling protection will take place very gradually and cautiously because as world grain stocks are lowered, then market shocks such as the uncertain buying intentions by the USSR and China or the greenhouse effect, could have a much greater impact on prices than might otherwise be the case.

The cereals sector in the USSR
and eastern Europe

The recent political trends in eastern Europe prompt discussion on the future developments in the cereal sectors of the USSR and the five major eastern European countries, Poland, Hungary, Czechoslovakia, Bulgaria and Rumania plus East Germany. The question follows: how will these events influence food production and in particular the production of cereals at the farm level? Perhaps more importantly, from the viewpoint of UK and EC cereal producers, are there likely to be any further export possibilities in these countries? If some of these countries were to seek membership of the EC how would this affect the EC cereal balance? How likely is it that what is now generally known as the USSR could within the next 5 or 10 years become self-sufficient in cereals? One might ask, is there any possibility that some eastern European countries, keen to acquire foreign exchange, would seek to export cereals onto the world market or into the EC Community?

The USSR, as it is presently structured, has ample land resources per head; at least twice that of the rest of Europe. Much of it (55%) is grassland prone to drought, and much of the arable land, 37·5%, has an average mean temperature of no more than 5°C. Yields achieved are therefore low. Although cereal yields have almost doubled in 20 years and that for wheat has increased from just over 1 t/ha in 1975 to 1·8 t/ha in 1989, a rate of growth somewhat faster than that of the USA, wheat yields are still rather low in a world context. The average world yield of wheat is now in excess of 2·35 t/ha. Indeed by way of further contrast, the yield of winter wheat is now close to 3·4 t/ha in China (International Wheat Council 1987). Yields for most other crops in the USSR are also low: potatoes 12·1 t/ha, sugar beet 26·6 t/ha and milk yields per cow in the region of 2552 kg (Table 5.5).

The USSR produces about 200 million tonnes of grain annually, making it one of the world's largest producers. Over the last 20 years this level of production has altered little, although yields per hectare have been increasing, the area planted has declined. Since the mid 1930s the production of wheat has increased from about 30 million tonnes annually to about 85 million tonnes in 1987–1990. Most of this increase took place in the 1960s, but since then production has tended to stagnate. In contrast, over the last 60 years world wheat production has expanded over three-fold, but since the 1960s has more than doubled to over 580 million tonnes in the late 1980s. By 1990/91 it will have reached over 592 million tonnes exposing yet again the flawed arithmetic of the Malthusian pessimists. To take yet another example the production of Soviet wheat changed but little between 1933 (22·5 million tonnes) and 1969 (27 million tonnes). Thereafter it steadily increased and is now about 85–90 million tonnes per year, yet the USSR is still one of the world's largest single importers of wheat and coarse grain, averaging over 35 million tonnes over the last decade. This was about three times the level of imports by China, which is a net importer of

TABLE 5.5. World distribution of yields in 1990 (t/ha, except for milk: kg/cow per year)

	World	USSR	W. Europe	EC
Wheat	2·35	1·77	4·59	6·9
Barley	2·21	1·57	3·97	5·2
Sugar beet	34·53	26·6	43·3	49·9
Potatoes	24·87	12·1	22·0	42·5
Milk	2110	2552	3758	5002

wheat of about 15 million tonnes per year. In the USSR, where wheat production has averaged about 85 million tonnes per year over the past decade, less than half is used for human consumption (estimated at between 120 and 130 kg per head). The bulk of imported wheat is used for milling. Yields of grain in the USSR, to be comparable with worldwide yields, probably need to be discounted by between 15 and 20% from the 'bunker weight' in which Soviet grain production is reported, which includes excess moisture and extraneous matter. Human consumption is probably closer to 100 kg per head. This compares with about 80 kg per head in the EC, which produced 79 million tonnes in 1989/90, of which about 31 million tonnes was used for human consumption (Table 5.6). There is no reason to believe that the Soviet Union consumes less wheat per head compared with other major world users of wheat, the EC, the USA or China.

In the six major eastern European countries wheat availability seems to be in the range 30–35 million tonnes per year, of which it is estimated 15 million tonnes are produced for human consumption, about 110 kg per head. The estimate for Poland is about 200 kg per head. It is not unreasonable to conclude that in the Soviet Union and eastern Europe, supplies of wheat per head are generally much higher than those available in the rest of the world. The USSR could satisfy its own demand from domestic supplies of wheat, assuming the quality was acceptable for bread-making, and confectionery.

If it is assumed that USSR supplies per hectare, after discounting for moisture and extraneous matter, are in the region of 180–190 million tonnes per year, this represents a per capita supply of well over 600 kg a year, including imports of a further 35 million tonnes per year (about 126 kg per head). The normal USSR feeding pattern for wheat is that high quantities are fed to livestock, estimated to approach 50 million tonnes per year. Soviet sources indicate that 8–10 million tonnes, or as much as a quarter of wheat fed, can be described as quality feed wheat.

Serious domestic price distortions are the primary reason for the heavy

TABLE 5.6. Wheat consumption in the USA, EC and USSR (million tonnes per year; means from 1986–1989/90). The waste/other category includes EC exports of 18.5 million tonnes and USA exports of 36 million tonnes. From FAO (1988)

	Human	Animal	Seed	Waste/Other
USSR	37	45	11	20
EC	31	21	3	28
USA	20	7	3	25

feeding of quality wheat and the apparent lack of interest in selling wheat to the State (Littman 1989). The price distortions arise out of the price gap between the State-regulated purchase price for wheat and the selling price for prepared feeding stuffs. The evidence seems to suggest that the low-cost wheat producing areas, which produce much of the high quality wheat grown in the USSR, ironically receive the lowest payments; between 70 and 80 roubles per tonne. In contrast the State-regulated price changed for prepared feeding stuffs are uniform throughout the country at about 200 roubles per tonne. Thus the cost of mixed feed is often over twice that of the wheat procurement price in the low-cost grain producing regions. The State prepared feeding stuffs are usually of poorer quality and lacking in adequate protein content. Farmers therefore prefer to feed wheat rather than sell it to the State and buy back inferior and overpriced compound feed. There also appears to be considerable wastage at docks and on railway sidings. A similar situation may be replicated throughout some of the other eastern European countries. This is not surprising since investigation of this nature is that the daily calorific intake through food and beverages is rather high in the USSR compared with other countries, at about 3500 kcal per day (World Bank 1987), and around 3350 kcal for much of eastern Europe. This is somewhat more than the average for western Europe, though the United States is slightly higher. The average Soviet diet consists of cereals and livestock products, including fish, but may be deficient in the consumption of fruit and vegetables and high quality processed products. In recent years food supplies have been enhanced by growing imports of livestock products, butter and meat. About 5% of meat consumption was imported in recent years. Total meat production is now approaching 20 million tonnes per year in the USSR (OECD 1989) In the late 1970s, in an effort to stimulate agriculture production, prices for several commodities were increased by between 13 and 30%.

The central problem with Soviet agriculture appears to be that despite considerable efforts on the part of government to allocate an adequate part of total investment to this sector, and the fact that producer prices are raised more frequently than consumer prices for food, increases in agricultural production may have been achieved more by increasing areas planted or livestock numbers rather than productivity. It could be summarized as being quantitative rather than qualitative growth and this syndrome is also evident in the pattern and style of food consumption. The linkages between farm prices, costs of production and consumer demand and preference are extremely weak and this partly explains the huge losses which consistently occur in the agricultural sector. More importantly in the context of this discussion, conversion rates in the animal feed sector seem to be extremely poor because of inadequate housing and inefficient technology. Twice as much grain is required to produce one pound of liveweight grain in the USSR as in western Europe and North America. Consequently the feed requirements of the livestock are well over twice those of the EC.

Though food consumption levels appear to be satisfactory in the Soviet Union, quality, presentation, consistency of supply to consumers, convenience of purchase all require improvement. If the present difficulties in the Soviet food sector can be resolved by the mid 1990s, and management and technical performance on farms improved, then net grain imports could be much smaller than they are at present. Better use of the vast grasslands might mean less reliance on cereals as a source of animal feed, resulting in a lowering of net meat and butter imports. By the 21st century Soviet grain exports could become a reality.

The cereals sector in China

A discussion on world cereal production which excludes China would be less than complete. Although China is the world's largest producer of grain, in excess of 400 million tonnes, less than half is attributable to wheat and coarse grain. In recent years China has been producing close to 90 million tonnes of wheat and about 95 million tonnes of coarse grain. If imports of wheat are included, annual consumption is around 105 million tonnes, but annual consumption of grain is somewhat lower, at just over 90 million tonnes. China is a modest exporter of coarse grains, approaching 5 million tonnes in recent years. The combined area of wheat and coarse grain harvested over the last decade has remained more or less constant at just under 60 million hectares. In that time the yield per hectare has increased by about 1 tonne and by world standards China's farmers have now achieved fairly high wheat yields; certainly higher than those in India, Pakistan and countries cultivating large areas of dry land wheat such as Canada, Australia, the USSR and the USA. On the other hand wheat yields in Mexico, the EC and some parts of eastern Europe and Japan are higher than those in China.

Given sufficient economic incentives, China's farmers can still boost wheat yields with existing physical resources. A constraining factor will almost certainly be irrigation, and expanding production may require considerable investment in irrigation systems. Soil erosion may also be a problem in some areas. If past experience is anything to go on greater access to foreign plant-breeding techniques and supplies of genetic materials will enable China's plant breeders to develop still higher yielding varieties. Increased application of agrochemicals and fine tuning of irrigation and disease control should boost yields further. China's farmers could raise yields of wheat by 2–3% per year up to the year 2000 to in excess of 120 million tonnes.

Consumption trends in China

Wheat is a keenly sought after commodity in rural and urban markets. The government frequently purchases one quarter to one third of total wheat output, principally for transfer to urban areas, compared with only 15–21% of total rice, and much lower for other grains. For example, between 1979 and 1986 wheat made up over 28% of all grains purchased by the government under contract prices (procurement prices by the Grain Procurement Boards). Rice accounts for over 35%, and other grains 37% of total cereals output. Farmers prefer to consume wheat and are reluctant to sell it either to government boards or other outlets. Farmers in the past have sold a very small proportion of their wheat under negotiated procurement prices. Unlike wheat farmers, rice farmers are willing to sell a larger proportion of their output to government boards. The rapid increase in grain production since 1978 has permitted substantial shifts in consumption patterns. Coarse grains such as corn, sorghum, barley, oats and other grains have decreased substantially, while consumption of wheat and rice have increased. Recent rural survey data show that grain consumption per head rose from 248 kg in 1978 to 260 kg in 1987. In the same period what is known as fine grain consumption leaped from 123 to 210 kg per head. Coarse grain consumption in the same period fell from 125 to 48 kg per head. By international standards this level of grain consumption is extremely high. In addition urban household income and expenditure survey data available from 1981 to 1987 suggest that urban consumers reduced their grain consumption from 145 to 132 kg per head. Over the last thirty years wheat consumption per head has risen dramatically from only 29 kg in 1960 and throughout the 'reform' period it soared to 88 kg per head (1978–88).

International wheat trade and China

China has become a major wheat importer since the emergence from the disturbances of the 'great leap forward' when millions of people are reputed to have died from malnutrition. From 1960 to 1987 imports averaged only 6·9 million tonnes per year but reached 15·5 million tonnes in 1988/89. Since 1972 the USA has been the major supplier, frequently accounting for over 60% of total imports. It is reasonable to assume that wheat consumption per head will be maintained or possibly increased slightly up to the year 2000. Given the limited potential for expanding the cultivated area and despite sustained yield improvements, domestic supplies may not match demand. It is highly probable that up to the year 2000 China could expand imports of wheat up to 20 million tonnes per year.

THE FUTURE PROSPECTS FOR LAND USE IN THE EC

The biggest threats to EC farming are the continued pressure to keep output closer to consumption levels and the surge of output from the East will be a new challenge. The future for cereal production in the UK/EC and the USA will be one of economic austerity until such time as effects of price distortions are eased or eliminated. There will be renewed pressure for policy makers to encourage farmers to think again on the merits of 'set-aside' as a measure of taking land out of cereal production. Though this would increase the cost of production per tonne of grain produced, it would help to increase world prices. Such trends could be reinforced with greater incentives to reduce inputs, by rationing fertilizer application through a system of permits or emissions/pollution licenses; these could be tradable. There could also be incentives through grants or other income aids for those farmers, who opt to move towards organic farming methods. This would have the twin merits of reducing output and inducing the withdrawal of chemical inputs, thus controlling or limiting agricultural pollution of rivers and water supplies. However inspection of pollution maps for various parts of the EC will suggest that by far the worst pollution occurs near towns and industrialization sites. Farming is but a small contributor to pollution.

Organic farming

One of the major stumbling blocks towards the transition from conventional to organic farming is the adjustment of costs of production. Recent evidence suggests that as farmers move towards organic management, savings in the cost of chemical and other inputs are usually less than the decline in total gross output per farm. This results in a lower farm gross margin (gross output minus variable costs) and unless there is a compensating decline in fixed costs, farm income falls. Consequently if organic farming is to be adopted there would have to be swingeing reductions in rents/land values or mortgage charges.

Savings would also have to be made in labour and machinery costs — plainly very difficult. Indeed if these costs were to be cut, husbandry methods would deteriorate, possibly resulting in lower yields (Table 5·7) and output. Furthermore organic farming is best undertaken if there are adequate supplies of farmyard and other sources of organic manure. Specialization in farming over the past 40 years has resulted in the disappearance of livestock from many farms, particularly in the arable regions of the UK/EC. If farmers in these areas were to attempt organic farming, output would fall to levels which could result in insolvency in many cases.

Annual support would have to be in the region of £300–500 per hectare to tide farmers over the transition stage. It would not be reasonable to advise farmers to adopt organic farming if it were to result in abject poverty or bankruptcy following large reductions in gross output. The argument that integrated farming is the answer looks attractive (El Titi, this volume, pp. 399–411) but is not widely substantiated and may depend upon adequate supplies of manure in the right places. We have an unevenly distributed surplus of livestock products at present and policy makers are seeking a reduction in the output from this section.

In the longer term the answer may rest with extensive methods for the production of all livestock (including free-range hens and outdoor pigs). Indeed as the effects of GATT and further measures to limit production of all farm produce begin to take effect farming may become extensive (Murphy 1991). This would be especially so if farmers seek to stay in business at those levels of output which are profit maximizing. It is likely that by the 21st century farms will be larger, and farmers will be more sophisticated in management and possibly more secret about their business and farming methods.

There is growing concern that unless policy makers put in place stringent controls on the use of chemicals and other environmentally harmful agents used in the production of industrial and agricultural pro-

TABLE 5.7. Cereal yields (t/ha) on 'organic' farms in England, 1989. W = winter crop; S = spring crop

	W. wheat	S. wheat	W. barley	S. barley	Oats
Conventional	6·05	4·95	5·27	3·44	4·72
Organic	3·77	3·50	3·14	3·00	3·65

ducts, the quality of life in the 21st century may be unnecessarily impaired. Consumers are the problem not the solution, and they will ultimately bear the brunt of the cost, under the umbrella of the 'Polluter Pays' approach, now widely argued and discussed. The environment is now an issue which farming and the supply industry cannot ignore. If for whatever reason consumers wish to consume green food and savour a greener countryside, then farmers should be ready to stoke the thirst of green and hopefully better informed consumers and taxpayers.

REFERENCES

The figures used in this paper are often combined from a variety of official sources; it is not practical to itemize every one. Only reviews, articles and major sources are referred to here.

Becker, H. (1990). Influencing environmental quality by induced innovations in agricultural production systems. *European Agriculture, In Search of New Strategies, Theme 3, New Strategies in Agriculture and Social Policy*, pp. 49–62. European Association of Agricultural Economists, The Hague.

Dietz, F. & Hoogervorst, N. (1990). The economics of Dutch manure policy. *European Agriculture, In Search of New Strategies, Theme 3, New Strategies in Agriculture and Social Policy*, p. 70. European Association of Agricultural Economists, The Hague.

Dunne, N. (1990). Grassroots anxiety in agriculture. *Financial Times*, 2 November 1990.

FAO (1988). *Yearbook, 1988*, Vol. 42. Food & Agriculture Organization, Rome.

International Wheat Council (1987). *World Wheat Statistics*. London.

International Wheat Council (1990). Report, September 1990. London.

Laan A. & Thiyssen, G. (1990). Towards an environmental accounted agriculture — direct or indirect regulation? *European Agriculture, In Search of New Strategies, Theme 3, New Strategies in Agriculture and Social Policy*, pp. 63–76. European Association of Agricultural Economists, The Hague.

Littman, E.L. (1989). Agricultural policies in the USSR, problems, trends and prospects. *Journal of Agricultural Economics*, **40**, 84.

Murphy, M. (1991). *The Impact of GATT Reforms on Arable Farmers' Incomes in England*. Agricultural Economics Conference Discussion Paper, April 1991. Nottingham.

OECD (1989). *Agricultural Policies of Market and Trade; Monitoring and Outlook*. Organization for Economic Co-operation & Development, Paris.

Russell, N.P. (1990). *Efficiency of Four Conservation and Output Reduction Policies*. Department of Agricultural Economy Working Paper 1990.02. University of Manchester, Manchester.

USDA (1990). *World Grain Situation and Outlook*, January 1990. FG1–90. United States Department of Agriculture.

PART 2
THE SOIL

Session Chairman & Organizer: J.F. DARBYSHIRE

At the microscopic level, soils usually consist of complex mixtures of mineral and organic particles surrounded by an intricate assortment of pores filled with either the soil atmosphere or solution. As a result, the soil micro-environment is a difficult environment to investigate. It is opaque, heterogeneous and contains a multitude of micro-organisms, plant roots and animals as well as their residues. To make matters worse, some of the biological component will be growing and decomposing throughout the life of the cereal crop. A comprehensive coverage of all aspects of the soil is obviously not possible in three papers, but some of the more important aspects for the growth of cereal crops are addressed. The microbial biomass of the soil, that community of micro-organisms less than a thousand cubic microns (e.g. fungi, bacteria, actinomycetes and protozoa), are responsible for many of the transformations of plant and animal residues in the soil. Some of these transformations lead to the release of inorganic nutrients for plants. The microbial biomass also contains a significant reserve of nutrients within its own protoplasm. Both these characteristics make the microbial community particularly important for the nutrition of cereal plants under low input farming.

In the first paper, Phil Brookes and colleagues consider the changes that can occur in the soil microbial biomass of cereal fields, particularly with regard to cereal straw amendments. Next, Gerhard Jagnow reviews what is known about the region of soil close to plant roots, the so-called rhizosphere, and the microbial interactions that occur in this region. The rhizosphere is obviously an important part of the soil for cereals, because it is the region where plant roots assimilate nutrients and where many pathogens attempt to invade root tissues.

Apart from earthworms, soil animals are sometimes neglected in cereal-field ecology. In the concluding paper, Kor Zwart and Lijbert Brussaard attempt to redress the balance with a review of soil animals, especially the secondary decomposers of the microflora. They discuss recent data from current long-term Dutch experiments concerned with low input farming systems.

93

6. SOIL MICROBIAL BIOMASS DYNAMICS FOLLOWING THE ADDITION OF CEREAL STRAW AND OTHER SUBSTRATES TO SOIL

P.C. BROOKES,* J. WU† AND J.A. OCIO‡

*Soil Science Department, AFRC Institute of Arable Crops Research, Rothamsted Experimental Station, Harpenden, Hertfordshire AL5 2JQ, UK; †Changsha Institute of Agricultural Modernization, Changsha, China; ‡Granja Modelo, Arkaute, Alava, Spain

INTRODUCTION

The soil microbial biomass comprises all soil organisms less than about $5 \times 10^3 \ \mu m^3$ (e.g. fungi, bacteria, actinomycetes, yeasts, protozoa), other than living plant material. It can thus be considered as a living fraction of soil organic matter. It generally comprises about 1–3% of the total organic matter in soil, which actually means it is a very large population.

The amount of biomass in soil depends on many factors but the two most important are: (i) the annual inputs of fresh substrates (usually mainly plant residues, e.g. roots, stubble and often straw in cereal fields, and (ii) soil texture (especially clay content). Invariably clay soils will contain at least two to five times as much biomass as sandy soils under similar management.

Soil management affects the amount of biomass in soil quite markedly. For example, the biomass in four Rothamsted soils (two arable, one woodland and one grassland) of the Batcombe soil series ranged from about 400 kg C/ha in a soil which has never been fertilized, to more than 2000 kg C/ha in a permanent grassland soil (Fig. 6.1), reflecting the larger input of plant-derived carbon in the grassland soil. Figure 6.1 also shows the large quantities of nitrogen and phosphorus in the cells of the microbial biomass in these soils. During biomass turnover these nutrients are mineralized. However, some will be transferred directly to the next generation of micro-organisms, so are not available to plants. The magnitude of this effect is not known.

The biomass decomposes all dead plants and animals that enter the soil, ultimately degrading them to carbon dioxide, water and simple inorganic salts. It is thus an essential component of the global nutrient cycles of

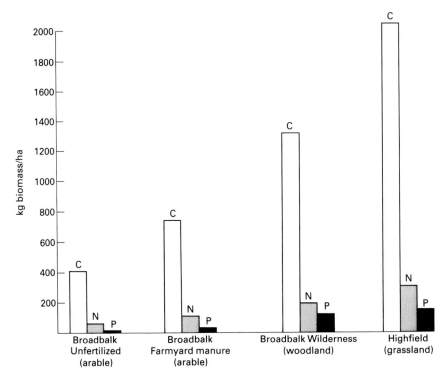

FIG. 6.1. Effects of different soil managements on amounts of microbial biomass carbon (C), nitrogen (N) and phosphorus (P) to a soil depth of 23 cm in Rothamsted soils of the Batcombe series. Adapted from Brookes, Powlson & Jenkinson (1984).

carbon, nitrogen and phosphorus. Therefore, an understanding of the properties and physiology of the biomass is essential if we are to understand the functioning of natural and agricultural ecosystems.

Treating the micro-organisms in soil as a single undifferentiated compartment (collectively the microbial biomass) is clearly an oversimplification. To better understand, for example, soil carbon and nitrogen dynamics, the separate contributions of the different microbial groups are needed. So far, however, attempts to split the 'black box' in this way have not generally been very successful. Indeed, we contend that appropriate laboratory methods to do so generally do not yet exist. Some success in estimating the fungal and bacterial biomasses by direct microscopy has been achieved (e.g. Jenkinson, Powlson & Wedderburn 1976; Schnürer *et al.* 1986), but the methodology is tedious and not suitable for routine use. Similarly, the protozoan component may be estimated by the most probable number technique (Darbyshire *et al.* 1974). However, in order to make reliable routine measurements of the principal groups in the biomass (i.e. fungi, bacteria, yeasts, actinomycetes and protozoa) in relation to nitrogen and carbon cycling we require new techniques to extract the separate microbial groups from soil in sufficient quantities (and free of contamination from other soil components) to permit accurate chemical analysis. Some progress has already been made in separating the bacterial biomass from soil (Macdonald 1986), but more needs to be done.

The biomass has been considered as 'the eye of the needle' through which all dead plant and animal remains pass as they are broken down to their simple inorganic components (Jenkinson 1977). In soils growing cereals, substrates such as dead roots and other plant tissue and root exudates will be released into the soil throughout the life of the crop. However, there will also be a larger pulse of substrate entering the soil as the roots and stubble are ploughed back after harvest or left to decompose on the soil surface in direct-drill systems. In addition, over the last few years there has been a significant increase in the proportion of UK cereal straw that is incorporated into soil, as environmental pressure to reduce straw burning increases. For example, in 1983 only about 2% of total UK cereal straw production was incorporated at harvest (Attwood 1984). By 1987 this had increased to 18% (MAFF 1987). Straw burning will probably be completely banned in the UK within the next 3 years.

The amounts of cereal residues entering soil each year obviously depend on the final total yield of the crop and are thus variable. Powlson *et al.* (1990) reported that a 8·9 t/ha grain yield gave 8·2 tonnes straw, chaff and stubble at harvest, while a grain yield of 0·6 t/ha gave only 2·4 t/ha of total above-ground crop residues. The total straw weights at harvest for these two crops were 5·3 and 1·4 t/ha, respectively. Roots, in addition, would add about another 1–1·5 and 0·5–1·0 t/ha dry matter at harvest (P.B. Barraclough, personal communication). Eventually, all this material will be mineralized to water, carbon dioxide and simple inorganic salts by the microbial biomass.

In this paper, we will describe the recently developed fumigation–extraction method to measure the microbial biomass (Vance, Brookes & Jenkinson 1987a). Its great advantage over earlier methods is that it can be used to measure the biomass in soils with intense microbial activity, e.g. immediately after incorporation of fresh substrates. This was hitherto impossible. We are using fumigation–extraction and other methods to investigate the role of the biomass in the decomposition of straw and other substrates. This is discussed by reference to three recent experiments.

In the first, the role of the biomass in straw decomposition and soil nitrogen dynamics was investigated to determine the potential of straw incorporation to decrease autumn and winter nitrate leaching losses.

The second experiment investigated soil priming effects. Following substrate incorporation into soil an accelerated rate of mineralization of soil organic matter may be observed (the priming effect). Sometimes, the reverse situation develops and the mineralization rate is depressed. The origin of this effect has long been controversial and we offer an explanation.

The third experiment investigated the effect of soil texture on biomass turnover times. Simply measuring the 'standing crop' of biomass tells us little about the dynamics of the microbial population. However, a simple procedure has recently been developed to measure biomass turnover in the laboratory and possibly under field conditions, although this needs to be tested. We have studied the relationships between biomass turnover times, soil organic matter contents and substrate inputs in soils of different texture.

METHODS

Measuring the soil microbial biomass

Several different methods are available to measure the total amount of microbial biomass in soil. These include direct microscopic counting (Jenkinson, Powlson & Wedderburn 1976), the fumigation–incubation method (Jenkinson & Powlson 1976), the fumigation–extraction method (Vance, Brookes & Jenkinson 1987a), the substrate-induced respiration method (Anderson & Domsch 1978) and the soil adenosine 5'-triphosphate (ATP) method (Jenkinson & Oades 1979). Jenkinson (1988) reviewed these methods and assessed their relative merits and demerits. The fumigation–incubation method is usually the standard procedure against which the others are calibrated. However, for many purposes the recently developed fumigation–extraction method offers many advantages over fumigation–incubation or the other methods, and it is now the usual method to measure biomass carbon and nitrogen at Rothamsted. It has the advantage that it is rapid, simple and is suitable for automated analysis.

The fumigation–extraction method

Chloroform fumigation of soils lyses the cells of the microbial biomass and makes them partially extractable to 0.5 M K_2SO_4 (or other aqueous re-agents). The cytoplasm is probably the main microbial fraction made extractable by $CHCl_3$, although this has not yet been proved. The basic procedure to measure biomass by this approach (fumigation–extraction) is as follows. The moist field soil is sieved (usually <2 mm) and portions (usually at about 40% of full water-holding capacity and containing 50 g soil on an oven-dry basis) are fumigated with $CHCl_3$ for 24 h. The $CHCl_3$ is then removed and the soil extracted for 30 min with 0·5 M K_2SO_4. Other, non-fumigated fractions are extracted at the start of fumigation. The soil extracts are then filtered and either analysed immediately or stored frozen at −15 °C. Microbial biomass carbon (biomass C), biomass nitrogen (biomass N) and biomass ninhydrin N may then be estimated from the *increases* in organic C, N and ninhydrin N in the K_2SO_4 extracts of fumigated soils compared to the unfumigated soil extracts. Full analytical details for measuring biomass C (Vance, Brookes & Jenkinson 1987a), biomass N (Brookes *et al.* 1985) and biomass ninhydrin N (Amato & Ladd 1988) are given in those papers.

Using fumigation–extraction, biomass C and N measurements can be made across the whole pH range (Vance, Brookes & Jenkinson 1987b) in soils containing actively decomposing substrates (Ocio & Brookes 1990) and in freshly sampled soils. Reliable biomass measurements by fumigation–extraction have also been reported in paddy (i.e. waterlogged) soils (Inubushi, Brookes & Jenkinson 1989). For various reasons, this is mainly not possible with the other methods.

Fumigation–extraction also has the big advantage that the labelled biomass that develops during early substrate decomposition can be measured (Wu, Brookes & Jenkinson 1989; Ocio, Martinez & Brookes 1991). This is impossible with the other methods.

RESULTS AND DISCUSSION

Soil biomass nitrogen

About 3–5% of total soil N is found in the soil microbial biomass (Brookes *et al.* 1985). However, the biomass has a much shorter turnover time (about 1·5–2 years) than total soil organic matter, some fractions of which may have turnover times of thousands of years (Jenkinson & Ladd 1981). Thus the N held in the microbial biomass is much more labile than total soil N. The biomass can therefore be considered as a pool of N, which may become available to plants during biomass turnover. This N pool can be very large: arable soils may contain 30–150 kg biomass N/ha in the plough layer and grassland considerably more (Brookes *et al.* 1985).

We do not yet know all the factors involved in the turnover of soil microbial biomass N. The turnover time of the biomass in temperate soils is about 2 years (Jenkinson & Ladd 1981). For a biomass containing 200 kg N/ha this would imply an annual flux of 100 kg N/ha through the biomass. However, not all of this N will enter the soil inorganic N pool as plant-available N, because some of the microbial N will be incorporated directly into new microbial cells and not released to the soil (Jenkinson 1985). It seems likely that protozoa play an important role in the turnover and mineralization of biomass N by grazing on the bacterial biomass (e.g. Elliott, Coleman & Cole 1979; Clarholm 1985). Kuikman, Van Vuuren & Van Veen (1989) reported that predation by protozoa increased the plant uptake of bacterial N by 15% and 40% in microcosms with stable and fluctuating soil moisture conditions, respectively. Darbyshire (1976) showed that the activity of protozoa was suppressed in dry soils, which is in agreement with the observation that mineralization of soil organic C and N are also decreased under these conditions (Shen, Brookes & Jenkinson 1987). Factors such as soil temperature, air-drying, rewetting, freeze–thaw and alterations in substrate supply are also likely to be important in regulating biomass turnover.

The C:N ratio of the soil microbial biomass is remarkably constant (about 5–7) in unamended soils differing widely in texture, organic matter content and management (Jenkinson 1988). Thus, the biomass is much richer in N than soil organic matter as a whole, which in temperate arable soils usually has a C:N ratio around 10.

Soil nitrogen dynamics following straw incorporation

Previously, we could not measure changes in total biomass C or N as it developed on the decomposing straw, due to limitations in the methodology. However, the fumigation–extraction method permits biomass N determinations in soils containing recently added substrates (Ocio & Brookes 1991). This was first tested in the following experiment. Chopped wheat straw (1–2 mm diameter pieces) was added (2% wt/wt) to moist, sieved (<6·25 mm) portions of two arable soils, which were then adjusted to 50% of full water-holding capacity (WHC) and incubated at 25 °C. Soil 1 (a sandy-loam soil with 9% clay) contained 1·30% total organic C and had a pH of 7·1. Soil 2 (a clay-loam soil with 40% clay) contained 2·76% C and had a pH of 6·6. Biomass C and biomass N were determined by fumigation–extraction after 13 and 35 days of incubation. Straw addition caused large increases in both biomass C and N, between which there was a close linear relationship (r = 0·98) irrespective of soil type, straw addition or sampling date (Fig. 6.2). Averaged over both soils and treatments, the

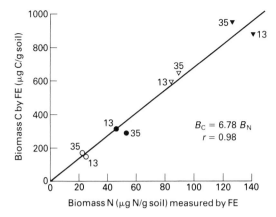

F IG. 6.2. The relationship between biomass C and biomass N both measured by fumigation–extraction in soil 1 and soil 2, amended and unamended with 2% straw. Measurements made 13 and 35 days after straw addition. See text for details.

biomass C : N ratio was virtually identical to the value of 6·8 in unamended soils given by Jenkinson (1988). The ability to make reliable biomass measurements in soils containing actively decomposing substrates has proved a real breakthrough in this work.

In UK arable soils, inorganic N is at considerable risk of leaching between the autumn and early spring (Macdonald *et al.* 1989). The N demands of an autumn-sown crop are very small during this period, and in the autumn there can be high N mineralization rates as the still-warm soils become wetter. Nitrogen held in the biomass is protected from leaching. Increasing the amount of biomass N at this time may help decrease N leaching losses. One way of doing this may be to incorporate straw after harvest. The biomass that develops in response to this input with a large C : N ratio has a large demand for N, at just the time when plant N demands are small or non-existent.

We tested this hypothesis in a field experiment at Rothamsted on a fallow clay-loam soil of the Batcombe series (Ocio, Brookes & Jenkinson 1991). Wheat straw (10 t/ha) was hand dug to 15 cm depth in 1 m^2 plots with or without an application of 100 kg N/ha as NH_4NO_3. Other (control) plots received neither treatment. The experiment commenced in October 1987 and continued until the autumn of 1988. Straw incorporation roughly doubled the amounts of biomass C and N (measured by fumigation–extraction) within 2 weeks, and thereafter they slowly declined. The biomass in the straw-amended soils contained about 50 kg N/ha more than in the control soil. There was only a small extra increase in biomass N, of about 10–20 kg (Fig. 6.3), in the straw+N soils.

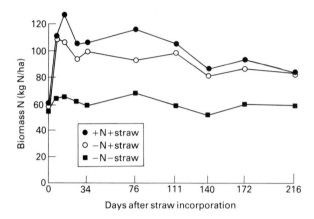

FIG. 6.3. Soil microbial biomass N in field soils following incorporation of 10 t/ha wheat straw to 15 cm, with or without 100 kg N/ha as NH₄NO₃.

At the time of maximum biomass N in the straw+N treatment (14 days after straw incorporation) we could still measure the 100 kg/ha of added inorganic N as NH_4-N and NO_3-N (Fig. 6.4). Also, the initial amounts of inorganic N in the previously fallow, control soils were very small (<10 kg N/ha) at the time of straw incorporation, and remained so throughout. Thus, the extra biomass N synthesized following straw addition did not come mainly from soil inorganic N reserves. Other possible sources for this N are soil organic N, biological N_2-fixation (unlikely) and straw-derived N. The straw itself contained about 90 kg N/ha so it could have

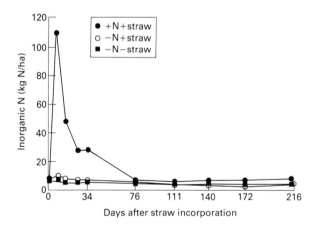

FIG. 6.4. Total soil inorganic N in field soils following incorporation of 10 t/ha straw to 15 cm, with or without 100 kg N/ha as NH₄NO₃.

been a major N source for the biomass. In the above experiment the straw was not ^{15}N-labelled so this could not be tested directly. However, further laboratory work (Ocio, Martinez & Brookes 1991) using ^{15}N-labelled straw, confirmed that within a few days of adding straw to soil there was rapid incorporation of ^{15}N-labelled straw−N into the biomass, reaching a maximum 5 days after straw addition.

If the biomass utilizes straw−N to a large extent, this lessens its demands for N from the large inorganic N pool usually present in arable soils at harvest. This, in turn, limits the potential of straw incorporation to decrease winter nitrate leaching. We are investigating this further in another field experiment, using ^{15}N-labelled straw and ^{15}N-labelled in-organic N, so that the contribution of the various possible N sources to the biomass can be better assessed.

The priming effect in soils

The fate of the carbon in substrates added to soil can theoretically be conveniently monitored if they are isotopically labelled with ^{14}C, because during the subsequent decomposition phase the [14]CO_2 evolved must be derived ultimately from the substrate and the non-labelled CO_2 from soil organic matter. However, it is often observed that *more* unlabelled CO_2 is evolved from soils containing labelled substrates than from similarly in-cubated soils without labelled substrate additions. Thus the substrate has in some way accelerated the decomposition of the native soil organic matter. The origin of this '*accelerated decomposition*' or '*priming effect*' has been debated for many years (Jenkinson, Fox & Rayner 1985). As atmospheric CO_2 concentrations are increasing with uncertain environmental con-sequences, a better understanding of the global carbon cycle is essential. It is important that we fully understand both the mechanisms which can produce priming effects and their magnitude in soil.

Soil organic matter can be considered to exist as two separate pools (excluding plant roots and organisms larger than about $5 \times 10 \, \mu m^3$): a pool of non-living complex polymeric organic molecules and a living pool — the soil microbial biomass. True priming effects, operating on soil organic matter itself, could therefore be caused by: (i) increased mineralization of C in the non-living soil organic matter pool; (ii) increased turnover of the C in the cells of the biomass; or (iii) both processes operating simultaneously.

The following experiment was designed to determine *if* true priming effects occur and, if so, which of these mechanisms operate (Wu 1991). A permanent grassland soil from the Highfield Ley-Arable experiment at Rothamsted (Johnston 1973) was sampled from the 0−10 cm horizon and was sieved to <2 mm. The soil (3·95% organic C, 0·4% total N) had a pH of 5·3. Two substrates, finely ground uniformly ^{14}C-labelled ryegrass, and ^{14}C-labelled glucose in aqueous solution were added separately to moist (50% WHC) portions of the soils (40 g oven-dry weight) to give 5000 µg C/g soil. A solution of $(NH_4)_2SO_4$ was then added, to give 160 µg added N/g soil and the soils were then incubated at 25 °C for 100 days (glucose) or 145 days (ryegrass). Carbon dioxide evolved during the incubations was trapped in 1·0 M NaOH and biomass C was measured by fumigation–extraction. Total and labelled C in both the evolved CO_2 and in the biomass were measured by chemical analysis and by scintillation counting.

Amounts of unlabelled CO_2 evolved from the grassland soil, with and without glucose and ryegrass, are shown in Figs 6.5 and 6.6, respectively. Both substrates produced priming effects, but of strikingly different natures. Consider first the unlabelled CO_2–C evolved following glucose addition (Fig. 6.5). A significant priming effect rapidly occurred, equivalent to about 60% of the CO_2 evolved from the control at day 10.

Now consider the rather more complex situation with ryegrass, where there was an initial, very slow, priming effect. Up to about day 30 the rate of unlabelled CO_2 evolution was slow and constant and at this time, it was about 16% of the CO_2 evolved from the control (Fig. 6.6). However, the rate then increased rapidly, reaching about 50% of the CO_2 evolved by the control by day 100.

In summary, both glucose and ryegrass produced priming effects, but of very different natures and presumably by different mechanisms. That produced by glucose began immediately after addition and was effectively

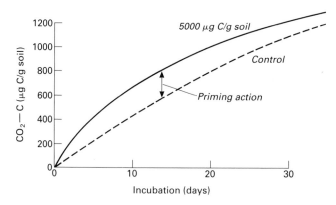

FIG. 6.5. Unlabelled CO_2 evolved from Highfield Grassland soil during the first 30 days of incubation, with or without 5000 µg ^{14}C-labelled glucose/g soil.

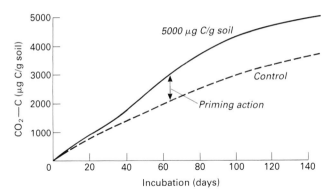

FIG. 6.6. Unlabelled CO_2 evolved from Highfield Grassland soil during 145 days of incubation with or without 5000 µg ^{14}C-labelled ryegrass/g soil.

over about 10 days later. That produced by ryegrass was much more complex, being apparently biphasic and had an appreciable lag phase.

Part of the explanation for the longer lag phase with ryegrass than glucose may have been because the glucose was completely soluble and thus totally available immediately, whereas the powdered ryegrass would have been colonized by micro-organisms after some delay.

The lag phase with the plant material may have been even longer, if intact roots or larger pieces of plant residue had been used. However, even when quite large pieces of straw are added to soil the initial growth of biomass is rapid. Thus, Ocio & Brookes (1990, and this paper) showed that 13 days after addition of wheat straw, sieved to 1–2 mm thickness, the biomass that grew on the straw had developed to its maximum size, then remained constant until the next sampling date (35 days after straw addition).

The total and labelled C in the biomasses of the soils with added glucose and ryegrass at different times after substrate addition are shown in Figs 6.7 and 6.8, respectively. Assuming the extraction efficiencies by K_2SO_4 were the same, the distribution of the label, like the priming effect itself, was quite different between the two substrates. This, we believe, helps explain the difference between the priming effects with glucose and ryegrass. The glucose amendment increased total biomass C by about 25% at day 20 and this increase was maintained until day 60, declining to little more than the amount in the control by day 100. However, the relatively small *net* increase in *total* biomass actually represents a considerable replacement of the original soil biomass with new ^{14}C-labelled biomass derived from the added glucose within 20 days (Fig. 6.7). This, we suggest, is the origin of the priming effect following glucose addition. The extra

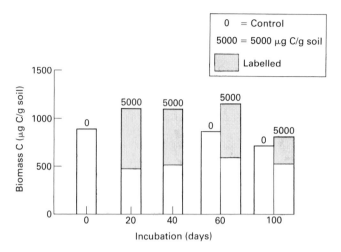

FIG. 6.7. Changes in unlabelled and labelled soil biomass following addition of 5000 µg ^{14}C-labelled glucose/g soil.

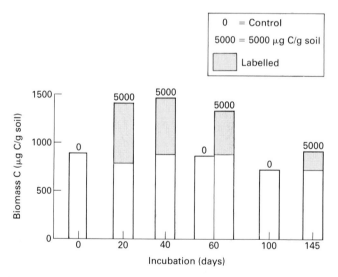

FIG. 6.8. Changes in unlabelled and labelled soil biomass following amendment with 5000 µg ^{14}C-labelled ryegrass/g soil.

unlabelled CO_2 evolved following glucose addition is mainly derived from the native biomass that the new biomass displaces. The approximate parallelism of the CO_2 respiration curves of unlabelled C, with and without the glucose amendment strongly suggests that the 'new' biomass does not increase the rate of mineralization of non-biomass soil organic matter.

The increase in total biomass (>50%) with ryegrass was rather larger than with glucose. More importantly, there was very little *replacement* of the native biomass with labelled biomass. Instead the ryegrass-derived biomass was simply added to the total amount of biomass (Fig. 6.8). Clearly, the priming effect due to ryegrass was caused by a different mechanism from that produced by glucose. We suggest that it was mainly caused by increased mineralization of non-biomass soil organic matter rather than by the ryegrass-derived biomass replacing the original biomass (as was the case with glucose). The shape of the CO_2 respiration curves following ryegrass addition gives some indication of the possible mechanism. During the first 40 days, very little extra unlabelled CO_2 was evolved due to ryegrass addition, unlike glucose. The priming effect mainly occurred after the rapid decomposition phase of the ryegrass was over (about day 30), and before the total biomass declined. After this time, there was a rapid increase in unlabelled CO_2 evolution. Thus, the initial long 'lag phase' was because the labelled biomass that developed on the ryegrass did not initially degrade soil organic matter; rather it 'switched' to metabolism of soil organic matter after the ryegrass was largely exhausted. Why glucose and ryegrass produced priming effects by these two different mechanisms is, as yet, unknown. It may be that different micro-organisms, with different properties, develop on glucose and ryegrass. It is also possible that the biomass that develops on the ryegrass first attacks the more readily available part of the tissue (e.g. the cytoplasmic component) and then, when the fraction is exhausted, switches to the more complex and resistant plant material and soil organic matter. We also do not know if labile plant components (e.g. root exudates) produce priming effects by mechanisms similar to those produced by glucose or ryegrass. Perhaps their priming effect operates by both mechanisms.

The turnover of the soil microbial biomass

Soil organic matter is produced by microbial degradation of the remains of plants and animals which enter soil. Ultimately, the degradation process converts organic matter to CO_2 and simple inorganic salts (e.g. nitrate) which, by becoming available again to plants, are recycled through the biosphere. Thus, soil organic matter can be considered as a large pool of potentially plant-available C, N and other nutrients, although the half-life of some of its components may be more than a thousand years (Jenkinson & Ladd 1981). Because much of the soil organic matter is so stable, changes in total amounts in response to changing soil management can often take many years before they can be detected by classical chemical analysis. In contrast, changes in total soil microbial biomass often occur much more quickly and it has been suggested that biomass measurements can provide an earlier 'warning' of changes in soil conditions long before changes in total soil organic matter can be detected (e.g. Powlson & Jenkinson 1976; Saffigna *et al.* 1989).

It seems likely that changes in total soil organic matter should also be related at an early stage to changes in biomass turnover. This is because soil organic matter itself is mainly composed of microbial metabolites with only small amounts of very resistant undecomposed plant and animal remains. Thus the balance between rates of microbial processes producing organic matter accumulation and those involved in organic matter degradation largely determines the amount of organic matter in soils whose total annual C inputs are similar.

We have developed an experimental procedure to measure biomass turnover in soil. It is being used to understand the relationships between soil physical properties (e.g. clay content), substrate inputs, soil organic matter contents and biomass turnover times. Our approach is firstly to label the biomass with ^{14}C-glucose (1000 µg ^{14}C/g soil) and incubate the soil at 25 °C in the laboratory. This causes a ^{14}C-labelled biomass to develop, which then slowly declines. After about 20 days the turnover rate of the ^{14}C-derived biomass is the same as the turnover rate of the entire biomass. Thus biomass C turnover rate (R_t) can be estimated by measuring the decline of ^{14}C-labelled biomass between 20 and 100 days incubation. Biomass C is measured by the fumigation–extraction method and biomass ^{14}C by scintillation counting of the K_2SO_4 soil extracts. Biomass C turnover time (T, in days) is then calculated from: $1/R_t$, where R_t is defined as:

$$R_t = \frac{\text{input to (or loss from) biomass C compartment per day}}{\text{biomass C in compartment}}$$

under steady-state conditions.

The difference in biomass turnover times in a sandy (9% clay) soil and a clay (40% clay) soil are shown in Table 6.1. Clay soils contain more soil organic matter and have a larger biomass (with a slower rate of respiration) than sandy soils under the same management. Thus clays can be considered to stabilize both microbial biomass and soil organic matter, although the mechanisms are not fully understood (e.g. Jenkinson & Ladd 1981). The biomass in the sandy soil had a much shorter turnover time (125 days) than that in the clay soil (383 days). Thus, the rate of transformation of fresh substrates (e.g. plant residues) and soil organic matter by the biomass in the sandy soil was several times faster than in the clay soil, presumably leading to the smaller amounts of total soil organic matter in the sandy soil (Table 6.1).

TABLE 6.1. Biomass C turnover time in a sandy and clay soil, both given NPK fertilizer

Soil	Texture	Soil organic C (%)	Estimated total annual C input (t C/ha per year)	Biomass C soil (μg C/g)	Biomass C turnover time (days)
Woburn	Sandy (9% clay)	1·10	1·1	125	125 ± 5
Northfield	Clay (40% clay)	3·39	1·9	788	383 ± 35

Although the estimated total annual C inputs to the clay soil (1·9 t C/ha were greater than to the sandy soil (1·1 t C/ha), this difference (Wu 1991) was far too small to account for the difference in total soil organic matter (1·1% C for the sandy soil and 3·4% C for the clay soil).

The measurements of biomass turnover times were done at 25 °C. This was partly to be able to make the measurements within the reasonable time period of 100 days. At lower soil temperatures (as in the field) the turnover time would obviously be longer. If we assume that the Q_{10} rule applies (i.e. there is a doubling in the rate for a 10 °C rise in temperature), this would give biomass turnover times at 10 °C of around 0·8 years for the sandy soil and 2·6 years for the clay soil. The validity of this extrapolation needs to be tested before it can be used with confidence. However, these turnover times are of the order of other turnover times estimated for the biomass under field conditions (e.g. Jenkinson & Ladd 1981). We hope to use the concept of biomass turnover time to explain accumulation rates of soil organic matter, for example during straw incorporation or conversion of arable land to grassland. Similarly, it may be helpful in understanding soil organic matter dynamics where organic matter is declining, for example when grassland is ploughed back into arable.

ACKNOWLEDGMENTS

J. Wu thanks the Chinese Government and J.A. Ocio thanks the Department of Agriculture and Fisheries of the Basque Government for their financial support. The authors also thank D.S. Jenkinson for his contribution to this research and for useful discussions.

REFERENCES

Amato, M. & Ladd, J.N. (1988). Assay for microbial biomass based on ninhydrin-reactive nitrogen in extracts of fumigated soils. *Soil Biology and Biochemistry*, **20**, 107–114.

Anderson, T.H. & Domsch, K.H. (1978). A physiological method for the quantitative measurement of microbial biomass in soil. *Soil Biology and Biochemistry*, **10**, 207–213.

Attwood, P. (1984). Straw incorporation — the background story. *Straw Incorporation, the Methods and Machines*, pp. 1–16. *Proceedings of ADAS/NAC Conference*, Stoneleigh, 16 February 1984.

Brookes, P.C., Landman, A., Pruden, G. & Jenkinson, D.S. (1985). Chloroform fumigation and the release of soil nitrogen: a rapid direct extraction method to measure microbial biomass nitrogen in soil. *Soil Biology and Biochemistry*, **17**, 837–842.

Brookes, P.C., Powlson, D.S. & Jenkinson, D.S. (1984). Phosphorus in the soil microbial biomass. *Soil Biology and Biochemistry*, **16**, 169–175.

Clarholm, M. (1985). Interactions of bacteria, protozoa and plants leading to mineralization of soil nitrogen. *Soil Biology and Biochemistry*, **17**, 181–187.

Darbyshire, J.F. (1976). Effects of water suctions on the growth in soil of the ciliate *Colpoda steinii* and the bacteria *Azotobacter chroococcum*. *Journal of Soil Science*, **27**, 369–376.

Darbyshire, J.F., Wheatley, R.E., Greaves, M.P. & Inkson, R.H.E. (1974). A rapid micromethod for estimating bacterial and protozoan populations in soil. *Revue d'Ecologie et de Biologie du Sol*, **11**, 465–475.

Elliott, E.T., Coleman, D.C. & Cole, C.V. (1979). The influence of amoebae on the uptake of nitrogen by plants in gnotobiotic soil. *The Soil–Root Interface* (Ed. by J.L. Harley & R. Scott Russell), pp. 221–229. Academic Press, London.

Inubushi, K., Brookes, P.C. & Jenkinson, D.S. (1989). A comparison of the fumigation–extraction, fumigation–incubation and ATP methods for measuring microbial biomass in waterlogged soils. *Proceedings of the 5th International Symposium on Microbial Ecology*, p. 69. Kyoto, Japan, 27 August–1 September 1989.

Jenkinson, D.S. (1977). The soil biomass. *New Zealand Soil News*, **25**, 213–218.

Jenkinson, D.S. (1985). The supply of nitrogen from the soil. *The Nitrogen Requirements of Cereals*, pp. 79–94, ADAS Conference September 1982. HMSO, London.

Jenkinson, D.S. (1988). The determination of microbial biomass carbon and nitrogen in soil. *Advances in Nitrogen Cycling in Agricultural Ecosystems* (Ed. by J.R. Wilson), pp. 368–386. CAB International, Wallingford.

Jenkinson, D.S. & Ladd, J.N. (1981). Microbial biomass in soil: measurement and turnover. *Soil Biochemistry* (Ed. by E.A. Paul, & J.N. Ladd), Vol. 5, pp. 415–471. Marcel Dekker, New York.

Jenkinson, D.S. & Oades, J.M. (1979). A method for measuring adenosine 5'-triphosphate in soil. *Soil Biology and Biochemistry* **11**, 193–199.

Jenkinson, D.S. & Powlson, D.S. (1976). The effects of biocidal treatments on metabolism in soil. V. A method for measuring soil biomass. *Soil Biology and Biochemistry*, **8**, 209–213.

Jenkinson, D.S., Fox, R.H. & Rayner, J.H. (1985). Interactions between fertilizer nitrogen and soil nitrogen — the so called 'priming effect'. *Journal of Soil Science*, **36**, 425–444.

Jenkinson, D.S., Powlson, D.S. & Wedderburn, R.W.M. (1976). The effects of biocidal treatments on metabolism in soil. III. The relationship between soil biovolume, measured by optical microscopy, and the flush of decomposition caused by fumigation. *Soil Biology and Biochemistry*, **8**, 189–202.

Johnston, A.E. (1973). The effects of ley and arable cropping systems on the amount of soil organic matter in the Rothamsted and Woburn Ley-Arable experiment. *Rothamsted Experimental Station, Report for 1972*, Part 2, pp. 131–159.

Kuikman, P.J., Van Vuuren, M.M.I. & Van Veen, J.A. (1989). Effects of soil moisture regime on predation by protozoa of bacterial biomass and the release of bacterial nitrogen. *Agricultural Ecosystems and Environment*, **27**, 271–279.

Macdonald, A.J., Powlson, D.S., Poulton, P.R. & Jenkinson, D.S. (1989). Unused fertilizer nitrogen in arable soils — its contribution to nitrate leaching. *Journal of the Science of Food and Agriculture*, **46**, 407–419.

Macdonald, R.M. (1986). Sampling soil microfloras: dispersion of soil by ion exchange and extraction of specific microorganisms from suspension by elutriation. *Soil Biology and Biochemistry*, **18**, 399–406.

MAFF (1987). *Straw Survey 1987 — England and Wales*. Government Statistical Service, Ref. No. Stats 325/87. HMSO, London.

Ocio, J.A. & Brookes, P.C. (1990). An evaluation of methods for measuring the microbial biomass in soils following recent additions of wheat straw and the characterization of the biomass that develops. *Soil Biology and Biochemistry*, **22**, 685–694.

Ocio, J.A., Brookes, P.C. & Jenkinson, D.S. (1991). Field incorporation of straw and its effects on soil microbial biomass soil inorganic N. *Soil Biology and Biochemistry*, **23**, 171–176.

Ocio, J.A., Martinez, J. & Brookes, P.C. (1991). Contribution of straw-derived N to total microbial biomass N following incorporation of cereal straw to soil. *Soil Biology and Biochemistry*, **23** (in press).

Powlson, D.S. & Jenkinson, D.S. (1976). The effects of biocidal treatments on metabolism in soil. II. Gamma irradiation, autoclaving, air-drying and fumigation. *Soil Biology and Biochemistry*, **8**, 179–188.

Powlson, D.S., Hart, P.B.S., Poulton, P.R., Johnston, A.E. & Jenkinson, D.S. (1990). The influence of soil type, crop management and weather on the recovery of ^{15}N-labelled fertilizer applied to winter wheat in spring. *Journal of Agricultural Science, Cambridge* (in press).

Saffigna, P.G., Powlson, D.S., Brookes, P.C. & Thomas, G.A. (1989). Influence of sorghum residues and tillage on soil organic matter and soil microbial biomass in an Australian vertisol. *Soil Biology and Biochemistry*, **21**, 759–765.

Schnürer, J.M., Clarholm, M., Boström, S. & Rosswall, T. (1986). Microbial biomass and activity in an agricultural soil with different organic matter contents. *Soil Biology and Biochemistry*, **17**, 611–618.

Shen Shan-Min, Brookes, P.C. & Jenkinson, D.S. (1987). Soil respiration and the measurement of microbial biomass C by the fumigation technique in fresh and in air-dried soils. *Soil Biology and Biochemistry*, **19**, 153–158.

Vance, E.D., Brookes, P.C. & Jenkinson, D.S. (1987a). An extraction method for measuring soil microbial biomass C. *Soil Biology and Biochemistry*, **19**, 703–707.

Vance, E.D., Brookes, P.C. & Jenkinson, D.S. (1987b). Microbial biomass measurements in forest soils: the use of the chloroform fumigation–incubation method in strongly acid soils. *Soil Biology and Biochemistry*, **19**, 697–702.

Wu, J. (1991). *The turnover of organic C in soil*. PhD thesis, University of Reading.

Wu, J., Brookes, P.C. & Jenkinson, D.S. (1989). Measuring the dynamics of the turnover of biomass carbon in soil. *Proceedings of the 5th International Symposium on Microbial Ecology*, p. 148. Kyoto, Japan, 27 August–1 September 1989.

7. MICROBIAL INTERACTIONS IN THE CEREAL RHIZOSPHERE

G. JAGNOW

Bundesforschungsanstalt für Landwirtschaft,
Institut für Bodenbiologie, Bundesallee 50, 3300 Braunschweig, Germany

INTRODUCTION

Living root systems of cereals provide ample substrates and habitats for a complex community of soil micro-organisms and small animals. Even dry seeds are colonized by dormant cells of bacteria and fungi, which may in some instances be inside the seed coat and make it difficult during experiments to sterilize cereal seeds. As the seeds swell after sowing, soluble compounds leak into the adjacent soil and specific communities of micro-organisms are stimulated to grow in this region, the so-called spermosphere. Subsequently, the seedlings grow and additional secondary roots soon overtake the primary roots in their rate of growth and biomass. The roots penetrate the soil by the force of elongation of the new cells. Micro-organisms from the spermosphere and the surrounding soil colonize the young epidermal and root-hair cells, utilizing root exudates and root mucilage, as well as senescent or dead epidermal, root-hair and cortical cells. The root surfaces and surrounding soil affected by root components constitute the rhizosphere.

In order to understand interactions between micro-organisms and roots in the rhizosphere we have to consider root morphology and modes of microbial colonization. We also have to look at the total mass, length, distribution and decomposition of roots during cereal development. This in turn determines the possible microbial biomass in the rhizosphere and its turnover. Root exudates play a dominant role in determining the size and composition of rhizosphere microflora. The quantities of root exudates and their constitution differ between cereal cultivars and also between different plant growth stages. The rhizosphere microflora, including their functions and interactions, the role of protozoa and small animals are discussed in this contribution. Various attempts have been made to manipulate the rhizosphere by inoculation of seeds or seed-beds with bacteria or fungi capable of colonizing roots and promoting plant growth with varying levels of success. Future prospects for microbial inoculants in biological control of root diseases and plant growth promotion are reviewed in the context of changes occurring in cereal production in Europe and other temperate regions of the world.

ROOT BIOMASS AND COMPOUNDS RELEASED BY ROOTS

The main sources of carbon in the soil are plant roots and litter. In the ecosystem of cereal fields, the carbon inputs from photosynthesis tend to compensate for losses of soil carbon due to microbial decomposition. Rhizosphere micro-organisms are the first to decompose and transform this continual flow of primary production from cereal roots. To understand the flux of carbon in the soil, detailed knowledge is required about root development, root biomass and the compounds released by cereal roots.

Root biomass

The roots of temperate cereals reach a considerable length and biomass over the growing season with the maximal length and depth in the flowering stage. Recent field estimates of root biomass for winter wheat in England were 173 and 900 kg/ha at booting and flowering stages, corresponding to total root lengths of 1·6 and 9·8 km/m^2, respectively (Vincent & Gregory 1989). At tillering, heading and grain filling of spring barley near Nottingham even higher estimates of 500, 2500 and 1300 kg/ha were made by Wood (1987), resulting in a maximal root density of about 25 km/m^2 and a 50% loss of the root mass by the grain-filling stage. Laboratory estimates of decay rates of roots by Barber & Lynch (1977) in unsterile soil were 35 mg/g per day for the same spring barley cultivar (Proctor). Taking an average root weight of 1430 kg/ha for the 84 days between tillering and grain filling, this results in a biomass input of about 4200 kg/ha. This estimate is in agreement with results obtained by Sauerbeck & Johnen (1976) with young wheat plants grown in $^{14}CO_2$ atmospheres. These latter authors concluded that the total amounts of carbon transferred to the soil during the growth of a wheat crop is three to four times larger than the root carbon left at the harvest. Estimates by various authors of root carbon transferred to sterile or unsterile soil by young cereal plants growing in $^{14}CO_2$ for 3 weeks range between 10 and 20% of root carbon (Barber & Martin 1976; Helal & Sauerbeck 1989). In a $^{14}CO_2$ pulse-labelled experiment in large containers until harvest, root losses by maize were found to be smaller. Extrapolation to field conditions of maize with a top growth of 10 t/ha would result in a total of 1230 kg/ha of organic matter released by roots and only 140 kg/ha root biomass left at harvest. Accordingly, frequent maize cultivation can lead to decreases of soil organic matter (Davenport & Thomas 1988).

Insoluble root losses

As the roots grow, root cap and epidermal cells of the elongation zone secrete a viscous heteropolysaccharide, termed the mucigel, which lubricates the passage of roots through the soil. Root mucigels are mainly heteropolysaccharides consisting of 5–8 different sugar moieties. Rovira, Foster & Martin (1979) classified the organic materials in the rhizosphere and Oades (1978) described root mucigels in detail. The ultrastructure of cereal rhizosphere was studied by Foster & Rovira (1976) and rhizosphere microhabitats are also described by Foster (1988). Epidermal cells of wheat form curved microfibrils on their surface that aid attachment to soil particles or micro-organisms (Leppard 1974). These microfibrils consist of polygalacturonic acid (Hale & Moore 1979). Root cap cells are continuously sloughed off and their turnover time is about 6–9 days in wheat and barley (Feldman 1984). Balandreau & Knowles (1978) gave an estimate of 10^4 dead cells lost from a root cap of maize per day. Even in young oat plants, 57% of their root losses consist of dead cells and mucigels and only 43% of soluble compounds when the plants are grown without micro-organisms for 4 weeks (Trofymow, Coleman & Cambardella 1987). In maize, most of the unbranched seminal and nodular roots are surrounded by soil sheaths held together by root hairs and mucigel (Vermeer & McCully 1982; Fyson *et al.* 1988). Due to their rapid growth of about 5 mm per day, the zones of elongation are only sparsely colonized by rhizosphere microorganisms with very few dead cells present. Even detached root-cap cells of maize within the mucilage of soil sheaths have been seen to exhibit active cytoplasmic streaming (Vermeer & McCully 1982).

Other root losses

The major soluble constituents in root exudates of young barley and maize are sugars and organic acids, which are released in milligram quantities per'gram of roots. Individual amino acids were limited to a few μg/g root. Using root exudate data reported for micro-organism-free barley (Barber & Gunn 1974), maize (Kraffczyk, Trolldenier & Beringer 1984) and also data of Lambert *et al.* (1987), relatively large C:N ratios of about 100 or more can be estimated. With maize, the dominant sugars are glucose, arabinose, fructose and saccharose, while fumaric and oaxalacetic acid are the dominant acids. Nevertheless, some qualitative and quantitative differences in root exudates from different cereal cultivars and plant growth stages exist.

Apart from particulate and soluble compounds released from roots, volatile compounds may originate from cereal roots. Helal & Sauerbeck (1989) showed that the 'priming effect' of ^{14}C-labelled material from maize roots on the decomposition of unlabelled soil organic matter extended to a distance of 20 mm from the root and suggested that volatile compounds such as ethylene or undefined substances were involved.

RHIZOSPHERE MICRO-ORGANISMS

The nature of the rhizosphere

The interface zone between roots and soil contains many microhabitats for rhizosphere micro-organisms, as has been clearly demonstrated in electron-micrographs (Foster & Rovira 1976; Foster 1988). These authors showed that the surfaces of young wheat roots between the root cap and the root-hair zone were colonized only sparsely with micro-organisms. By the time of flowering, however, many different types of micro-organisms were found in the rhizosphere and the outer cortical cells of the roots showed all stages of microbial decay. Single cells or small groups of bacteria were found embedded in cavities in the secondary cellulose layers of cell walls of roots. Actinomycetes were also present in the outer cortex. Lignified primary walls were more resistant to decay. Foster & Rovira (1976) found some bacterial cocci and rods as well as fungal hyphae in the cereal root cortex and the lytic action of these micro-organisms on the cell walls of older root zones was evident. The increasing bacterial colonization from younger to older parts of roots has also been studied in soil suspension cultures of 5-day-old maize seedlings, where the area colonized by bacteria increased from 4 to 20% between the root-hair formation zone and the oldest zone below the seed, with the root surface population reaching $1\cdot2 \times 10^9$/g root (Schönwitz & Ziegler 1986). On 10-day-old seminal wheat roots in soil, van Vuurde & Schippers (1980) found two maxima of bacterial colonization at root ages of 3–5 and 8–9 days with densities up to 3×10^5/cm. These dense populations were associated with areas of accelerated decomposition of epidermal cells. Newman & Bowen (1974) also discussed pattern and distribution of bacteria on root surfaces and Newman & Watson (1977), using data for substrate flow from roots and the efficiency of microbial utilization, proposed a suitable computer model for microbial abundance in the rhizosphere.

Attachment mechanisms to the root

Fluorescent pseudomonad bacteria become attached to maize roots by means of cell wall projections or fimbriae (Vesper 1987), but more specific adsorption is mediated by root lectins which bind to specific sugar components of the microbial cell walls. Such mechanisms are essential for successful host–parasite or symbiotic interactions (Whatley & Sequeira 1981). Less specific root surface compounds (agglutinins) have been shown to be involved in the bacterial–fungal interactions in the rhizosphere of vegetables (Tari & Anderson 1988). These heat-labile proteinaceous substances were able to agglutinate with *Pseudomonas putida* cells antagonistic to *Fusarium* in liquid cultures. Spontaneous agglutinin-negative mutants of *P. putida* were adsorbed much less to roots and had lost their capacity to protect the roots against *Fusarium* wilt. Chaopongpang *et al.* (1989) demonstrated similar agglutinins in rice roots, which promoted the adsorption of N_2-fixing *Klebsiella* bacteria. It is not known if agglutinins are involved, however, in the preferential microbial adsorption with temperate cereals. Certainly, some rhizosphere bacteria show more affinity for the root surface, while others colonize the intercellular spaces between cortical cells of wheat seedlings more readily (Lindberg, Granhall & Tomenius 1985).

Distribution of root surface microflora in soil

The bacteria most frequently dominant on and in roots of the major temperate cereals belong to Gram-negative genera, such as *Pseudomonas, Enterobacter, Klebsiella* and *Xanthomonas*, but some Gram-positive bacteria like *Bacillus* and coryneform species are always present. The ability of specific bacteria and fungi to colonize roots in competition with other micro-organisms in field soils may be determined by many factors. Such rhizosphere specificity has been most reliably determined by seed inoculation experiments in the field. For example, Juhnke, Mathre & Sands (1987) screened 537 isolates from wheat roots and found that these isolates varied considerably in their ability to colonize wheat roots in subsequent years. Recently, these authors showed that root colonization was largely independent of the size of the inoculum (Juhnke, Mathre & Sands 1989). Specific saprophytic fungi are also present in the cereal rhizosphere, e.g. species of *Trichoderma* and *Penicillium*, and may reach levels of biomass and colonization similar to rhizosphere bacteria (Campbell 1985).

Motile bacteria may be attracted to plant roots in soil by chemotaxis and may in some instances be specific for roots of one plant species. For example, Mandimba *et al.* (1986) found that strains of *Azospirillum lipoferum* and *Enterobacter cloaceae* isolated from the maize rhizosphere, were attracted by specific components of the maize mucigel, but strains of the same species of bacteria isolated from rice roots were not. Differences in colonization patterns in different areas of roots, however, are controlled by other mechanisms. Howie, Cook & Weller (1987) found that motile and non-motile strains of *P. fluorescens* bacteria were distributed on the root with progressively lower densities with increasing distances from the point of inoculation irrespective of the motility of the bacterial strains. Distribution apparently occurred by passive transport on growing roots from the source of inoculation. A wide range of bacteria colonize cereal roots in soil in discontinuous patches. The distribution would seem to be a much more random process than expected from present knowledge. The distribution of root-surface bacteria amongst maize plants in a field was studied in detail by Lambert *et al.* (1987). The proteins of 1366 isolates were characterized and separated into 236 different protein patterns, which are probably equivalent to bacterial species or sub-species. Lambert *et al.* (1987) found that most of these protein groups occurred very infrequently on the plants and most could not be completely identified by current systematics. Clearly, we need to know far more about the environment of the rhizosphere at the microscopic level, large-scale improvements are required in microbial systematics and more information is required about the physiological needs of rhizosphere micro-organisms before we can predict rhizosphere colonization in soil.

Environmental factors affecting root colonization by micro-organisms

The composition of cereal rhizosphere microflora is also dependent on plant age, physical factors and microbial interactions. Marked shifts in the predominance of bacterial groups in the wheat rhizosphere between tillering and flowering were reported recently (Ruppel 1988). Patterns of bacterial distribution on developing roots were also described by Newman & Bowen (1974) and Lawley, Campbell & Newman (1983). Rhizosphere microbial populations on wheat roots were markedly different at 15 and 32 °C with more pseudomonad bacteria present at the lower temperature. In contrast, the largest bacterial numbers in root-free soil were found at the higher temperature (Rouatt, Peterson & Katznelson 1963). This inverse temperature effect in the rhizosphere is likely to be a repercussion of the effect of temperature on root exudation. Organic manuring (Deavin, Horsgood & Rusch 1981) and foliar N applications (Vrany & Macura 1971) also increase the density of rhizosphere microflora. Turner, Newman & Campbell (1985) showed shifts of microbial populations in ryegrass rhizosphere due to different concentrations of nitrogen and phosphorus. Potassium fertilization of the soil decreases the leakage of carbohydrates from roots and reduces the numbers of rhizosphere micro-organisms in the wheat rhizosphere (Trolldenier 1971). In the densely colonized rhizosphere, microbial interactions are likely to be more intense than in root-free soil. Microbial antagonisms between actinomycetes, fungi and bacteria are common, e.g. the destruction of root pathogens by *Trichoderma* spp. and the importance of predation of rhizosphere bacteria and fungi by protozoa (Clarholm 1981, 1989), nematodes and small arthropods is becoming more widely appreciated (Hunt *et al.* 1987). Predator–prey interactions are also common between certain members of the micro- and mesofauna. Predatory bacteria, such as *Bdellovibrio* (Stolp & Starr 1963) and bacteriolytic bacteria (Casida 1988), are known to attack specific groups of rhizosphere bacteria. Furthermore, bacteriophages can specifically suppress the populations of host bacteria to low levels (Germida 1986). Genetic changes and plasmid transfer by conjugation have also been shown to occur in wheat rhizosphere (van Elsas, Trevors & Starodub 1988), contributing to the image of the rhizosphere as a region of marked change and interaction.

Nitrogenous transformations in the rhizosphere

Plant nutrient release and immobilization are important activities of the rhizosphere community. Nitrogen and phosphorus transformations are particularly enhanced in the rhizosphere. Release of NH_4^+—N from dead cells or soil organic matter is generally accelerated in the root zone, but ammonium oxidation to nitrite and nitrate by the autotrophic bacteria, *Nitrosomonas* and *Nitrobacter* spp., may be reduced by the excretion of inhibitory substances from wheat roots (Moore & Waid 1971). Also, maize roots contain compounds inhibitory to *Nitrosomonas*, but not to *Nitrobacter* spp. (Molina & Rovira 1964). A slight inhibition of nitrification may be beneficial by retarding or preventing nitrate loss from leaching, if this is not counteracted by plant uptake. Nitrogen-fixing bacteria are widespread in cereal rhizosphere, among them *Enterobacter, Bacillus* and *Azospirillum* spp. However, many [15]N- and acetylene reduction based estimates of fixed nitrogen show that N inputs to temperate cereal ecosystems by rhizosphere bacteria are negligible and are more useful as a strategy for bacterial survival than as N source for cereal nutrition (Lethbridge & Davidson 1983; Lamb, Doran & Peterson 1987; Jagnow 1989).

Many rhizosphere bacteria use dissimilatory nitrate reduction for respiration, if O_2 concentrations are limiting. The reduced molecules, N_2O or N_2, are lost to the atmosphere (Haller & Stolp 1985). This loss of soil or fertilizer N was measured in wheat and barley fields near Braunschweig in different years using the acetylene inhibition technique to recover denitrification products as N_2O. The resulting estimates for cereals receiving 150 kg fertilizer N/ha were only 9–15 kg/ha or 6–10% of fertilizer N, which is far less than previously indicated by N or ^{15}N balances (Benckiser et al. 1987; Schneider, Haider & Mahmoud 1990). In fact, denitrifiers do not compete with crop plants for nitrate. Christensen & Tiedje (1988) showed that NO_3 uptake by barley stops at 1·6% O_2 saturation from depressed respiration. Denitrification by Pseudomonas was already completely inhibited at 0·04% O_2-saturation and reached only 10% of its maximal anaerobic rate at 0·01% O_2-saturation. Plant, rhizosphere and micro-organisms, however, must be viewed as a functional entity. According to model simulations of Robinson et al. (1989), the N mineralization–immobilization sequences in the rhizosphere that are initiated by the consumption of dead root cells and root exudates, ultimately result in a higher uptake of mineral N by the roots than would take place in the absence of these processes. Activation of rhizosphere micro-organisms by available plant carbon also leads to an increased N mineralization of soil organic matter, depending on its C:N ratio and relative availability. Predation of bacteria and fungi by protozoa and nematodes also leads to an intensified N mineralization by a quicker release of mineral N from consumed microbes and from predator cells (Hunt et al. 1987; Hendrix et al. 1989). Since nitrogen is more often a limiting plant nutrient than carbon, which is freely available from photosynthesis, plants may benefit even if a relatively high proportion of root carbon is sacrificed for an increased return of available nitrogen (Robinson et al. 1989).

Phosphorus transformations in the rhizosphere

Phosphorus would be a limiting plant nutrient to cereals in temperate regions, if it was not supplied regularly by super or rock phosphate fertilizers. However, due to the continuous immobilization of phosphorus in Ca, Fe or Al phosphates with extremely low solubility and low availability, the soil solution contains very low concentrations of 0·1–1 p.p.m. phosphorus (Paul & Clarke 1988). Consequently, diffusion alone often cannot satisfy plant demands and again rhizosphere micro-organisms play a vital function for the plant root. Phosphorus uptake rates of 1–2 µmol/g of root per day are required for adequate supply during rapid plant growth of 7–10% per day. This flux needs a turnover of dissolved P 50–250 times per day and rhizosphere micro-organisms contribute to this by accelerated solubilization of inorganic and organic phosphorus fractions (Paul & Clarke 1988). Solubilization of phosphorus in the rhizosphere of maize was demonstrated by Helal & Sauerbeck (1984). Inorganic P is mainly solubilized by microbial production of bicarbonate anions from respiration and by organic acids of the citric acid cycle, gluconic, lactic and other hydroxy-acids, acting as chelators mainly on calcium phosphates (Swaby & Sperber 1959; Duff, Webley & Scott 1963). Organic P fractions normally constitute 30–50% of total soil phosphorus. They are mainly phytates, nucleic acids, nucleotides and phospholipids. Phytin is inositol hexaphosphate formed by both plants and micro-organisms. It is the most stable fraction and may range between 3 and 52% of the total organic phosphorus in different soils. Microbial phosphatases are probably responsible for converting phytin into the lower substituted inositol mono- to pentaphosphates, which normally occur together with the hexaphosphate (Greaves, Vaughan & Webley 1967). The chemical nature of about half of the organic phosphorus in soils is unknown (Paul & Clarke 1988). Nucleic acids and nucleotides constitute only about 1% (Greaves, Vaughan & Webley 1970) and phospholipids from 1 to 2% of the organic phosphorus. Beneficial effects of inoculation with the so-called 'phosphobacteria', which have been claimed to increase P mineralization, have never been demonstrated conclusively with cereals (Smith, Allison & Soulides 1961). Solubilization of mineral phosphates by formation of the above-mentioned acids frequently has been shown in pure cultures, but whether these are formed to any extent in soil is uncertain. Nevertheless organic phosphates can be readily decomposed by the majority of rhizosphere micro-organisms (Paul & Clarke 1988). Therefore, screening soil micro-organisms for the ability to solubilize phosphorus rapidly for subsequent use in inoculation trials, does not seem to be a promising research initiative. In contrast, symbiotic VA mycorrhizal fungi, which are frequently very active in improving phosphorus nutrition of cereals, may prove more rewarding for use as inocula.

GROWTH-PROMOTING CAPABILITIES OF RHIZOSPHERE MICRO-ORGANISMS

Numerous research workers within the last decade have investigated how rhizosphere micro-organisms can promote crop growth and plant health. This is a large and expanding interdisciplinary research topic, involving laboratory, glasshouse and field inoculation experiments, plant physiology, plant pathology and molecular genetic analysis. Only a few examples of these experiments can be briefly described.

Results from cereal crops

Most of the field inoculation experiments with cereals have been carried out with *Azospirillum* spp. in subtropical or tropical climates with the expectation that N_2 fixation by these bacteria would increase crop yields. Cereal inoculation experiments with *Azotobacter* have also been reported, mainly from India, with marginal responses in grain yield and half as much as obtained with *Azospirillum* spp. Earlier inoculation experiments in Russia provided inconsistent responses with *Azotobacter* spp. (Mishustin 1966). The inconsistency of these responses to inoculants with increasing levels of N fertilizers suggests that N_2 fixation is not involved in these yield increases (Jagnow 1987), because N_2 fixation is repressed by mineral nitrogen. Out of fifteen field experiments, each with several levels of N fertilizer, eight showed an increase and seven a decrease in grain yield with increments of N fertilizer. Furthermore, straw and grain yields sometimes responded to N increments in dissimilar ways. Cereal yield increases obtained by inoculating seeds with different rhizosphere bacteria in European soils are summarized in Table 7.1. Apart from the example with bacterial antagonists to take-all fungus, harmful micro-organisms do not seem to be involved in these growth enhancements and phytohormones are the most probable agents (Brown 1975). The increases in wheat yields after inoculation with *Rhizobium* spp. (Table 7.1) seem to be surprising, but were also reported recently with wheat from Indian and Egyptian field experiments (Kavimandan 1986; Ishac 1989). Furthermore, high *Rhizobium* populations in soil under continuous wheat monoculture for more than 150 years suggest that rhizobia survive and multiply also in the cereal soils (Werner 1987).

Growth promotion by phytohormones

Cereals inoculated with *Azospirillum* spp. produce more root hairs and lateral roots than control plants (Okon 1984). Pure cultures of *A. brasilense* during the early stationary phase also produced the plant growth hormone indoleacetic acid (IAA) without amendment with tryptophane (Horemans & Vlassak 1985). Kolb & Martin (1985) found that a mutant of *A. brasilense* producing excessive amounts of IAA also increased root extension and lateral root production in unsterile soil. Using monoclonal anti-

TABLE 7.1. Yield increases of temperate cereals after bacterial inoculation

Crops	Bacteria	Soil	Country	Yield increase (%)	Reference
Winter wheat	*Azospirillum brasilense*	Albic luvisol	Belgium	15	Reynders & Vlassak (1982)
Spring wheat	*Azospirillum lipoferum*	Gleysol	FRG	22	Mertens & Hess (1984)
Spring wheat	*Agrobacterium tumefaciens*	Leptic podzol	GDR	9	Ruppel (1987)
Spring barley	*Azospirillum brasilense*	Gleysol	USSR	8–16	Vasjuk et al. (1989)
Spring barley	*Azospirillum lipoferum*	Gleysol	USSR	21–41	Vasjuk et al. (1989)
Spring barley	*Rhizobium trifolii*	Albic luvisol	GDR	6	Höflich (1989)
Maize	*Azotobacter chroococcum*	Calcaric fluvisol	Spain	10	Martinez-Toledo et al. (1988)
Spring wheat	*Bacillus pumilus*	Loamy humisol	England	105	Capper & Campbell (1986)

bodies in an enzyme immunoassay, Müller, Deigele & Ziegler (1989) showed that IAA, abscisic acid and three cytokinins were the major phytohormones produced by pure and mixed cultures of rhizosphere bacteria from maize. Bacterial cultures with populations comparable to those in the maize rhizosphere produced these phytohormones in similar concentrations to those detected in the rhizosphere of maize crops. Inoculation of maize seedlings growing in mineral nutrient solution with *Azotobacter chroococcum* or *Pseudomonas fluorescens* resulted in synergistic increases of hormone concentrations above the sums of hormone production from bacterial cultures and sterile plants. The inhibitory effects of rhizosphere pseudomonads on growth of sterile barley seedlings reported by Lynch & Clarke (1984) may have been partly due to inhibitory IAA concentrations. Gibberellin production by *A. brasilense* has also been reported (Umali-Garcia *et al.* 1980). Inbal & Feldman (1982) transformed a dwarf mutant of wheat into a normal-size plant by *Azospirillum* inoculation. O'Hara, Davey & Lucas (1981) also increased the growth of maize by inoculation with a nitrogenase-negative *Azospirillum* mutant.

Recently, Zimmer, Roeben & Bothe (1988) reported stimulatory effects with nitrite on root elongation of wheat seedlings similar to IAA amendments. Apparent differences reported in $^{15}NO_3$ uptake after inoculation with *A. brasilense* or a N_2-fixing *Bacillus* sp. between two Canadian wheat cultivars were also shown to be due to phytohormonal effects (Kucey 1988). Phytohormonal interactions with *Azospirillum* inoculants can affect the development of wheat seedlings in the first 3 weeks, resulting in an increased root mass and root surface area, an increased number of fertile tillers and an accelerated heading and flowering (Kapulnik *et al.* 1981; Kapulnik, Gafny & Okon 1985). Improvements of crop yield can also result from increasing the rooting depth after inoculation, if the crop is liable to suffer from drought (Sarig, Blum & Okon 1988). Root metabolism as well as root development may be affected by bacterial inoculation of maize plants. Specific activities of acid phosphatase, glutamine synthase, isocitrate, malate and shikimate dehydrogenase and pyruvate kinase were increased by *A. brasilense* inoculants, but not those enzymes normally stimulated in response to microbial infections, phenylalanine-ammonia-lyase and glucose-6-phosphate dehydrogenase (Fallik, Okon & Fischer 1988). With *Azospirillum*-inoculated maize roots, the uptake of nitrate, potassium and phosphate was increased by 30–50% after 2 weeks. Thin sections of the inoculated roots showed a loosening of the outer four to five layers of cortical cells and Lin, Okon & Hardy (1983) suggested that the increased intercellular free space encouraged nutrient uptake.

Growth promotion by suppression of root pathogens
or rhizosphere saprophytes

Biocontrol of root diseases is a rapidly expanding field. Recent reviews dealing with bacteria were given by Weller (1988) and Kloepper, Lifshitz & Zablotowicz (1989). Similar antagonistic fungi were reviewed by Baker (1989). Apart from soilborne root pathogenic fungi, crop growth can be reduced by rhizosphere bacteria producing toxic compounds, e.g. Enterobacteriaceae, *Arthrobacter* and *Pseudomonas* spp. (Suslow & Schroth 1982). Non-fluorescent pseudomonads produce a toxic polypeptide, which retards the growth of wheat roots. When inoculated on to non-sterile wheat straw, these bacteria multiplied up to 10^{10}/g within 6 days and became the dominant bacteria, but only at low temperatures of $5-15\,°C$. When inoculated on to straw in the field, the same bacteria maintained high populations during winter with a decline in the spring, but multiplication was resumed with the start of crop growth. No-till plots had ten-fold higher populations than ploughed treatments (Elliott & Lynch 1985; Stroo, Elliott & Papendick 1988). Some *Pseudomonas* species from the wheat rhizosphere are able to inhibit root growth of host plants by the production of cyanide (Aström & Gerhardson 1989). Probably the most significant root disease of temperate cereals is take-all caused by the fungus *Gaeumannomyces graminis* and *Rhizoctonia solani*. Infected root and straw residues persist during winter to attack the following crop and the severity of take-all increases with the number of preceding susceptible crops. Sumner & Minton (1989) give examples from maize field in Georgia, USA, where plots with low and high inoculum levels of *R. solani* suffered percentage yield losses of 15, 19, 1 and 47, 42, 8, respectively, during 3 consecutive years in comparison with uninfected plots. The declining severity of yield losses is probably indicative of the gradual increase of natural microbial antagonists, which may convert disease-conducive soils into disease-suppressive soils for a particular root pathogen. This was shown also by artificial cycles of infection of wheat with *G. graminis* in a glasshouse study by Charigkapakorn & Sivasithamparam (1987). The severity of disease produced by the inocula decreased significantly with each generation. Root disease index fell from 100% in the first generation to 80, 50 and 8% in the second, third and fourth generations, respectively. Diseased roots had higher populations of fluorescent pseudomonads and 33% of them from the second generation of infection were able to inhibit *G. graminis in vitro*.

Antagonists to fungal pathogens and growth-inhibiting bacteria can employ different strategies of suppression. These may involve either substrate competition, competition for essential elements, antibiotics, lysis and hyperparasitism or combinations of these. Substrate competition is most likely to succeed when the competitors utilize the same substrates, e.g. when a non-toxic mutant of the detrimental strain is involved. This type is one of the commonest interactions observed in biocontrol phenomena. Some soils of neutral or alkaline pH have a very low availability of soluble iron and competition for this essential element involves the production of specific compounds (siderophores) with a high affinity for iron. These compounds make the iron unavailable to the fungal pathogen or the toxigenic bacterium. Siderophores from fluorescent pseudomonads have a high affinity for iron and several examples of biocontrol of root pathogens by such action are quoted by Weller (1988). A major role in the suppression of root pathogenic fungi is also played by antibiotics produced by bacterial antagonists. For instance, an effective biocontrol strain of *Pseudomonas fluorescens* is active against *G. graminis*, because of its phenazine-type antibiotic suppressing the fungus at concentrations less than 1 µg/ml. Mutants without antibiotic production lost their disease suppressiveness (Weller & Cook 1983; Thomashow & Weller 1988). Other strains of *P. fluorescens* produced the antibiotic pyoluteorin effective against *Pythium ultimum* (Howell & Stipanovic 1980). Another example of bacterial antagonism may be the lysis of chitinaceous fungal cell walls, e.g. *Enterobacter cloacae* lysing hyphae of the root-colonizing fungus *Pythium ultimum* (Nelson *et al.* 1986). Good examples of microbial hyperparasitism in the biocontrol of fungal root pathogens are provided by strains of the soil and rhizosphere fungi *Trichoderma viride* and *T. harzianum*, whose delicate hyphae parasitize the large hyphae of root pathogens like *Pythium* or *Rhizoctonia* spp. (Baker 1989). Pseudomonads, most frequently active against pathogens by combinations of these strategies, may also promote plant growth in the absence of pathogens under sterile conditions by hormonal interactions (Elad, Chet & Baker 1987).

Benefits of vesicular–arbuscular mycorrhizae

Symbiotic infection with a special group of Zygomycete fungi, which form vesicular–arbuscular mycorrhizae (VAM) frequently occurs with all cereals and most other crops. For details of this increasingly important topic see the reviews of Harley & Smith (1983) and Safir (1987). Unlike other symbiotic fungus–root associations, no visible changes in root morphology develop after infection. Only after clearing and staining the roots can the extent of VAM infections in the roots be observed microscopically.

These fungi require a living host plant and cannot at present be grown in pure cultures. Depending on their inoculum potential, VAM fungi invade varying proportions of roots of several host plants in the intercellular spaces of the root cortex. Cortical cells are invaded and filled by extensively branched haustoria through which the fungus is supplied with carbohydrates. The plant in turn receives mineral nutrients, mainly phosphorus from inorganic and organic fractions of the soil inaccessible to roots, because the fungal hyphae can penetrate micropores in soil aggregates inaccessible to plant roots (Powell & Daniel 1978; Reid 1984; Pacovsky, Bethlenfalvay & Paul 1986). Although crops benefit most from VAM symbiosis in soils with low concentrations of available P, other beneficial effects of this symbiosis may result in increased growth and health of cereal crops. Fungal root pathogens can be reduced in VA mycorrhizal crops by competitive exclusion of the fungi and by the stimulation of plant defence reactions such as intensified phenol and lignin synthesis (Dehne & Schönbeck 1979; Dehne 1982). Several years of field trials with winter wheat and maize have shown that inoculation with efficient VAM fungal spores grown on roots in a mineral substrate consistently increased yields and % VAM infection of roots. These effects with winter wheat were most noticeable in patches of the crops under temporary waterlogging. The state of health of wheat roots at flowering was markedly improved by VAM infection. While the percentage of healthy, senescent and diseased roots was 9, 28 and 63%, respectively, with non-mycorrhizal plants, it was 30, 55 and only 15% with mycorrhizal plants (Dehne 1986). VAM infection of maize can be increased by simultaneous *Azospirillum* inoculation by 20%. Such dual inoculated plants had similar size, N content and a higher P content than uninoculated plants given N + P fertilizer (Barea, Bonis & Olivares 1983). A similar, very strong interaction of the same symbionts was also reported with wheat in India (Subba Rao, Tilak & Singh 1985). *Glomus* inoculation increased the Cu and Zn contents of *Sorghum* (Pacovsky 1988). Increased root length and improved water status was also found after VAM inoculation of *Sorghum* growing under drought stress (Sieverding 1986).

THE INFLUENCE OF CHANGING FARMING PRACTICES ON RHIZOSPHERE MICRO-ORGANISMS

Up to now, most of the knowledge about the capabilities of cereal rhizosphere micro-organisms has been obtained from intensive, conventional agriculture, which uses mouldboard ploughing, restricted rotations involving many cereal crops, liberal applications of mineral N fertilizers, little or no use of organic or green manuring and large applications of pesticides. However, the overproduction of grain in Europe, the low economic return, the reduction in subsidies from taxpayers and not least the growing environmental destruction have encouraged the transposition of existing agriculture into organic or integrated farming systems (Murphy, this volume, pp. 69–91). Integrated farming involves less tillage, less N fertilizer to produce c. 80% of maximal yields, the inclusion of organic and green manures, reintegration of animal husbandry into crop production and the limited use of pesticides to prevent impending economic losses from pests and diseases (e.g. El Titi, this volume, pp. 399–411). Although some experience already exists with respect to root diseases, the likely effects of integrated farming on cereal rhizosphere micro-organisms remains speculative at present. Nevertheless, some attempt will be made to predict the effects of changes in crop rotation, fertilizer use, soil tillage and pesticide use on the mineralization of nutrients, the incidence of root diseases, on growth-promoting and growth-inhibiting bacteria, on VAM infection and on prospects of successful seed and VAM inoculation to promote cereal growth.

The numbers of root infections with take-all fungus are strongly influenced by the sequence of preceding crops, as discussed earlier. No yield reductions were observed when the previous crops were unsuitable as hosts for the fungus, but even in these situations a small infection rate of 5–20% of plants prevailed in some instances. If the previous crop was a host plant, plant infections of 70–100% could result in yield reductions of cereals up to 50% and this was more severe when the two preceding crops were host plants (Steinbrenner & Obenauf 1986). Fifteen years of continuous winter wheat monoculture in Bavaria resulted in yield reductions of 11–19% from reduced tillering and grain weight and in a 58% higher incidence of root diseases compared with wheat in a 3-year crop rotation. Wheat root mass was reduced by 50% on continuous culture plots, possibly indicating higher activities of growth-inhibitory micro-organisms. Moreover, the fraction of soil organic carbon present as microbial biomass carbon was only 2·64% with continuous culture, while it was 3·04% in crop rotation (Pommer, Beck & Borchert 1989). Crop residues from different crops seem to be used by a greater variety of micro-organisms more efficiently for cell synthesis than those from a single crop (Anderson & Domsch 1989). The incidence and degree of infection of winter barley with symbiotic VAM fungi, on the other hand, was not affected by continuous culture compared with barley in rotation with winter wheat, rye and oats (Baltruschat & Dehne 1989). Therefore, it can be expected that reduction of cereal cropping frequency with integrated crop production will enhance natural microbial disease control and improve crop health and yields.

Reduced N fertilization will have important consequences for rhizosphere micro-organisms as well. Since spores of parasitic fungi need mineral soil N to germinate and grow far enough to infect a host root, decreased N fertilization and N immobilization by decomposing crop residues will decrease the inoculum potential of pathogens and incidence of root disease. The level of VAM mycorrhiza infection also responds favourably. Increasing N fertilizer from 100 to 200 kg/ha or using an annual application of green manure decreased VAM colonization of winter barley (Baltruschat & Dehne 1989). The effects of mineral N fertilizers on natural VAM infection of winter wheat were influenced by soil texture. While 180 kg/ha N reduced VAM root infection from 52% without N to 22% on a sandy soil, 200 kg/ha N fertilizer had no detrimental effect on a heavy loam soil, where root length of fertilized plants was increased from 28 to 64 m/1000 ml soil and VAM infection remained unimpaired at 61% (Dehne 1986). Reduced applications of N fertilizer may decrease the production of root-inhibitory microbial secondary metabolites, which are often products of unbalanced N metabolism. Zvyagintsev & Guzev (1987) detected increasing numbers of fungi and bacteria with toxic secondary metabolites when soils were amended with increasing doses of mineral N. Microbial interactions are also more likely to control detrimental micro-organisms at suboptimal N concentrations. Seed inoculation with growth promoting rhizosphere bacteria is more likely to enhance crop growth, when yields have not reached the maxima possible with N fertilizers. For instance, green manuring or straw incorporation altered rhizosphere colonization of wheat and rye. In these situations, the rhizosphere contained higher proportions of fluorescent pseudomonads and more *Bacillus* spp. than crops given mineral N fertilizer only (Höflich & Steinbrenner 1988). Thus, integrated agriculture probably will allow a better use of growth-promoting and/or disease-controlling microbial inoculants.

A likely agricultural trend in the future is more direct drilling of seeds by reduced or zero tillage. In the absence of soil mixing by mouldboard ploughing, the crop residues, soil organic matter, plant nutrients, moisture and crop roots will be concentrated in the upper 10 cm of soil. No-till cereal soils have about 40% more aerobic micro-organisms and twice as many facultatively anaerobic and denitrifying bacteria as well as larger soil enzymatic activities than ploughed soils (Doran 1980). Large water contents and more frequent anaerobiosis of no-till cereal soils resulted in small nitrifier populations, cool soil temperatures and reduced mineralizable N (Broder *et al.* 1984). Field denitrification rates from no-till maize fields were always higher than from ploughed fields, sometimes up to seventy-seven times (Rice & Smith 1982). The effects of soil cultivation on disease incidence and microbial inoculants are hard to predict, because decreased soil cultivation also favours the soil micro- and mesofauna. The increased faunal–microfloral interactions would probably increase turnover and mineralization of microbial biomass and may aid in controlling soil pathogens. In a microbial survey of cereal soils with different incidences of take-all, Feest & Campbell (1986) found a negative correlation between numbers of dictyostelid slime moulds and giant amoebae with the disease index of take-all.

Another feature of integrated agriculture is the reduced use of pesticides. There are only a few examples of significant pesticide side effects on rhizosphere microbes when the pesticides are used at recommended rates of application. Nitrifiers and rhizobia are relatively susceptible, but most active ingredients of pesticides are degraded by some soil micro-organisms and this will limit their persistence in the soil (Domsch, Jagnow & Anderson 1983). There are a few examples of beneficial effects of fungicides on microbial interactions with cereal roots. The control of powdery mildew in winter wheat, for instance, can prevent reductions of VAM root infection, root length and yield. Plots without fungicide control of mildew had reductions of 80, 50 and 30%, respectively, in VAM infection, root length and yield (Dehne 1986). For root disease control, antagonists selected for resistance against the fungicide benomyl may be used, such as resistant, avirulent *Gaeumannomyces graminis* strains against take-all (Baker 1989).

Many promising new research topics can be suggested to improve our manipulations of the cereal rhizosphere, such as the mechanisms for controlling root colonization, growth promotion and disease control. The improvement of VAM inoculants also offers great promise, if techniques can be integrated with agricultural practice. The smaller and versatile companies, who are alert to new developments in soil biotechnology, should be well placed to provide such improvements.

REFERENCES

Anderson, T.H. & Domsch, K.H. (1989). Ratios of microbial biomass carbon to total organic carbon in arable soils. *Soil Biology and Biochemistry*, **21**, 471–479.

Aström, B. & Gerhardson, B. (1989). Wheat cultivar reactions to deleterious rhizosphere bacteria under gnotobiotic conditions. *Plant and Soil*, **117**, 157–165.

Baker, R. (1989). Improved *Trichoderma* spp. for promoting crop productivity. *Trends in Biotechnology*, **7**, 34–38.

Balandreau, J. & Knowles, R. (1978). The rhizosphere. *Interactions between Nonpathogenic Microorganisms and Plants* (Ed. by Y.R. Dommergues & S.V. Krupa), pp. 243–268. Elsevier, Amsterdam.

Baltruschat, H. & Dehne, H.W. (1989). The occurrence of vesicular-arbuscular mycorrhiza in agro-ecosystems II. Influence of nitrogen fertilization and green manure in continuous monoculture and in crop rotation on the inoculum potential of winter barley. *Plant and Soil*, **113**, 251–256.

Barber, D.A. & Gunn, K.E. (1974). The effect of mechanical forces on the exudation of organic substances by the roots of cereal plants grown under sterile conditions. *New Phytologist*, **73**, 39–45.

Barber, D.A. & Lynch, J.M. (1977). Microbial growth in the rhizosphere. *Soil Biology and Biochemistry*, **9**, 305–308.

Barber, D.A. & Martin, J.K. (1976). The release of organic substances by cereal roots into soil. *New Phytologist*, **76**, 69–80.

Barea, J.M., Bonis, A.F. & Olivares, J. (1983). Interactions between *Azospirillum* and VA mycorrhiza and their effects on growth and nutrition of maize and ryegrass. *Soil Biology and Biochemistry*, **15**, 705–709.

Benckiser, G., Gaus, G., Syring, K.M., Haider, K. & Sauerbeck, D. (1987). Denitrification losses from an inceptisol field treated with mineral fertilizer or sewage sludge. *Zeitschrift für Pflanzenernährung und Bodenkunde*, **150**, 241–248.

Broder, M.W., Doran, J.W., Peterson, G.A. & Fenster, C.R. (1984). Fallow tillage influence on spring populations of soil nitrifiers, denitrifiers, and available nitrogen. *Soil Science Society of America Journal*, **48**, 1060–1067.

Brown, M.E. (1975). Rhizosphere microorganisms — opportunists, bandits or benefactors. *Soil Microbiology — a Critical Review* (Ed. by N. Walker), pp. 21–38. Butterworths, London.

Campbell, R. (1985). *Plant Microbiology.* Blackwell Scientific Publications, Oxford.

Capper, A.L. & Campbell, R. (1986). The effect of artificially inoculated antagonistic bacteria on the prevalence of take-all disease of wheat in field experiments. *Journal of Applied Bacteriology*, **60**, 155–160.

Casida, L.E. (1988). Minireview: Nonobligate bacterial predation of bacteria in soil. *Microbial Ecology*, **15**, 1–8.

Chaopongpang, S., Limpananont, J., Chaisiri, P. & Boonjawat, J. (1989). Adhesion of nitrogen-fixing *Klebsiella* on the root surface of rice via lectin. *Proceedings of the 5th International Symposium on Microbial Ecology*, Kyoto, Japan.

Charigkapakorn, N. & Sivasithamparam, K. (1987). Changes in the composition and population of fluorescent pseudomonads on wheat roots inoculated with successive generations of root-piece inoculum on the take-all fungus. *Phytopathology*, **77**, 1002–1007.

Christensen, S. & Tiedje, J.M. (1988). Oxygen control prevents denitrifiers and barley roots from directly competing for nitrate. *FEMS Microbiology Ecology*, **53**, 217–221.

Clarholm, M. (1981). Protozoan grazing of bacteria in soil — impact and importance. *Microbial Ecology*, **7**, 343–350.

Clarholm, M. (1989). Effects of plant–bacterial–amoebal interactions on plant uptake of nitrogen under field conditions. *Biology and Fertility of Soils*, **8**, 373–378.

Davenport, J.R. & Thomas, R.L. (1988). Carbon partitioning and rhizo-deposition in corn and bromegrass. *Canadian Journal of Soil Science*, **68**, 693–701.

Deavin, A., Horsgood, R.K. & Rusch, V. (1981). Rhizosphere microflora in relation to soil conditions. II: Rhizosphere and soil 'coliform bacteria'. *Zentralblatt für Bakteriologie II*, **136**, 619–627.

Dehne, H.W. (1982). Interactions between vesicular-arbuscular mycorrhizal fungi and plant pathogens. *Phytopathology*, **72**, 1115–1119.

Dehne, H.W. (1986). Improvement of the VA mycorrhiza status in agriculture and horticulture. *Transactions of the 13th Congress of the International Society of Soil Science*, Hamburg, **VI**, 817–825.

Dehne, H.W. & Schönbeck, F. (1979). Untersuchungen zum Einfluss der endotrophen Mycorrhiza auf Pflanzenkrankheiten, II. Fremdstoffwechsel und Lignifizierung. *Phytopathologische Zeitschrift*, **95**, 210–216.

Domsch, K.H., Jagnow, G. & Anderson, T.H. (1983). An ecological conception for the assessment of side-effects of agrochemicals on soil micro-organisms. *Residue Reviews*, **86**, 65–105.

Doran, J.W. (1980). Soil microbial and biochemical changes associated with reduced tillage. *Soil Science Society of America Journal*, **44**, 765–771.

Duff, R.B., Webley, D.M. & Scott, R.O. (1963). Solubilisation of minerals and related materials by 2-ketogluconic acid-producing bacteria. *Soil Science*, **95**, 105–114.

Elad, Y., Chet, I. & Baker, R. (1987). Increased growth response of plants induced by rhizobacteria antagonistic to soilborne pathogenic fungi. *Plant and Soil*, **98**, 325–330.

Elliott, L.F. & Lynch, J.M. (1985). Plant growth-inhibitory pseudomonads colonizing winter wheat (*Triticum aestivum* L.) roots. *Plant and Soil*, **84**, 57–65.

Fallik, E., Okon, Y. & Fischer, M. (1988). The effect of *Azospirillum brasilense* inoculation on metabolic enzyme activity in maize root seedlings. *Symbiosis*, **6**, 17–28.

Feest, A. & Campbell, R. (1986). The microbiology of soils under successive wheat crops in relation to take-all disease. *FEMS Microbiology Ecology*, **38**, 99–111.

Feldman, L.J. (1984). Regulation of root development. *Annual Review of Plant Physiology*, **35**, 223–342.

Foster, R.C. (1988). Microenvironments of soil microorganisms. *Biology and Fertility of Soils*, **6**, 189–203.

Foster, R.C. & Rovira, A.D. (1976). Ultrastructure of wheat rhizosphere. *New Phytologist*, **76**, 343–352.

Fyson, A., Kerr, P., Lott, J.N.A. & Oaks, A. (1988). The structure of the rhizosphere of maize seedling roots, a cryogenic scanning electron microscopy study. *Canadian Journal of Botany*, **66**, 2431–2435.

Germida, J.J. (1986). Population dynamics of *Azospirillum brasilense* and its bacteriophage in soil. *Plant and Soil*, **90**, 117–128.

Greaves, M.P., Vaughan, D. & Webley, D.M. (1967). The hydrolysis of inositol phosphates by *Aerobacter aerogenes*. *Biochimica Biophysica Acta*, **132**, 412–418.

Greaves, M.P., Vaughan, D. & Webley, D.M. (1970). The degradation of nucleic acids by *Cytophaga johnsonii*. *Journal of Applied Bacteriology*, **33**, 380–389.

Hale, M.G. & Moore, L.D. (1979). Factors affecting root exudation II: 1970–1978. *Advances in Agronomy*, **31**, 93–124.

Haller, H. & Stolp, H. (1985). Quantitative estimation of root exudation of maize plants. *Plant and Soil*, **86**, 207–216.

Harley, J.L. & Smith, S.E. (1983). *Mycorrhizal Symbioses*. Academic Press, London.

Helal, H.M. & Sauerbeck, D. (1984). Influence of plant roots on carbon and phosphorus metabolism in soil. *Plant and Soil*, **76**, 175–182.

Helal, H.M. & Sauerbeck, D. (1989). Carbon turnover in the rhizosphere. *Zeitschrift für Planzenernährung und Bodenkunde*, **152**, 211–216.

Hendrix, P.F., Parmelee, R.W., Crossely, D.A., Coleman, D.C., Odum, E.P. & Groffman, P.M. (1989). Detritus food webs in conventional and no-tillage agro-ecosystems. *Bioscience*, **36**, 374–380.

Höflich, G. (1989). Einfluss einer Inokulation mit Rhizobium-Baketerien auf das Wachstum von Getreide. *Zentralblatt für Mikrobiologie*, **144**, 73–79.

Höflich, G. & Steinbrenner, K. (1988). Einfluss acker- und pflanzen-baulicher Massnahmen auf einige bodenbiologische Faktoren. *Zentralblatt für Mikrobiologie*, **143**, 611–620.

Horemans, S. & Vlassak, K. (1985). Production of indole-3-acetic acid by *Azospirillum* *brasilense. Azospirillum III, Genetics, Physiology, Ecology* (Ed. by W. Klingmüller), pp. 98–108. Springer, Berlin.

Howell, C.R. & Stipanovic, R.D. (1980). Suppression of *Pythium ultimum*-induced damping-off of cotton seedlings by *Pseudomonas fluorescens* and its antibiotic, Pyoluteorin. *Phytopathology,* **70**, 712–715.

Howie, W.J., Cook, R.J. & Weller, D.M. (1987). Effects of soil matric potential and cell motility on wheat root colonization by fluorescent pseudomonads suppressive to take-all. *Phytopathology,* **77**, 286–292.

Hunt, H.W., Coleman, D.C., Ingham, E.R., Ingham, R.E., Elliott, E.T., Moore, J.C., Rose, S.L., Reid, C.P.P. & Morley, C.R. (1987). The detrital food web in a short grass prairie. *Biology and Fertility of Soils,* **3**, 57–68.

Inbal, E. & Feldman, M. (1982). The response of a hormonal mutant of common wheat to bacteria of the genus *Azospirillum. Israel Journal of Botany,* **31**, 257–263.

Ishac, Y.V. (1989). Application of Biofertilizers in Egypt. *Proceedings of the 5th International Symposium on Microbial Ecology,* Kyoto, Japan.

Jagnow, G. (1987). Inoculation of cereal crops and forage grasses with nitrogen-fixing rhizosphere bacteria: Possible causes of success and failure with regard to yield response — a review. *Zeitschrift für Pflanzenernnährung und Bodenkunde,* **150**, 361–368.

Jagnow, G. (1989). Contributions of nitrogen-fixing rhizosphere bacteria to nitrogen uptake and plant growth. *Recent Advances in Microbial Ecology* (Ed. by T. Hattori, Y. Ishida. Y. Maruyama, R.Y. Morita & A. Uchida), pp. 201–205. *Proceedings of the 5th International Symposium on Microbial Ecology,* Kyoto, Japan.

Juhnke, M.E., Mathre, D.E. & Sands, D.C. (1987). Identification and characterization of rhizosphere-competent bacteria of wheat. *Applied and Environmental Microbiology,* **53**, 2793–2799.

Juhnke, M.E., Mathre, D.E. & Sands, D.C. (1989). Relationship between bacterial seed inoculum density and rhizosphere colonization of spring wheat. *Soil Biology and Biochemistry,* **21**, 591–595.

Kapulnik, Y., Gafny, R. & Okon, Y. (1985). Effect of *Azospirillum* spp. inoculation on root development and NO_3 uptake in wheat (*Triticum aestivum* c. Miriam) in hydroponic systems. *Canadian Journal of Botany,* **63**, 627–631.

Kapulnik, Y., Kigel, J., Okon, Y., Nur, I. & Henis, Y. (1981). Effect of *Azospirillum* inoculation on some growth parameters and N-content of wheat, *Sorghum* and *Panicum. Plant and Soil,* **61**, 65–70.

Kavimandan, S.K. (1986). Influence of *Rhizobium* inoculation on yield of wheat (*Triticum aestivum* L.). *Plant and Soil,* **95**, 297–300.

Kloepper, J.W., Lifshitz, R. & Zablotowic, R.M. (1989). Free-living bacterial inocula for enhancing crop productivity. *Trends in Biotechnology,* **7**, 39–44.

Kolb, W. & Martin, P. (1985). Response of plant roots to inoculation with *Azospirillum brasilense* and to application of indole acetic acid. *Azospirillum. III. Genetics, Physiology, Ecology* (Ed. by W. Klingmüller), pp. 215–221. Springer, Berlin.

Kraffczyk, I., Trolldenier, G. & Beringer, H. (1984). Soluble root exudates of maize: influence of potassium supply and rhizosphere micro-organisms. *Soil Biology and Biochemistry,* **16**, 315–322.

Kucey, R.M.N. (1988). Plant growth altering effects of *Azospirillum brasilense* and *Bacillus* C-11-25 on two wheat cultivars. *Journal of Applied Bacteriology,* **64**, 187–196.

Lamb, J.A., Doran, J. W. & Peterson, G.A. (1987). Nonsymbiotic dinitrogen fixation in no-till and conventional wheat-fallow systems. *Soil Science Society of America Journal,* **51**, 356–361.

Lambert, B., Leyns, F., van Rooyen, L., Gossele, F., Papon, Y. & Swings, J. (1987).

Rhizobacteria of maize and their antifungal activities. *Applied and Environmental Microbiology*, **53**, 1886–1871.

Lawley, R.A., Campbell, R. & Newman, E.I. (1983). Composition of the bacterial flora of the rhizosphere of three grassland plants grown separately and in mixture. *Soil Biology and Biochemistry*, **15**, 605–607.

Leppard, G.G. (1974). Rhizoplane fibrils in wheat: demonstration and derivation. *Science*, **185**, 1066–1067.

Lethbridge, G. & Davidson, M.S. (1983). Root-associated nitrogen-fixing bacteria and their role in the nitrogen nutrition of wheat estimated by ^{15}N isotope dilution. *Soil Biology and Biochemistry*, **15**, 365–374.

Lin, W., Okon, Y. & Hardy, R.W.F. (1983). Enhanced mineral uptake by *Zea mays* and *Sorghum bicolor* roots inoculated with *Azospirillum brasilense*. *Applied and Environmental Microbiology*, **45**, 1775–1779.

Lindberg, T., Granhall, U. & Tomenius, K. (1985). Infectivity and acetylene reduction of diazotrophic rhizosphere bacteria in wheat (*Triticum aestivum*) seedlings under gnotobiotic conditions. *Biology and Fertility of Soils*, **1**, 123–129.

Lynch, J.M. & Clark, S.J. (1984). Effects of microbial colonization of barley (*Hordeum vulgare* L.) roots on seedling growth. *Journal of Applied Bacteriology*, **56**, 47–52.

Mandimba, G., Heulin, T., Bally, R., Guckert, A. & Balandreau, J. (1986). Chemotaxis of free-living nitrogen-fixing bacteria towards maize mucilage. *Plant and Soil*, **90**, 129–139.

Martinez-Toledo, M.V., Gonzalez-Lopez, J., de la Rubia, T., Moreno, J. & Ramos-Cormenzana, A. (1988). Grain yield response of *Zea mays* (hybrid AE 703) to *Azotobacter chroococcum* H 23. *Biology and Fertility of Soils*, **6**, 352–353.

Mertens, T. & Hess, D. (1984). Yield increases in spring wheat (*Triticum aestivum* L.) inoculated with *Azospirillum lipoferum* under greenhouse and field conditions of a temperate region. *Plant and Soil*, **82**, 87–99.

Mishustin, E.M. (1966). Action d'*Azotobacter* sur les vegetaux superieurs. *Annales de l'Institut Pasteur*, **III**, Suppl. 3, 121–135.

Molina, J.A.E. & Rovira, A.D. (1964). The influence of plant roots on autotrophic nitrifying bacteria. *Canadian Journal of Microbiology*, **10**, 249–257.

Moore, D.R.E. & Waid, J.S. (1971). The influence of washings of living roots on nitrification. *Soil Biology and Biochemistry*, **3**, 69–83.

Müller, M., Deigele, C. & Ziegler, H. (1989). Hormonal interactions in the rhizosphere of maize (*Zea mays* L.) and their effects on plant development. *Zeitschrift für Pflanzenernährung und Bodenkunde*, **152**, 247–254.

Nelson, E.B., Wei-Liang Chao, Norton, J.M., Nash, G.T. & Harman, G.E. (1986). Attachment of *Enterobacter cloaceae* to hyphae of *Pythium ultimum*: possible role in the biological control of *Pythium* pre-emergence damping-off. *Phytopathology*, **76**, 327–335.

Newman, E.I. & Bowen, H.J. (1974). Patterns of distribution of bacteria on root surfaces. *Soil Biology and Biochemistry*, **6**, 205–209.

Newman, E.I. & Watson, A. (1977). Microbial abundance in the rhizosphere: a computer model. *Plant and Soil*, **48**, 17–56.

Oades, J.M. (1978). Mucilages at the root surface. *Journal of Soil Science*, **29**, 1–16.

O'Hara, G.W., Davey, M.R. & Lucas, J.A. (1981). Effect of inoculation of *Zea mays* with *Azospirillum brasilense* strains under temperate conditions. *Canadian Journal of Microbiology*, **27**, 871–877.

Okon Y. (1984). Response of cereal and forage grasses to inoculation with N_2-fixing bacteria. *Advances in Nitrogen Fixation Research* (Ed. by C. Veeger & W.E. Newton), pp. 303–309. Martinus Nijhoff, The Hague.

Pacovsky, R.S. (1988). Influence of inoculation with *Azospirillum brasilense* and *Glomus fasciculatum* on sorghum nutrition. *Plant and Soil*, **110**, 283–287.

Pacovsky, R.S., Bethlenfalvay, G.J. & Paul, E.A. (1986). Comparisons between phosphorus-fertilized and mycorrhizal plants. *Crop Science*, **26**, 151–156.

Paul, E.A. & Clarke, F.E. (1988). Phosphorus transformations in soil. *Soil Microbiology and Biochemistry* (Ed. by E.A Paul & F.E. Clark), pp. 222–232. Academic Press, New York.

Pommer, G., Beck, T. & Borchert, H. (1989). 15 jähriger Vergleich von Daueranbau und Fruchtwechsel bei Winterweizen — Auswirkungen auf Ertrag, Ertragsbildung, Wurzelwachstum, Krankheitsbefall und Merkmale der Bodenfruchtbarkeit. *Kali-Briefe (Büntehof)*, **19**, 663–675.

Powell, C.L. & Daniel, J. (1978). Mycorrhizal fungi stimulate uptake of soluble and insoluble phosphate fertilizer from a phosphate deficient soil. *New Phytologist*, **80**, 351–358.

Reid, C.P.P. (1984). Mycorrhiza: A root/soil interface in plant nutrition. *American Society of Agronomy Special Publication*, **47**, 29–50.

Reynders, L. & Vlassak, K. (1982). Use of *Azospirillum brasilense* as bio-fertilizer in intensive wheat cropping. *Plant and Soil*, **66**, 217–223.

Rice, C.W. & Smith, M.S. (1982). Denitrification in no-till and plowed soils. *Soil Science Society of America Journal*, **46**, 1168–1173.

Robinson, D., Griffiths, B.S., Ritz, K. & Wheatley, R. (1989). Root-induced nitrogen mineralisation: A theoretical analysis. *Plant and Soil*, **117**, 185–193.

Rouatt, J.W., Peterson, E.A. & Katnelson, H. (1963). Microorganisms in the root zone in relation to temperature. *Canadian Journal of Microbiology*, **9**, 227–236.

Rovira, A.D., Foster, R.C. & Martin, J.K. (1979). Origin, nature and nomenclature of the organic materials in the rhizosphere. *The Soil-Root Interface*. (Ed. by J.L. Harley & R.S. Russell), pp. 1–4. Academic Press, London.

Ruppel, S. (1987). *Isolation diazotropher Bacterien aus der Rhizosphare von Winterweizen und Charakterisierung ihrer Leistungsfahigkeit.* PhD thesis, Academy of Agriculture of the GDR, Muncheberg.

Ruppel, S. (1988). Vorkommen diazotropher Bakterien in der Endorhizosphare und Rhizoplane von Winterweizen auf einem Standort mit gemassigtem Klima. *Zentralblatt für Mikrobiologie*, **143**, 621–629.

Safir, G.R. (1987). *Ecophysiology of VA Mycorrhizal Plants.* CRC Press, Boca Raton, Florida.

Sarig, S., Blum, K. & Okon, Y. (1988). Improvement of the water status and yield of field-grown sorghum (*Sorghum bicolor*) by inoculation with *Azospirillum brasilense*. *Journal of Agricultural Science, Cambridge*, **110**, 271–277.

Sauerbeck, D. & Johnen, B.G. (1976). Der Umsatz von Pflanzenwurzeln im Laufe der Vegetationsperiode und dessen Beitrag zur 'Bodenatmung'. *Zeitschrift für Pflanzenernährung und Bodenkunde*, **139**, 315–328.

Schneider, U., Haider, K. & Mahmood, T. (1990). Preliminary results by comparing gaseous denitrification losses with ^{15}N-balance losses in a wheat and a barley field. *Mitteilungen der Deutschen Bodenkundlichen Gesellschaft*, **60**, 37–44.

Schönwitz, R. & Ziegler, H. (1986). Quantitative and qualitative aspects of a developing rhizosphere microflora of hydroponically grown maize seedlings. *Zeitschrift für Pflanzenernährung und Bodenkunde*, **149**, 623–634.

Sieverding, E. (1986). Influence of soil water regimes on VA mycorrhiza IV. Effect on root growth and water relations of *Sorghum bicolor*. *Journal of Agronomy and Crop Science*, **157**, 36–42.

Smith, J.H., Allison, F.E. & Soulides, D.A. (1961). Evaluation of phosphobacterin as a soil inoculant. *Soil Science Society of America Proceedings*, **25**, 109–111.

Steinbrenner, K. & Obenauf, U. (1986). Untersuchungen zum Einfluss der Vorfrucht und Vorvorfrucht auf den Ertrag der Wintergetreidearten und den Befall durch

Gaeumannomyces graminis. Archiv für Acker- und Pflanzenbau und Bodenkunde, **30**, 773–779.

Stolp, H. & Starr, M.P. (1963). *Bdellovibrio bacteriovorus* gen. et sp. n. a predatory, ectoparasitic, and bacteriolytic microorganism. *Antonie van Leeuwenhoek*, **29**, 217–248.

Stroo, H.F., Elliott, L.F. & Papendick, R.I. (1988). Growth, survival and toxin production of root-inhibitory pseudomonads on crop residues. *Soil Biology and Biochemistry*, **20**, 201–207.

Subba Rao, N.S., Tilak, K.V.B.R. & Singh, C.S. (1985). Synergistic effect of vesicular-arbuscular mycorrhizas and *Azospirillum brasilense* on the growth of barley in pots. *Soil Biology and Biochemistry*, **17**, 934–941.

Sumner, D.R. & Minton, N.A. (1989). Crop losses in corn induced by *Rhizoctonia solani* AG-2-2 and nematodes. *Phytopathology*, **79**, 934–941.

Suslow, T.V. & Schroth, M.N. (1982). Role of deleterious rhizobacteria as minor pathogens in reducing crop growth. *Phytopathology*, **72**, 111–115.

Swaby, R.J. & Sperber, J. (1959). Phosphate dissolving microorganisms in the rhizosphere of legumes. *Nutrition of Legumes* (Ed. by E.D. Hallsworth), pp. 289–294. Butterworths, London.

Tari, P.H. & Anderson, A.J. (1988). *Fusarium* wilt suppression and agglutinability of *Pseudomonas putida*. *Applied and Environmental Microbiology*, **54**, 2037–2041.

Thomashow, L.S. & Weller, D.M. (1988). Role of a phenazine antibiotic from *Pseudomonas fluorescens* in biological control of *Gaeumannomyces graminis* var. *tritici*. *Journal of Bacteriology*, **170**, 3499–3508.

Trofymow, J.A., Coleman, D.C. & Cambardella, C. (1987). Rates of rhizo-deposition and ammonium depletion in the rhizosphere of axenic oat roots. *Plant and Soil*, **97**, 333–344.

Trolldenier, G. (1971). Der Einfluss der Kalium- und Stickstoffernährung des Weizens auf die bakterielle Besiedelung der Rhizosphare. *Sonderheft Landwirtschaftliche Forschung*, **26**, 37–46.

Turner, S.M., Newman, E.I. & Campbell, R. (1985). Microbial populations of ryegrass root surfaces: Influence of nitrogen and phosphorus supply. *Soil Biology and Biochemistry*, **17**, 711–715.

Umali-Garcia, M., Hubbell, D.H., Gaskins, M.H. & Dazzo, F.B. (1980). Association of *Azospirillum* with grass roots. *Applied and Environmental Microbiology*, **39**, 219–226.

van Elsas, J.D., Trevors, J.T. & Starodub, M.E. (1988). Bacterial conjugation between pseudomonads in the rhizosphere of wheat. *FEMS Microbiology Ecology*, **53**, 299–306.

van Vuurde, J.W.L. & Schippers, B. (1980). Bacterial colonization of seminal wheat roots. *Soil Biology and Biochemistry*, **12**, 559–565.

Vasjuk, L.F., Borovkov, A.V., Chalcickii, A.E., Ionkova, C.V. & Cmeleva, Z.V. (1989). Bacterii roda *Azospirillum* i ich vjijanie na produktivnost nebobovych rastenii. *Mikrobiologija*, **58**, 642–648.

Vermeer, J. & McCully, M.E. (1982). The rhizosphere in *Zea*: new insight into its structure and development. *Planta*, **156**, 45–61.

Vesper, S.J. (1987). Production of pili (fimbriae) by *Pseudomonas fluorescens* and correlation with attachment to corn roots. *Applied and Environmental Microbiology*, **53**, 1397–1405.

Vincent, C.D. & Gregory, P.J. (1989). Effects of temperature on the development and growth of winter wheat roots. II. Field studies of temperatures, nitrogen and irradiance. *Plant and Soil*, **119**, 99–110.

Vrany, J. & Macura, J. (1971). Changes in bacterial population during colonization of wheat rhizosphere, following aseptic cultivation and foliar application of urea. *Zeitschrift für Bakteriologie II*, **126**, 399–408.

Weller, D.M. (1988). Biological control of soilborne plant pathogens in the rhizosphere with bacteria. *Annual Review of Phytopathology*, **26**, 379–407.

Weller, D.M. & Cook, R.J. (1983). Suppression of take-all of wheat by seed treatments with fluorescent pseudomonads. *Phytopathology,* **73**, 463–469.

Werner, D. (1987). Die Rhizobium/Bradyrhizobium-Fabales-Symbiose: Konkurrenz und Überlebensfähigkeit im Boden. *Pflanzliche und Mikrobielle Symbiosen* (Ed. by D. Werner), pp. 48–50. Thieme, Stuttgart.

Whatley, M.H. & Sequeira, L. (1981). Bacterial attachment to plant cell walls. *Phytochemistry of Cell Recognition and Cell Surface Interactions* (Ed. by F.A. Loewus & C.A. Ryan), pp. 213–240. Plenum Press, New York.

Wood, M. (1987). Predicted microbial biomass in the rhizosphere of barley in the field. *Plant and Soil,* **97**, 303–314.

Zimmer, W., Roeben, K. & Bothe, H. (1988). An alternative explanation for plant growth promotion by bacteria of the genus *Azospirillum. Planta,* **176**, 333–342.

Zvyagintsev, D.G. & Guzev, V.S. (1987). Effect of mineral fertilizers on microbiological processes in the soil. *Proceedings of the 9th International Symposium on Soil Biology and Conservation of the Biosphere, Budapest* (Ed. by J. Szegi), pp. 3–13. Akademia Kiado, Budapest.

8. SOIL FAUNA AND CEREAL CROPS[*]

K.B. ZWART[†] AND L. BRUSSAARD[‡]

†*Institute for Soil Fertility, Department of Soil Biology, PO Box 30003,*
9750 RA Haren, The Netherlands; ‡Institute for Soil Fertility, Haren,
and Agricultural University, Department of Soil Science and Geology,
PO Box 37, 6700 AA Wageningen, The Netherlands

INTRODUCTION

Cereals, in particular winter wheat and spring barley, are important crops in Dutch arable farming. Arable production in The Netherlands is among the most intensive in the world as a result of several environmental conditions and husbandry practices. Firstly, the natural soil fertility in The Netherlands is high, especially on reclaimed marine soils. Secondly, the use of large amounts of artificial fertilizers and animal manure by the farmers, prevents any deficiency of the basic nutrients during crop growth. Thirdly, the extensive use of chemicals suppresses weeds and pests. And finally, a good soil structure is created and maintained by intensive tillage.

During the last 10–15 years, the negative aspects related with this high production have begun to attract more attention. Several areas in The Netherlands are facing environmental problems due to: (i) contamination of groundwater by nitrate, originating from artificial fertilizers or animal manure; (ii) contamination of groundwater with agrochemicals; and (iii) deterioration of soil structure due to the use of heavy machinery. Apart from causing environmental problems, this so-called conventional farming (CF) is associated with high costs of agrochemicals and machinery, constituting a high demand for energy. Vereijken (1986) proposed integrated farming (IF) as a new way to overcome many problems associated with CF. In integrated farming, reduction of input in terms of artificial fertilizers, agrochemicals and soil tillage will result in lower crop production, but in economic terms this will, to a greater or lesser extent, be offset by lower costs (Vereijken 1989). A cost-effective and long-term environmentally sound arable farming may stimulate farmers to change farm management.

For integrated farming to become successful, basic knowledge about the mechanisms involved in the functioning of the soil-crop ecosystem is needed (Brussaard *et al.* 1988). The objective of the Dutch Programme on

[*] Communication No. 22 of the Dutch Programme on Soil Ecology of Arable Farming Systems.

Soil Ecology of Arable Farming Systems is to obtain such knowledge with respect to: (i) matching of nutrient supply by the soil with nutrient demand by crops; and (ii) enhancement of the contribution of soil organisms to soil structure formation (Brussaard *et al.* 1988). Research within this programme is performed at laboratory and field level. Field research is carried out at the experimental farm Lovinkhoeve in the north-east Polder, which was reclaimed in 1942. The soil is a marine silt loam, with a pH-KCl of 7·5, an organic matter (OM) content of 2·3–2·8% and a total nitrogen (N) content of 0·09–0·14%, both depending on the history of the various trial fields. The annual rainfall is 650–800 mm. The trial includes three different types of management: (i) CF, (ii) IF (both since 1985); and (iii) IF with minimum tillage (since 1986). The third management system will not be discussed here. The site and farm management characteristics of CF and IF were described in detail by Kooistra, Lebbink & Brussaard (1989) and are summarized for winter wheat in Table 8.1. Field studies include collection of data on:

1 biological characteristics;
    (a) biomass of micro-organisms (bacteria and fungi),
    (b) biomass of fauna (protozoa, nematodes, collembola, mites, enchytraeids, earthworms and ground beetle larvae);
2 chemical/physical characteristics;
    (a) soil organic matter, total N, mineral N, N mineralization or immobilization,
    (b) total C and N input,
    (c) description of macropore structure of the soil,
    (d) quantification of soil macro- and microporosity, aggregate size distribution and aggregate stability.

Laboratory studies include:
1 studies on the effects of biotic and abiotic factors on the dynamics of the soil biota, related to the C and N turnover;
2 assessment of the impact of microflora and fauna interactions on C and N flows;
3 studies on the effects of enchytraeids and earthworms on soil structure.

Studies in the Dutch Research Programme are linked to earlier and ongoing studies on the ecology of arable soils performed in the Swedish Arable Land Project (e.g. Andrén *et al.* 1989b), the North American Detrital Food Web Project (e.g. Ingham *et al.* 1986), the North American Horse Shoe Bend studies (e.g. Hendrix *et al.* 1986) and the work of the Canadian group at Edmonton, Alberta (e.g. Rutherford & Juma 1989). The Dutch programme started in 1985 and the data presented in this paper relate to winter wheat in 1986. Since 1986 no extensive data on the soil

TABLE 8.1. Husbandry practices under conventional and integrated farming on the Lovinkhoeve site of a winter wheat crop in 1986. From Kooistra, Lebbink & Brussaard (1989)

Management	Conventional farming (CF)	Integrated farming (IF)
Tillage		
October	Plough	Fixed-tine cultivator
Depth	20 cm	12 cm/8 cm
Seed-bed	Spring-tine cultivator	Spring-tine cultivator
Depth	Superficial	Superficial
Fertilizer		
N	200 kg/ha*	155 kg/ha*
Crop protection		
Weed control	Mainly chemical	Mainly mechanical
Pest control	Recommended dosages of pesticides (EPIPRE system)	Less pesticides
	Soil fumigation after harvest	No soil fumigation

* Including mineral N amount in the soil (0–100 cm) in early spring.

fauna of cereal crops have been collected with the exception for winter wheat in 1990. These 1990 data, however, have not been completely analysed and will not be discussed here.

The focus of this paper is on soil fauna as related to C and N dynamics and to soil structure under cereals. Soil microflora will also be discussed. The paper is not meant to be a comprehensive review of the subject. Ample reference, however, will be made to the above-mentioned projects and other relevant literature.

SOIL BIOTA AND CARBON AND NITROGEN DYNAMICS

Matching of the nutrient supply by the soil and the nutrient demand by plants requires knowledge of the kinetics of biological processes such as decomposition of organic matter and mineralization, especially when organic nutrient sources are important, as is the case in integrated farming (IF). Therefore, factors that determine decomposition and mineralization receive particular attention in the Dutch Programme, in which the situation in conventional farming (CF) is compared with IF.

The role of soil fauna in mineralization

Mineralization is the process by which dead organic matter (DOM) is converted from labile complex biopolymers into stable simple mineral elements. Simultaneously, complex humic compounds are formed (stabilized DOM). Physical, chemical and especially biological factors play an important role in this process. The importance of the soil fauna for the decomposition of organic matter and the release of nutrients is dealt with in recent synthesis and review papers, e.g. Anderson (1988), Crossley, Coleman & Hendrix (1989) and Verhoef & Brussaard (1990). In arable land, labile DOM originates from several sources: crop residues, decaying roots from (previous) crops, decaying micro-organisms, root exudates, animal manure, green manure, etc. The estimated labile DOM input at CF and IF at the Lovinkhoeve site associated with winter wheat cropping, is given in Table 8.2. Besides labile DOM, arable soil contains a large pool of previously stabilized DOM (Table 8.2), which, although slowly degradable, may be just as important as the labile DOM, in view of its total mass. DOM is mineralized by a range of organisms, which are connected in detritus-based food webs. In such below-ground food webs, the organisms are grouped functionally into different trophic levels related to their main food source and life history characteristics, rather than taxonomically. Moore & De Ruiter (1990) have described the webs for CF and IF, respectively, at the Lovinkhoeve site. Figure 8.1 shows the food web for IF.

Figures 8.2 and 8.3 show the food webs as described for agro-ecosystems

TABLE 8.2. Input of organic matter immediately prior to and during growth of winter wheat on the Lovinkhoeve site, under conventional and integrated farming in 1986 (kg/ha)

Input	Conventional farming (CF)	Integrated farming (IF)
Previous crop*	500	4500
Exudation from wheat[†]	1940	1900
Degradation of SDOM[‡]	1370	1750
Total	3810	8150
Total carbon (50%)	1905	4075

* Sugar beet, input was different for CF and IF: on CF roots only; on IF roots and leaves were ploughed into the soil.
[†] Assumed to be 10% of primary production (Woldendorp 1981).
[‡] Degradation of SDOM (stable dead organic matter) assumed to be 2% per year (Kortleven 1963); CF < IF, because of slightly higher organic matter content in the IF plots.

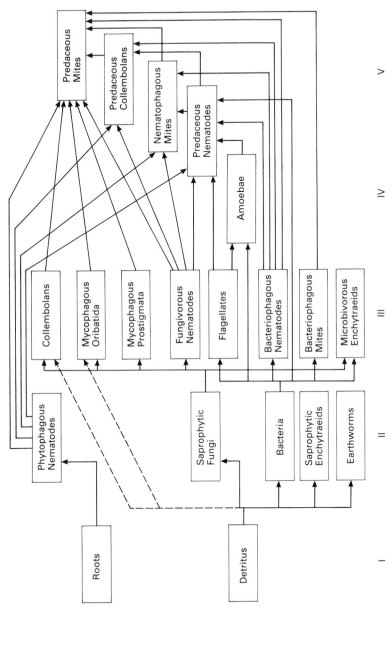

FIG. 8.1. Subterranean food webs for the Lovinkhoeve site under winter wheat with integrated farming. The roman numerals refer to trophic levels. Dashed vectors indicate potential feeding relationships, which have not been quantified. After Moore & De Ruiter (1990).

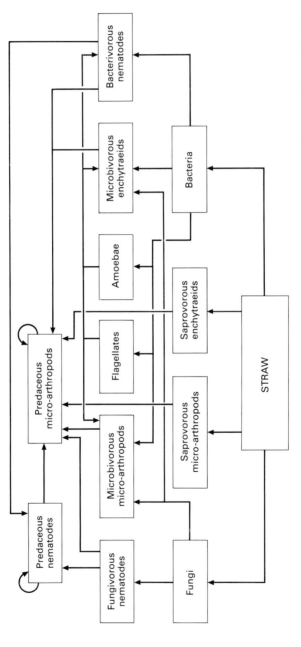

FIG. 8.2. Conceptual subterranean food web of decomposing barley-straw litter during decomposition. From Andrén, Paustian & Rosswall (1988).

in Sweden (Andrén, Paustian & Rosswall 1988) and North America (Hendrix *et al.* 1986, 1987). The role of different trophic levels of such a food web in mineralization is schematically presented in Fig. 8.4. In each degradation step, organic matter (OM) is converted into newly formed biomass, partly digested OM and mineral elements. The amount of mineral elements released by functional groups determines their impact on *direct* mineralization. The amount of newly formed biomass and partly decomposed OM determines their impact on *indirect* mineralization, since newly formed biomass may be further consumed (*i.e.* mineralized) by higher trophic levels, while the partly decomposed OM returns to the DOM pool and re-enters the process. Organisms of a functional group also have an *indirect* effect on mineralization if they affect the decomposition rates of lower trophic levels. The quantities and proportions of biomass, released OM and minerals produced vary widely among the functional groups within the trophic levels. Therefore, the role of different groups will show wide variation. The relative importance of organisms in mineralization will generally depend on their biomass (C and N pool), their population turnover rate and their assimilation efficiency (C and N fluxes).

The average biomass of several functional groups during the growing season of winter wheat under CF and IF on the Lovinkhoeve site is given in Table 8.3. Microbial biomass forms by far the largest pool under winter wheat, followed by earthworms (IF only), protozoa, micro-arthropods and enchytraeids. Within the microbial biomass, fungi are of minor importance. The irregular distribution of earthworms has not resulted from different management systems, but rather from the history of the fields. The polder where the Lovinkhoeve site is located, was reclaimed from the sea almost

TABLE 8.3. Average biomass (kg C/ha) in the top 25 cm of soil under winter wheat on the Lovinkhoeve site under conventional and integrated farming during the 1986 growing season

Organisms	Conventional farming (CF)	Integrated farming (IF)
Bacteria	339·80	436·70
Fungi	16·20	24·70
Protozoa	10·00	13·00
Nematodes	0·48	1·20
Micro-arthropods	0·79	0·64
Macro-arthropod larvae	0·30	0·00
Enchytraeids	0·78	0·55
Earthworms	0·00	15·00
Total biomass C	368·35	491·81

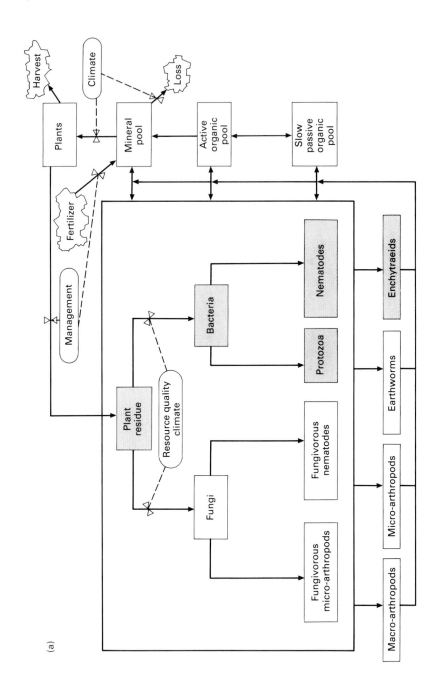

(a)

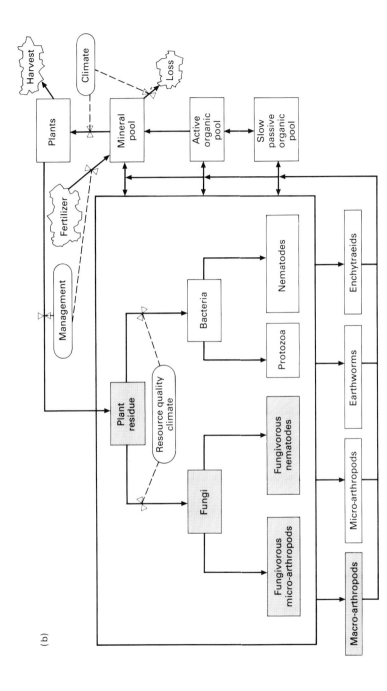

(b)

F IG. 8.3. Conceptual detritus based, subterranean food webs under conventional tillage (a) and no tillage (b) agro-ecosystems. Boxes = nutrient stores, 'clouds' = nutrient sources or sinks, arrows = nutrient transfer pathways, valve symbols on arrows indicate that nutrient transfers are influenced by factors connected by dotted lines. The shaded boxes denote the major nutrient pools. From Hendrix *et al.* (1987).

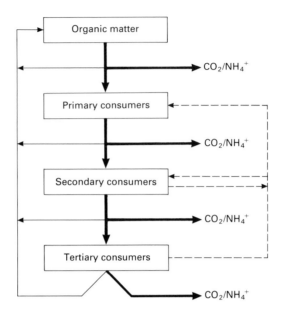

F I G. 8.4. The role of different trophic levels in a food web on mineralization of organic matter. Bold arrows indicate direct mineralization, fine arrows indicate recirculation of organic matter and dashed lines indicate possible effects on the activity of lower trophic levels.

50 years ago. Earthworms are still colonizing parts of the site. Both intensive tillage and the exclusive use of artificial fertilizers have probably prevented colonization of the conventionally managed trial fields (Marinissen 1990). The quantitative variations in some of these groups during the growing season are shown in Fig. 8.5 (Moore & De Ruiter 1990). Table 8.4 summarizes the biomass of several groups of organisms found in studies in Europe and North America. To allow comparison, the data are given as percentages of the total biomass. In all situations the microflora form the largest pool, followed by protozoa (if determined) and earthworms (if present). The role of each of these respective groups in mineralization are discussed successively in more detail.

Microflora

Although this paper is primarily concerned with soil fauna, their activities should not be isolated from the microflora. Bacteria and fungi are the major primary consumers and not only constitute by far the largest biomass pool, but also serve as food for many organisms in the below-ground food web.

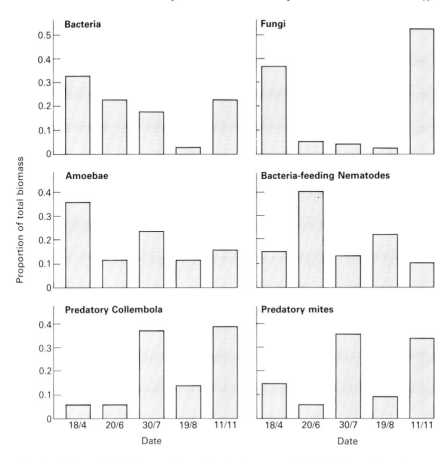

F IG. 8.5. Fluctuation in biomass of some functional groups at the Lovinkhoeve site under integrated farming. The columns represent the proportions of the total biomass of each group over the sampling period from April until November. From Moore & De Ruiter (1990).

In 1986 there was no significant difference in microbial biomass between CF and IF at the Lovinkhoeve site (Table 8.3). During the growing season of wheat, some variation in biomass was found (Fig. 8.5). This variation has also been found by others under cereals (Campbell & Biederbeck 1976; Carter & Rennie 1982; Elliott *et al.* 1984; Schnürer, Clarholm & Rosswall 1986; Granatstein *et al.* 1987) and other crops (Ross *et al.* 1981; Nannipieri 1984). It is explained by variation in temperature and moisture (Elliott *et al.* 1984; Schnürer *et al.* 1986; Granatstein *et al.* 1987), by addition of crop residues after harvesting (Granatstein *et al.* 1987) and by predation (Schnürer *et al.* 1986).

TABLE 8.4. Biomass of micro-organisms and fauna groups as a percentage of total carbon in arable soils under cereals of different locations in the world

Organisms	Lovinkhoeve[1]		Ellerslie[2]	Breton[3]	Horse Shoe Bend[4]		Kjettslinge[5]		Akron[6]	
	CF	IF			CT	NT	B0	B120	SM	NT
Total biomass (kg C/ha)	368	492	609	554	910	740	2360	3050	235	273
Bacteria	92·20	88·80	75·3*	47·6*	76	59·5	29·7	29·5	99*	99*
Fungi	4·40	5·00	24·6	52·3	16·5	21·5	63·6	68·9		
Protozoa	2·85	2·60	0·01	0·01	5·5	5·4	5·9	1·0	0·6	0·6
Nematodes	0·24	0·24	0·06	0·03	0·22	0·14	0·05	0·03	0·3	0·3
Micro-arthropods	0·19	0·13	NA	NA	0·08	0·27	0·02	0·01	0·06	0·09
Macro-arthropods	0·10	0·00	NA	NA	0·01	0·03	0·05	0·01	NA	NA
Enchytraeids	0·21	0·11	NA	NA	0·03	0·07	0·18	0·12	NA	NA
Earthworms	0·00	3·04	NA	NA	2·2	13·5	0·47	0·45	NA	NA

[1] Lovinkhoeve site, The Netherlands, Typic Fluvaquent, silt loam, 0–25 cm, spring/summer, winter wheat.
[2] Ellerslie, Alberta, Canada, Black Chernozem, silt clay loam, 0–10 cm, barley, summer/autumn (Rutherford & Juma 1989).
[3] Breton, Alberta, Canada, Gray Luvisol, silt loam, 0–10 cm, barley, summer/autumn (Rutherford & Juma 1989).
[4] Horse Shoe Bend site, Athens, Georgia, USA, Hiwassi loam, Typic Rhodudult, sandy clay loam, 0–15 cm, winter/spring, grain/rye (Hendrix et al. 1986, 1987).
[5] Kjettslinge site, Sweden, mixed, frigid Haplaquoll, loam, 0–27 cm, barley, 1982–83 (Andrén, Paustian & Rosswall 1988; Andrén et al. 1989a).
[6] Akron, Colorado, USA Mollisol, 0–10 cm, fallow/wheat, summer (Elliott et al. 1984).
* Including fungi.

CF, conventional farming. IF, integrated farming. NT, no tillage. CT, conventional tillage. B0, no N fertilizer. B120, 120 kg N fertilizer/ha. SM, stubble mulch. NA, data not available.

Fungal populations, as estimated by direct counts, are quantitatively unimportant in the Lovinkhoeve soil. This observation is in agreement with several other investigations; in Sweden, however, the fungi seem to be more important than bacteria (Table 8.4). The relatively low contribution of fungi to organic matter degradation at the Lovinkhoeve site was confirmed by the presence of low numbers of fungivorous animals. Moore & De Ruiter (1990) performed cluster analysis of functional groups at the Lovinkhoeve site using the functional link approach, as suggested by Moore *et al.* (1988). The functional link approach attempts to project species into niche space and to study the collective behaviour of species (assembled in functional groups) and their impact on ecosystem-level processes, such as cycling of elements. Moore & De Ruiter (1990) distinguished three clusters of functional groups under CF and four under IF. Fungi formed a distinct cluster under IF, whereas under CF they were grouped together with bacteria. This might suggest that fungi were relatively more important under IF. Hendrix *et al.* (1986, 1987; Fig. 8.3) suggested that changing from conventional tillage to no tillage would result in a relative increase in the function of fungi and fungal-feeding organisms in organic matter degradation. Numerical data from both the Lovinkhoeve site and the Horse Shoe Bend site, however, indicate that bacteria rather than fungi are the most important primary consuming micro-organisms under all types of management (Table 8.4).

Although microbial biomass is an important ecological parameter in relation to mineralization, microbial growth rate is even more important. Unfortunately, growth rate measurements in soil are very difficult to perform. Bloem & Brussaard (1989) used the frequency of dividing and divided cells (FDDC) as an indication of growth rates of bacteria in a field plot experiment during drying and rewetting of Lovinkhoeve soil under sugar beet. They found a non-significant increase in bacterial biomass (based on direct microscopic counts) shortly after rewetting, notwithstanding a FDDC indicating a generation time of approximately 10 h. This discrepancy was explained by assuming a high activity of predators. Based on total carbon input during 1985 and 1986 (Table 8.2), assuming a C:N ratio of 20 for this organic input and a bacterial C:N ratio of 5, one might expect a turnover rate for the bacterial biomass of 0·7 times per year on CF and 1·2 times on IF. In that case there was no net N mineralization by micro-organisms. In case of a very efficient remineralization of nutrients by the soil fauna, a turnover rate of 1·2 and 2·5 times per year was estimated for CF and IF, respectively. The turnover rate estimated for the Lovinkhoeve site is fairly close to the rate of 0·4–0·8 times per year found by Jenkinson, Ladd & Rayner (1980) for soil bacteria. Schnürer, Clarholm

& Rosswall (1985) reported an annual turnover rate of 0·5 for microbial biomass under fertilized barley. There is, however, an enormous discrepancy between growth rates resulting in a turnover rate of 0·7–1·2 per year and generation times of approximately 10 h. This may indicate that bacterial growth rates show wide variations with bursts of rapid growth followed by periods of very slow growth or dormancy. Therefore, a reliable and accurate method to establish bacterial production and consumption by predators in soil is badly needed.

Protozoa

Protozoa are the most predominant faunal group in arable land under cereals (Table 8.4). They constitute 1–6% of the total biomass, and Rutherford & Juma (1989) found even higher densities. However, the counting method used by Rutherford & Juma (1989) was different from the one of Darbyshire et al. (1974), which is employed most frequently in soil protozoan studies. This may explain, at least partly, the extremely high protozoan density in Canadian soils. At the Lovinkhoeve site, the protozoan pool in 1986 represented approximately 10 kg C; the amounts in CF and IF did not differ significantly (Table 8.3).

Protozoa probably are the most important bacterial consumers in soil, followed by bacterivorous nematodes. The major representatives at the Lovinkhoeve site are small naked amoebae and flagellates, which are well adapted to the thin waterfilms around soil particles. Ciliates are often used in model studies to test the effect of protozoa on soil processes (Darbyshire 1976; Griffiths 1986). The number of ciliates at the Lovinkhoeve site, however, was below the detection limit of the most probable number technique for enumeration of soil protozoa (Darbyshire et al. 1974). Clarholm (1983, 1989) considers naked amoebae as the most important protozoa in arable land under barley.

Initially, protozoa were assumed to have an adverse effect on soil fertility (Russell & Hutchinson 1909). Waksman (1916), however, considered them to have a beneficial effect or to be harmless. Since that time, evidence has been obtained for the positive effects of protozoa on soil processes. Nitrogen fixation by Azotobacter was stimulated by the presence of protozoa (Cutler & Bal 1926; Telegdy-Kovats 1932; Hervey & Greaves 1941; Nikoljuk 1969; Darbyshire 1972a,b) and protozoa also stimulate the mineralization of N and P in soil (Cole, Elliott & Coleman 1978; Elliott, Coleman & Cole 1979; Woods et al. 1982; Clarholm 1983; Griffiths 1986; Kuikman & Van Veen 1989). Elliott, Coleman & Cole (1979), Clarholm (1983) and Kuikman (1990) demonstrated that plant N uptake from an organic source is stimulated by protozoan activity.

Several explanations for the stimulatory effect of protozoa have been presented. One obvious explanation is that protozoa excrete excess N and/or P (phosphorus) from bacteria which they use as a source of energy. Furthermore, protozoa may excrete residues of their food (e.g. bacterial cell walls). Further digestion of these organic excretion products by bacteria may be facilitated, which will then lead to an increase in mineralization. Finally, several authors have suggested that grazing of protozoa on bacteria stimulates bacterial activity by preventing or releasing nutrient limitation (Barsdate, Prentki & Fenchel 1974; Cutler & Crump 1929; Fenchel & Harrison 1976; Meiklejohn 1930, 1932; Pussard & Rouelle 1986). These authors contend that, although grazing leads to a decrease in the bacterial population, the specific activity of bacteria will increase to such an extent that the total activity of a grazed population is higher than that of an ungrazed population.

Mathematical simulation models are used to estimate the fluxes of C and N through different functional groups (O'Neill 1969; Hunt *et al.* 1984; 1987). De Ruiter (personal communication) estimated the flux of nitrogen via animal consumption of bacteria at CF and IF under winter wheat in the top 25 cm of soil. The protozoan contribution was 40 and 65 kg N/ha per year in CF and IF respectively, which was over 70% of the total faunal flux. In Sweden protozoan contribution to the flux of N under fertilized barley was approximately 30 kg/ha per year, i.e. 16% of the total N flux, which was over 50% of the flux via the soil fauna (Andrén *et al.* 1989a). According to Hendrix *et al.* (1987) the protozoan contribution to soil respiration was 7.6 and 6.6% under CT and NT, respectively (Table 8.5), which was 56 and 18%, respectively, of the faunal respiration. From these figures it may be concluded that protozoa play a major role in mineralization of microbial biomass under cereals. The accuracy of simulation models with respect to protozoan activity may be further improved. For example, one should distinguish between active and encysted cells. Furthermore, the effects of abiotic factors on growth of protozoa need to be included to allow for the temporal dynamics of protozoa (Kuikman 1990). Little is known about the effect of moisture on small amoebae and flagellates. There is some information that protozoan growth ceases when pores with a diameter larger than 6 µm become dry (Darbyshire 1976; Alabouvette *et al.* 1981). Postma (1989) and Kuikman (1990) assumed that pores with neck diameters smaller than 3 µm are not accessible to protozoa. However, common soil flagellates of the genus *Cercomonas* showed almost maximum growth rates in sterilized Lovinkhoeve soil with a moisture content of 18·6%, which is equivalent in this soil with a water potential of −100 KPa, where only pores smaller than 3 µm in diameter are filled with water (Fig. 8.6). Moreover, also at lower moisture conditions, where only pores smaller

TABLE 8.5. Flows of carbon or nitrogen through micro-organisms and fauna of arable fields under cereals as a percentage of total annual flows/ha at the Horse Shoe Bend site (Hendrix *et al.* 1986) and the Kjettslinge site (Andrén *et al.* 1989)

Organisms	Horse Shoe Bend rye		Kjettslinge barley, 120 kg N fertilizer/ha
	CT	NT	
Bacteria	81.0	56.8	69.1*
Fungi	5.7	6.6	—
Protozoa	7.6	6.6	16.0
Nematodes	0.4	0.2	9.4
Micro-arthropods	0.08	0.25	2.2
Macro-arthropods	0.008	0.03	0.2
Enchytraeids	0.04	0.02	1.6
Earthworms	5.3	29.0	0.5
Total annual flow/ha	2630 kg C	2410 kg C	198 kg N

* Including fungi.
CT, conventional tillage. NT, no tillage.

than 1 μm were filled with water, the flagellates still showed maximum growth (Fig. 8.6). This may indicate that small flagellates, like *Cercomonas*, are able to consume bacteria even under dry conditions, when other bacterivorous organisms are no longer active. Their growth rates under

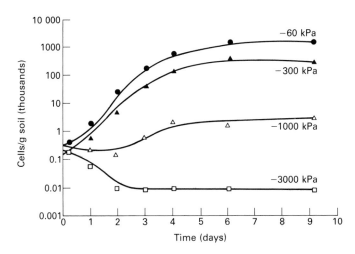

FIG. 8.6. Growth of the common soil flagellate *Cercomonas* sp. in microcosms of sterilized soil from the Lovinkhoeve site, in the presence of bacteria, *Pseudomonas fluorescens*, as food, under different moisture conditions at 20°C.

these conditions may indicate that these organisms have turnover rates which are higher than assumed thus far in simulation model studies.

Nematodes

Total nematode biomass at the Lovinkhoeve site was less than 0·15% of the total biomass (Table 8.4), representing approximately 1 kg C/ha. This fraction of 0·15% is in between the ones observed at other locations (Table 8.4). Under conventional tillage (CT) at Horse Shoe Bend and in Akron nematodes constitute between 0·25 and 0·3% of the total biomass and in Sweden and Canada the nematode fraction is 0·05% or less (Table 8.4). Biomass under CF was almost similar to that under IF at the Lovinkhoeve site. Stinner & Crossley (1982) found no differences in biomass under CT or no tillage (NT). Hendrix *et al.* (1986, 1987), however, reported somewhat higher numbers under CT than under NT. According to Boström & Sohlenius (1986), there was no difference in taxonomic diversity of nematodes under single or continuous crops of barley. Plant-parasitic nematodes were more abundant under IF than under CF due to diminished nematocides on CF (Bouwman, unpublished result).

Free-living nematodes under barley responded positively to manure and fertilizer (Andrén & Lagerlof 1983). The very low nematode biomass may indicate that the role of these organisms in mineralization will probably be small. Sohlenius, Boström & Sandor (1988) estimated that only 1·3–2% of the total C input on arable land was respired by nematodes. They also calculated the annual turnover rate of the nematode population under barley. In rotational cropping systems, the turnover rate was 9·5 times per year v. 9·1 times per year with continuous barley. Based on standing stock population and turnover rates the impact of free-living nematodes on mineralization appears to be small. This is confirmed by estimations of respiration. Carbon respiration of nematodes under CT and NT was only 0·4 and 0·2% of total respiration, respectively (Hendrix *et al.* 1986, 1987; Table 8.5). Andrén *et al.* (1989a), on the other hand, estimated that 9·4% of the total N flow through the biomass was via nematodes (Table 8.5). Microcosm studies with sterilized soil and an organic N or P source, have sometimes indicated higher mineralization rates in the presence of nematodes than in their absence (Griffiths 1986). Nevertheless, Woods *et al.* (1982) observed a stimulation of nitrogen mineralization only during the decline of the nematode population, while Cole, Elliott & Coleman (1978), on the other hand, found little effect of nematodes on phosphorus mineralization in microcosm studies. Although these results are rather conflicting, it could mean that the impact of nematodes on mineralization is larger than expected solely from standing stock and population turnover rates. Possibly, nematodes have considerable indirect effects on the activity of other groups of organisms. Cole, Elliott & Coleman (1978) observed a positive effect of nematodes on amoebal activity and Brussaard *et al.* (1991) explained part of the positive effect of bacterivorous nematodes on N mineralization by suggesting that nematodes carry commensal or epizoic flagellates. Finally, there are indications that nematode turnover rates are higher than the values given by Sohlenius, Boström & Sandor (1988). Nematophagous fungi, organisms not considered in most food web studies, may effectively keep the nematode population at a low level in the Lovinkhoeve soil (Bouwman, unpublished). So, possibly the role of nematodes is more likely to be underestimated than overestimated.

Micro-arthropods

Micro-arthropods (acari and collembola) represent only a minor fraction of the total biomass at the Lovinkhoeve site (Tables 8.3 and 8.4) and other sites (Table 8.4). Their overall densities at IF were similar to those at CF, but there was a large difference in abundance of species under both types of management (Brussaard *et al.* 1988). Andrén & Lagerlöf (1983) found a negative response of collembola and most of the acari to N fertilizer in various cropping systems. However, no effect of N fertilizer on micro-arthropods under barley was found (Lagerlöf 1987; Lagerlöf & Andrén 1988). Acari and collembola consumed less than 1% of the estimated microbial production (Lagerlöf 1987; Lagerlöf & Andrén 1988). Andrén & Schnürer (1985) found little effect on C mineralization of barley straw by the collembolan *Folsomia fimetaria* in pot experiments. Different numbers of grazing collembola did not result in differences in respiration, mass loss or microbial biomass. Based on these data, it would seem unlikely that micro-arthropods would have a great impact on mineralization. Indeed, less than 3% of the N flow under fertilized barley was via arthropods, macro-arthropods including (Andrén *et al.* 1989a; Table 8.5). Carbon respiration of micro-arthropods under CT and NT was 0·08 and 0·25%, respectively, of total respiration (Hendrix *et al.* 1986, 1987; Table 8.5). However, in other cases fungal-grazing collembola or mites played an important role as regulators of decomposition in microcosms (Morley *et al.* quoted by Elliott *et al.* 1986) and in the field, both under natural (Whitford *et al.* 1982) and agricultural (Parmelee, Beare & Blair 1989) conditions. In microcosm studies by Morley *et al.* (quoted by Elliott *et al.* 1986) the presence of mites resulted in a four-fold increase of the specific respiration rate of the microflora.

Besides microbivorous micro-arthropods, nematophagous mites have been reported to affect decomposition and mineralization rates. Lagerlöf & Andrén (1988) found that the consumption by predatory mites corresponded to 20–60% of the total nematode production. In terms of biomass this may be small, but it may affect bacterial production considerably. Despite their low direct contribution to carbon and nitrogen dynamics, micro-arthropods therefore may exert a stabilizing effect on microflora, nematode abundance, decomposition and mineralization rates (Moore, Walter & Hunt 1988). The available evidence (Coleman *et al.* 1984; Seastedt 1984; Verhoef & Brussaard 1990), however, suggests that the importance of such activities is greater in the case of more recalcitrant substrates, i.e. more important in natural systems than in agro-ecosystems (Crossley, Coleman & Hendrix 1989).

Enchytraeids

At the Lovinkhoeve site, the enchytraeids constituted approximately 0·2% of the total biomass under wheat in 1986 both under CF and IF (Tables 8.3, 8.5 and 8.6). In IF however, the organisms were concentrated in the upper 10 cm of the soil ($P < 0.01$, Friedman test), whereas in CF more animals occurred in the 10–40 cm layer ($P < 0.05$, Friedman test) (Didden 1991). Lagerlöf, Andrén & Paustian (1989) found that different cropping systems had no effect on enchytraeid densities. Using the values of Persson &

TABLE 8.6. Presence of enchytraeidae: abundance (10^6/ha), biomass (kg C/ha) and carbon flows (kg C/ha per year)

	Lovinkhoeve[1] (0–25 cm)		Kjettslinge[2] (0–20 cm)	
	CF	IF	B0	B120
Abundance	90	83	100	81
Biomass[3]	0·78	0·55	3·2	2·7
Respiration	8·75	7·75	25·9	20·7
Respiration, percentage of organic C input	0·46	0·19	1·73	1·15
Consumption	58·33	51·67	172·8	138·2
Consumption, percentage of organic C input	3·06	1·27	11·52	7·62

[1] Samples from wheat (January–December), Lovinkhoeve site (Didden 1991).
[2] Samples September 1980–1984, barley, Kjettslinge site (Lagerlöf, Andrén & Paustian 1989).
[3] Biomass and C-flow calculations according to Persson & Lohm (1977) (B0 and B120).
CF, conventional farming. IF, integrated farming. B0, no N fertilizer. B120, 120 kg N/ha.

Lohm (1977), they assumed that enchytraeids are 50% microbivorous and 50% saprovorous. Based on this assumption and figures of Heal & MacLean (1975), they used values 0·12 and 0·18 for production: consumption and respiration : consumption efficiencies, respectively. The estimates on enchytraeid abundance, biomass and carbon flows under barley and winter wheat are listed in Table 8.6. The numbers of enchytraeids under barley and wheat were in the same range, despite different farming methods (Table 8.6). Tables 8.5 and 8.6 clearly show that the contribution of enchytraeids to soil respiration is small. Lagerlöf, Andrén & Paustian (1989) reported an enchytraeid respiration under barley of 12–26 kg C/ha per year. Enchytraeids however, have a low assimilation : consumption ratio (Heal & MacLean 1975), and therefore consume a significant amount of organic matter. When all enchytraeids found in the 0–40 cm layer are taken into account, the respiration as a percentage of organic C input was 1·0% and 0·52% for CF and IF, respectively, while the values for consumption were 6·78% and 3·44%. The calculated amount of microbial C consumed by enchytraeids was 69–86 kg/ha per year under barley (0–25 cm) (Lagerlöf, Andren & Paustian 1989) and 36–47 kg/ha per year under winter wheat (0–40 cm). This is up to 10% of the total microbial production under barley and almost 6% under wheat. Andren & Lagerlöf (1983) and Lagerlöf, Andren & Paustian (1989) found a positive response of enchytraeids to organic matter addition. This was explained by the increase in microbial biomass after the addition of a fresh nutrient and energy source. The higher OM content under NT compared with CT probably also explains why more enchytraeids were found under NT (Hendrix et al. 1986, 1987; Table 8.4). Therefore, it is somewhat puzzling that House & Parmelee (1985) found exactly the opposite: in contrast to all other functional groups enchytraeids were more abundant under CT than under NT, whereas the OM content under CT was lower. In many arable soils, the enchytraeid biomass is considerably lower than earthworm biomass. House & Parmelee (1985) state that enchytraeids may have a more pronounced effect on OM decomposition than earthworms, because of their much higher respiration rate, but Ryszkowski (1985) demonstrated that this does not always outweigh the effect of lower numbers.

Assuming values from the literature for N contents of 11–13%, depending on species, Didden (1991) calculated the N turnover rate of enchytraeid tissue, using daily estimates of standing stock and productivity. In 1986 the N flux in the 0–40 cm layer was 1·9 and 2·7 kg/ha under IF and CF, respectively. Since these organisms have a very low assimilation efficiency, the total N flux is almost ten times higher.

Earthworms

The number of earthworms may vary considerably in arable soils. It has been shown in many studies that management has a predominant effect on earthworm abundancy. For instance, under CT their number was much lower than under NT (Gerard & Hay 1979; Edwards & Lofty 1982; House & Parmelee 1985). Andrén, Paustian & Rosswall (1988) found no effect of N addition on the numbers of earthworms under barley. At the Lovinkhoeve site previous and present management has affected the colonization of earthworms, causing them to be present in IF, but not in CF.

Hendrix *et al.* (1986, 1987) estimated an earthworm respiration of 14 and 700 kg C/ha per year under CT and NT, respectively, equivalent with 5·3 and 29% of total soil respiration (Table 8.5). These results suggest that, at least under NT, earthworms play a major role in direct mineralization. Parmelee (1987) and Parmelee & Crossley (1988) calculated a population turnover of earthworms of 3·3 times per year. Based on this figure and a standing stock population under NT of approximately 143·5 kg dry matter/ha, they calculated an N flux from earthworm tissue of 40 kg/ha per year. At the Lovinkhoeve site, under winter wheat (IF only) and in Sweden under barley (Table 8.4), earthworm biomass was much lower, and consequently the N flux from tissue was lower as well. Figures for N flux for a number of ecosystems were in the wide range of 10–225 kg/ha per year as mentioned by Lee (1985). Besides nitrogen turnover from tissue, nitrogen excreted as urine and mucoproteins may quantitatively be important. Depending on earthworm abundance, mineral N flux from these sources was estimated to be 18–50 kg/ha per year (Lee 1985). In contrast to urine, the amounts of mineral N released via casts is relatively insignificant. It ranges from less than 35–50 g (Lee 1985) to 3·5 kg/ha per year in pasture (Syers, Sharpley & Keeney 1979). In addition to their contribution to direct mineralization, earthworms may through mucus production and casting have a strong effect on indirect mineralization by stimulating the microflora (Scheu 1987). Earthworms have high consumption rates and low assimilation efficiencies, resulting in comminution of large amounts of organic matter, followed by mixing with mineral soil and excretion in casts. Microbial activity in casts and soil around burrows is often several-fold increased compared with the surrounding soil (Shaw & Pawluk 1986; Scheu 1987). Boström (1988) reported the nitrogen flow to the soil resulting from earthworms and cocoons and from the excretion of urine and mucus to range between 3 and 12 kg N/ha per year in four arable cropping systems, the lower figure being typical for barley. If indirect effects on the microflora as a consequence of comminuting organic matter are included, these figures increase to 13–52 kg N/ha per year, corresponding to 16–25% of net annual mineralization. Altogether, the impact of earthworms on mineralization in arable soils is significant.

Macro-arthropods

Very little is known about the effect of macro-arthropod activity on mineralization dynamics under cereals. Where investigated, their biomass is generally low. If one assumes rather low turnover rates, their effect on direct mineralization will probably be small. Similar to earthworms, macro-arthropods will have an indirect effect through comminution of organic matter and stimulation of microbial activity in excrements. The significance of this indirect effect is unknown due to a lack of quantitative data.

SOIL FAUNA AND SOIL STRUCTURE

There is extensive literature on the interactions between soil fauna and soil structure (e.g. Spence 1985; Mitchell & Nakas 1986; Edwards *et al.* 1988; Elliott & Coleman 1988; Lavelle 1988). Oades (1984) defined soil structure as the size and arrangement of particles and pores in soil. Soil fauna may create new pores by burrowing, enlarge existing pores, or reduce these pores as a result of lateral pressure elsewhere in the soil, or deposition of material from elsewhere. Soil aggregates may be created and stabilized by excretory products from the fauna. Existing aggregates may be reduced in size, or destroyed completely by passage through soil animals. Complex polysaccharides and gums from micro-organisms and plant roots may also play an important role in the creation and stabilization of aggregates and pores (Tisdall & Oades 1982; Oades 1984).

In this contribution we restrict ourselves to recent findings at the Lovinkhoeve site, where the significance of the soil fauna in the maintenance or improvement of soil structure is likely to be accentuated more in IF than in CF. The effects of the soil biota on soil porosity after 20 years of farming was studied by Kooistra, Lebbink & Brussaard (1989) in CF, a CF-ley rotation with regular farmyard manure addition (CFL) and in CF with minimum tillage (CFMT). Earthworms had only colonized CFL. In CF less than 10% of the macropores larger than 30 μm in diameter, in the tilled layer had been influenced by the soil biota, including roots, whereas over 90% of the macropores had been affected by tillage, traffic and internal slaking (Fig. 8.7). In CFMT, however, the fraction of macro-

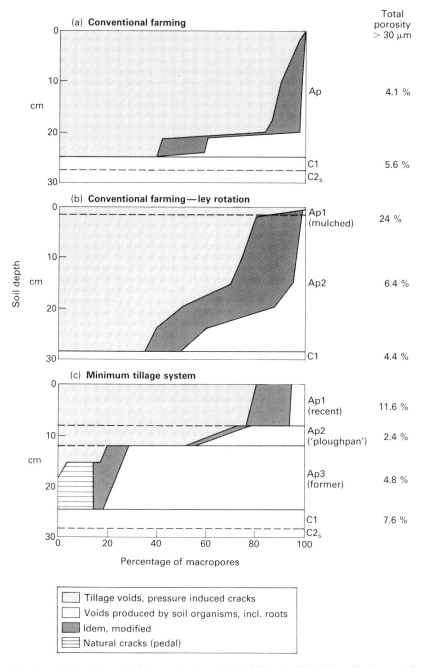

FIG. 8.7. Total soil porosity (>30 μm in diameter) at different soil depths and horizons at the Lovinkhoeve site under (a) conventional farming (CF), (b) conventional farming in a rotation with leys and regular farmyard manure addition (CFL) and (c) conventional farming with minimum tillage (CFMT). From Kooistra, Lebbink & Brussaard (1989).

pores affected by the soil biota rapidly increased below the shallow tilled Ap2 horizon to 80% at a depth of 25 cm. CFL was intermediate between CF and CFMT. Analysis of thin sections showed that the soil fauna, particulary enchytraeids and micro-arthropods, was the chief agent creating these voids or modifying existing voids by filling them with excrements.

Enchytraeids

The role of enchytraeids in soil structure evolution in agricultural fields under different management systems has been studied by Didden (1990). Enchytraeids transported only a very small part (0·001–0·01%) of bulk soil as mineral material. Their activity was somewhat higher in CF than in IF and their role in structure formation is probably small. However, in field-incubated soil cores they significantly ($P < 0.005$) increased the volume of pores with diameters of 50–200 µm, corresponding with their own diameter, and enhanced pore continuity. As a result, they increased air permeability up to two-fold depending on the time of the year. In addition, they increased the proportion of soil aggregates corresponding with the size of their fecal pellets (Didden 1990).

Earthworms

Earthworm activities have a clear effect on soil physical properties. Soil porosity and hydraulic conductivity are affected by burrowing. It is well known that the type of cultivation and cropping may affect earthworm populations dramatically (Edwards 1983). As casting may either increase or decrease soil aggregate stability (Lee 1985; Marinissen & Dexter 1990), earthworms may affect aggregate stability in different ways in different soils. In loamy sand or sandy clay loam soils of Georgia, USA, earthworm activity increased the average diameter of water stable aggregates, especially in the >2000 µm class. The effect in the loamy sand was stronger than in the sandy clay loam (Brussaard *et al.* 1990). In fluvial soils of the Netherlands, however, with textures ranging from sandy clay loam to clay loam and loam, earthworms caused a shift in diameter of water stable aggregates from 2000–8000 to 300–2000 µm (Brussaard *et al.* 1990).

CONCLUDING REMARKS

In terms of C and N mineralization only a few animals, notably protozoa and earthworms, seem to increase the amounts processed by the soil microflora substantially. A more detailed study of the activity of these groups of organisms is needed before a better understanding of the dynamics of mineralization can be gained, especially if one aims to match the supply of nutrients from organic matter and the demand of nutrients by plants. Enchytraeids, nematodes, micro- and macro-arthropods are less important in this respect. These groups may, however, exert an indirect effect on mineralization by stimulating (nematodes) or stabilizing (arthropods) the soil organisms that they consume. Also, their role may be more important in natural systems than in agricultural ecosystems (Crossley, Coleman & Hendrix 1989), but this warrants further research.

Moore *et al.* (1988) proposed the functional link approach as a means of integrating several ecological studies of soil fauna and of establishing the impact of functional groups within a food web on cycling of nutrients. Functional food webs have the capability to assess the regulatory capacity of any functional group and to represent interactions between the functional groups. This approach was applied to the data from the Lovinkhoeve site by Moore & De Ruiter (1990) and is currently being used to estimate the proportional contribution of the soil fauna to nitrogen mineralization in CF and IF. A preliminary analysis has yielded figures of 40% in CF and 50% in IF (P.C. De Ruiter, personal communication).

In terms of soil structure formation and soil aggregate stability, the meso- and macrofauna, especially earthworms, appear to be important but the contribution of the soil microfauna needs to be studied in more detail. Long-term studies with the soil fauna using labelled organic matter should provide a better understanding of the formation and maintenance of soil structure. Field and laboratory studies with known assemblages of organisms are needed with micromorphological and submicroscopical analysis of soil-thin sections.

With the current emphasis on soil conservation and reduced input agriculture, there is an increasing scope for research into the interplay between soil biota, soil structure and nutrient cycling. With the results of such research we aim to reach the point where we can, rather than by-passing the biology of soils by chemical and mechanical means, let the soil work for us (Elliott & Coleman 1988).

ACKNOWLEDGMENTS

Thanks are due to L.A. Bouwman and W.A.M. Didden for the supply of unpublished data and to P.C. De Ruiter and W.P. Wadman for helpful comments. This work was partly supported by the Netherlands Integrated Soil Research Programme.

REFERENCES

Alabouvette, C., Couteaux, M., Old, K.M., Pussard, M., Reisinger, O., & Toutain, F. (1981). Les protozoaires du sol: aspects ecologiques et methodologiques. *Annales Biologiques*, **20**, 253–303.

Anderson, J.M. (1988). Spatiotemporal effects of invertebrates on soil processes. *Biology and Fertility of Soils*, **6**, 216–227.

Andrén, O. & Lagerlöf, J. (1983). Soil fauna (Microarthropods, Enchytraeids, Nematodes) in Swedish agricultural cropping systems. *Acta Agriculturae Scandinavica*, **35**, 33–52.

Andrén, O. & Schnürer, J. (1985). Barley straw decomposition with varied levels of microbial grazing by *Folsomia fimetaria* (L.). *Oecologia (Berlin)*, **68**, 57–62.

Andrén, O., Paustian, K. & Rosswall, T. (1988). Soil biotic interactions in the functioning of agroecosystems. *Agriculture, Ecosystems and Environment*, **24**, 57–67.

Andrén, O., Lindberg, T., Boström, U., Clarholm, M., Hansson, A.-C., Johansson, G., Lagerlöf, J., Paustian, K., Persson, J., Pettersson, R., Sohlenius, B. & Wivstad, M. (1989a). Organic carbon and nitrogen flows. *Ecology of Arable Land — Organisms, Carbon and Nitrogen Cycling* (Ed. by O. Andrén, T. Lindberg, K. Paustain & T. Rosswall). *Ecological Bulletins*, **40**, 85–126.

Andrén, O., Lindberg, T., Paustian, K. & Rosswall, T. (Eds) (1989b). *Ecology of Arable Land — Organisms, Carbon and Nitrogen Cycling. Ecological Bulletins*, **40**, 1–222.

Barsdate, R.J., Prenkti, R.T. & Fenchel, T. (1974). Phosphorus cycle of model ecosystems: significance for decomposer food chains and effects of bacterial grazers. *Oikos*, **25**, 239–251.

Bloem, J. & Brussaard, L. (1989). Frequency of dividing-divided cells as an index of bacterial growth rate in soil. *Abstracts International Workshop on Modern Techniques in Soil Ecology*. Athens, Georgia, USA.

Boström, U. (1988). *Ecology of earthworms in arable land—population dynamics and activity in four arable cropping systems*. PhD thesis, Swedish University of Agricultural Sciences, Uppsala.

Boström, U. & Sohlenius, B. (1986). Short term dynamics of nematode communities in arable soil. Influence of a perennial and an annual cropping system. *Pedobiologia*, **29**, 345–357.

Brussaard, L., Coleman, D.C., Crossley, Jr., D.A., Didden, W.A.M., Hendrix, P.F. & Marinissen, J.C.Y. (1990). Impacts of earthworms on soil aggregate stability. *Proceedings of the 14th International Congress of Soil Science*, **3**, 100–105. Kyoto, Japan.

Brussaard, L., Kools, J.P., Bouwman, L.A. & De Ruiter, P.C. (1991). Population dynamics and nitrogen mineralization rates in soil as influenced by bacterial grazing nematodes and mites. *Proceedings of the 10th International Soil Zoology Colloquium*, Bangalore (in press).

Brussaard, L., Van Veen, J.A., Kooistra, M.J. & Lebbink, G. (1988). The Dutch programme on soil ecology of arable farming systems I. Objectives, approach and some preliminary results. *Ecological Bulletins*, **39**, 35–40.

Campbell, C.A. & Biederbeck, V.D. (1976). Soil bacterial changes as affected by growing season weather: a field and laboratory study. *Canadian Journal of Soil Science*, **56**, 293–310.

Carter, M.R. & Rennie, D.A. (1982). Changes in soil quality under a zero tillage farming system: distribution of microbial biomass and mineralizable C and N potentials. *Canadian Journal of Soil Science*, **62**, 587–597.

Clarholm, M. (1983). *Dynamics of soil bacteria in relation to plants, protozoa and inorganic nitrogen*. PhD thesis, University of Agricultural Sciences, Uppsala.

Clarholm, M. (1989). Effects of plant–bacterial–amoebal interactions on plant uptake of nitrogen under field conditions. *Biology and Fertility of Soils*, **8**, 373–378.

Cole, C.V., Elliott, E.T., Hunt, H.W. & Coleman, D.C. (1978). Trophic interactions in soils as they affect energy and nutrient dynamics V. Phosphorus transformations. *Microbial Ecology*, **4**, 381–387.

Coleman, D.C., Ingham, R.E., McClellan, J.F. & Trofymov, J.A. (1984). Soil nutrient transformations in the rhizosphere via animal–microbial interactions. *Invertebrate–Microbial Interactions* (Ed. by J.M. Anderson, A.D.M. Rayner & D.W.H. Walton), pp. 35–58. Cambridge University Press, Cambridge.

Crossley, Jr., D.A., Coleman, D.C. & Hendrix, P.F. (1989). The importance of the fauna in agricultural soils: research approaches and perspectives. *Agriculture, Ecosystems and Environment*, **27**, 47–55.

Cutler, D.W. & Bal, D.V. (1926). Influence of protozoa on nitrogen fixation by *Azotobacter chroococcum*. *Annals of Applied Biology*, **13**, 516–534.

Cutler, D.W. & Crump, L.M. (1929). Carbon dioxide production in sands and soils in the presence of amoebae. *Annals of Applied Biology*, **16**, 472–482.

Darbyshire, J.F. (1972a). Nitrogen fixation by *Azotobacter chroococcum* in the presece of *Colpoda steinii*. I. The influence of temperature. *Soil Biology and Biochemstry*, **4**, 359–369.

Darbyshire, J.F. (1972b). Nitrogen fixation by *Azotobacter chroococcum* in the presence of *Colpoda steinii*. II. The influence of agitation. *Soil Biology and Biochemistry*, **4**, 371–376.

Darbyshire, J.F. (1976). Effect of water suction on the growth in soil of the ciliate *Colpoda steinii* and the bacterium *Azotobacter chroococcum*. *Journal of Soil Science*, **27**, 369–376.

Darbyshire, J.F., Wheatley, R.E., Greaves, M.P. & Inkson, R.H.E. (1974). A rapid micro-method for estimating bacterial and protozoan populations in soil. *Revue d'Ecologie et de Biologie du Sol*, **11**, 465–475.

Didden, W.A.M. (1990). Involvement of Enchytraeidae (Oligochaeta) in soil structure evolution in agricultural fields. *Biology and Fertility of Soils*, **9**, 152–158.

Didden, W.A.M. (1991). *Population ecology and functioning of Enchytraeidae in some arable farming systems*. PhD thesis, Wageningen Agricultural University.

Edwards, C.A. (1983). Earthworm ecology in cultivated soils. *Earthworm Ecology from Darwin to Vermiculture*. (Ed. by J.E. Satchell), pp. 123–137. Chapman & Hall, London.

Edwards, C.A. & Lofty, J.R. (1982). The effect of direct drilling and minimal cultivation on earthworm populations. *Journal of Applied Ecology*, **19**, 723–734.

Edwards, C.A., Stinner, B.R., Stinner, D. & Rabatin, S. (Eds) (1988). Biological interactions in soil. *Agriculture, Ecosystems and Environment*, No. 1–3, 380 p.

Elliott, E.T. & Coleman, D.C. (1988). Let the soil work for us. *Ecological Bulletins*, **39**, 23–32.

Elliott, E.T., Coleman, D.C. & Cole, C.V. (1979). The influence of amoebae on the uptake of nitrogen by plants in gnotobiotic soils. *The Soil–Root Interface* (Ed. by J.L. Harley & R. Scott Russell), pp. 221–229. Academic Press, London.

Elliott, E.T., Horton, K., Moore, J.C., Coleman, D.C. & Cole, C.V. (1984). Mineralization dynamics in fallow dryland wheat plots, Colorado. *Plant and Soil*, **76**, 149–155.

Elliott, E.T., Hunt, H.W., Walter, D.E. & Moore, J.C. (1986). Microcosms, mesocosms and ecosystems: linking the laboratory to the field. *Perspectives in Microbial Ecology* (Ed. by F. Megusar & M. Gantar), pp. 472–480. *Proceedings of the 4th International Symposium on Microbial Ecology*, Slovene Society of Microbial Ecology, Ljubljana.

Fenchel, T. & Harrison, P. (1976). The significance of bacterial grazing and mineral cycling for the decomposition of particulate detritus. *The Role of Terrestrial and Aquatic Organisms in Decomposition Processes* (Ed. by J.M. Anderson & A. Macfadyen), pp. 285–299. Blackwell Scientific Publications, Oxford.

Gerard, B.M. & Hay, R.K.M. (1979). The effects on earthworms of ploughing, tined cultivation, direct drilling and nitrogen in a barley monoculture ecosystem. *Journal of Agricultural Science, Cambridge*, **93**, 147–155.

Granatstein, D.M., Bezdicek, D.F., Cochran, V.L., Elliott, E.T. & Hammel, J. (1987). Long-term tillage and rotation effects on soil microbial biomass, carbon and nitrogen. *Biology and Fertility of Soils*, **5**, 265–270.

Griffiths, B.S. (1986). Mineralization of nitrogen and phosphorus by mixed cultures of the ciliate protozoan *Colpoda steinii*, the nematode *Rhabditis* sp. and the bacterium *Pseudomonas fluorescens*. *Soil Biology and Biochemistry*, **18**, 637–641.

Heal, O.W. & MacLean, S.F. (1975). Comparative productivity in ecosystems — secondary productivity. *Unifying Concepts in Ecology* (Ed. by W.H. Van Dobben & R.H. Lowe-McConnel), pp. 88–109. PUDOC, Wageningen.

Hendrix, P.F., Crossley, Jr., D.A., Coleman, D.C., Parmelee, R.W. & Beare, M.H. (1987). Carbon dynamics in soil microbes and fauna in conventional and no-tillage agroecosystems. *INTECOL Bulletin*, **15**, 59–63.

Hendrix, P.F., Parmelee, R.W., Crossley, Jr., D.A., Coleman, D.C., Odum, E.P. & Groffman, P.M. (1986). Detritus food webs in conventional and no-tillage agroecosystems. *Bioscience*, **36**, 374–380.

Hervey, R.J. & Greves, J.E. (1941). Nitrogen fixation by *Azotobacter chroococcum* in the presence of soil protozoa. *Soil Science*, **51**, 85–100.

House, G.J. & Parmelee, R.W. (1985). Comparison of soil arthropods and earthworms from conventional and no-tillage agroecosystems. *Soil and Tillage Research*, **5**, 351–360.

Hunt, H.W., Coleman, D.C., Cole, C.V., Ingham, R.E., Elliott, E.T. & Woods, L.E. (1984). Simulation model of a food web with bacteria, amoebae and nematodes in soil. *Current Perspectives in Microbial Ecology* (Ed. by M.J. Klug & C.A. Reddy), pp. 424–434. American Society of Microbiology, Washington.

Hunt, H.W., Coleman, D.C., Ingham, E.R., Ingham, R.E., Elliott, E.T., Moore, J.C., Rose, S.L., Reid, C.P.P. & Morley, C.R. (1987). The detrital food web in a shortgrass prairie. *Biology and Fertility of Soils*, **3**, 57–68.

Ingham, E.R., Trofymow, J.A., Ames, R.N., Hunt, H.W., Morley, C.R., Moore, J.C. & Coleman, D.C. (1986). Trophic interactions and nitrogen cycling in a semi-arid grassland soil. *Journal of Applied Ecology*, **23**, 597–614 and 615–630.

Jenkinson, D.S., Ladd, J.N. & Rayner, J.H. (1980). Microbial biomass in soil: Measurement and turnover. *Soil Biochemistry* (Ed. by E.A. Paul & J.N. Ladd), Vol. 5. pp. 415–471. Marcel Dekker, New York.

Kooistra, M.J., Lebbink, G. & Brussaard, L. (1989). The Dutch Programme on Soil Ecology of Arable Farming Systems. II. Geogenesis, agricultural history, field site characteristics and present farming systems at the Lovinkhoeve experimental farm. *Agriculture, Ecosystems and Environment*, **27**, 361–387.

Kortleven, J.K. (1963). Kwantitatieve aspecten van humusopbouw en humusafbraak. *Verslag Landbouwkundig Onderzoek*, 69.1, 109 p.

Kuikman, P.J. (1990). *Mineralization of nitrogen by protozoan activity in soil*. PhD thesis, Wageningen Agricultural University.

Kuikman, P.J. & Van Veen, J.A. (1989). The impact of protozoa on the availability of bacterial nitrogen to plants. *Biology and Fertility of Soils*, **8**, 13–18.

Lagerlöf, J. (1987). *Dynamics and activity of microarthropods and enchytraeids in four cropping systems*. PhD thesis, University of Agricultural Sciences, Uppsala.

Lagerlöf, J. & Andrén, O. (1988). Abundance and activity of soil mites (Acari) in four cropping systems. *Pedobiologia*, **32**, 129–145.

Lagerlöf, J., Andrén, O. & Paustian, K. (1989). Dynamics and contribution to carbon flows of enchytraeidae (Oligochaeta) under four cropping systems. *Journal of Applied Ecology*, **26**, 183–199.

Lavelle, P. (1988). Earthworm activities and the soil system. *Biology and Fertility of Soils*, **6**, 237–251.

Lee, K.E. (1985). *Earthworms, their Ecology and Relationships with Soils and Land Use.* Academic Press, Sydney.

Marinissen, J.C.Y. (1991). Colonization of arable fields by earthworms in a newly reclaimed polder in The Netherlands. *Proceedings of the Xth International Soil Zoology Colloquium*, Bangalore (in press).

Marinissen, J.C.Y. & Dexter, A.R. (1990). Mechanism of stabilization of earthworm casts and artificial casts. *Biology and Fertility of Soils*, **9**, 163–167.

Meiklejohn, J. (1930). The relation between the numbers of a soil bacterium and the ammonium produced by it in peptone solutions; with some reference to the effect on this process of the presence of amoebae. *Annals of Applied Biology*, **17**, 614–637.

Meiklejohn J. (1932). The effect of *Colpidium* on ammonia production by soil bacteria. *Annals of Applied Biology*, **19**, 584–598.

Mitchell, M.J. & Nakas, J.P. (Eds) **(1986).** *Microfloral Interactions in Natural and Man-managed Ecosystems.* Nijhoff/Junk, Dordrecht.

Moore, J.C. & De Ruiter, P.C. (1990). Temporal and spatial heterogeneity of trophic interactions within belowground food webs: an analytical approach to understand multi-dimensional systems. *Agriculture, Ecosystems and Environment* (in press).

Moore, J.C., Walter, D.E. & Hunt, H.W. (1988). Arthropod regulation of micro- and mesobiota in below-ground detrital food webs. *Annual Review of Entomology*, **33**, 419–439.

Nannipieri, P. (1984). Microbial biomass and activity: ecological significance. *Current Perspectives in Microbial Ecology* (Ed. by M.J. Klug & C.A. Reddy), pp. 512–521. American Society of Microbiology, Washington, DC.

Nikoljuk, V.F. (1969). Some aspects of the study of soil protozoa. *Acta Protozoologica*, **7**, 99–101.

Oades, J.M. (1984). Soil organic matter and structural stability: mechanisms and implications for management. *Plant and Soil*, **76**, 319–337.

O'Neill, R.V. (1969). Indirect estimates of energy fluxes in animal food webs. *Journal of Theoretical Biology*, **22**, 284–290.

Parmelee, R.W. (1987). *The role of soil fauna in decomposition and nutrient cycling processes in conventional and no-tillage agroecosystems on the Georgia Piedmont.* PhD thesis, University of Georgia.

Parmelee, R.W., Beare, M.H. & Blair, J.M. (1989). Decomposition and nitrogen dynamics of surface weed residues in no-tillage agroecosystems under drought conditions: influence of resource quality on the decomposer community. *Soil Biology and Biochemistry*, **21**, 97–103.

Parmelee, R.W. & Crossley, Jr., D.A. (1988). Earthworm production and role in the nitrogen cycle of a no-tillage agroecosystem of the Georgia Piedmont. *Pedobiologia*, **32**, 353–361.

Persson, T., Baath, E., Clarholm, M., Lunkvist, H., Söderström, B.E. & Sohlenius, B. (1980). Trophic structure, biomass dynamics and carbon metabolism in a Scots pine forest. *Ecological Bulletins (Stockholm)*, **32**, 419–459.

Postma, J. (1989). *Distribution and population dynamics of Rhizobium sp. introduced into soil.* PhD thesis, Wageningen Agricultural University.

Pussard, M. & Rouelle, J. (1986). Predation de la microflore. Effect des protozoaires sur la dynamique de population bacterienne. *Protistologica*, **22**, 105–110.

Ross, D.J., Tate, K.R., Cairns, A., Meyrick, K.F. (1981). Fluctuations in microbial biomass indices at different sampling times in soils from tussock grasslands. *Soil Biology and Biochemistry*, **13**, 109–114.

Russell, E.J. & Hutchinson, H.B. (1909). The effect of partial sterilization of soil on the production of plant food. *Journal of Agricultural Science, Cambridge*, **3**, 111–144.

Rutherford, P.M. & Juma, N.G. (**1989**). Dynamics of microbial biomass and soil fauna in two contrasting soils cropped to barley (*Hordeum vulgare* L). *Biology and Fertility of Soils*, **8**, 144–153.

Ryszkowski, L. (**1985**). Impoverishment of soil fauna due to agriculture. *INTECOL Bulletin*, **12**, 7–17.

Scheu, S. (**1987**). Microbial activity and nutrient dynamics in earthworm casts (Lumbricidae). *Biology and Fertility of Soils*, **5**, 230–234.

Schnürer, J., Clarholm, M., Boström, S. & Rosswall, T. (**1986**). Effects of moisture on soil microorganisms and nematodes: a field experiment. *Microbial Ecology*, **12**, 217–230.

Schnürer, J., Clarholm, M. & Rosswall, T. (**1985**). Microbial biomass and activity in agricultural soil with different organic matter contents. *Soil Biology and Biochemistry*, **17**, 611–618.

Schnürer, J., Clarholm, M. & Rosswall, T. (**1986**). Fungi, bacteria and protozoa in soil from four arable cropping systems. *Biology and Fertility of Soils*, **2**, 119–126.

Seastedt, T.R. (**1984**). The role of microarthropods in decomposition and mineralization processes. *Annual Review of Entomology*, **29**, 25–46.

Shaw, C. & Pawluk, S. (**1986**). Faecal microbiology of *Octolasion tryteaum, Aporrectodea turgida* and *Lumbricus terrestris* and its relation to the carbon budgets of three artificial soils. *Pedobiologia*, **29**, 377–389.

Sohlenius, B., Boström, S. & Sandor, A. (**1988**). Carbon and nitrogen budgets of nematodes in arable soil. *Biology and Fertility of Soils*, **6**, 1–8.

Spence, J. (Ed.) (**1985**). Faunal influences on soil structure. *Quaestiones Entomologicae*, **21**, 371–700.

Stinner, B.R., & Crossley, Jr., D.A. (**1982**). Nematodes in no-tillage agroecosystems. *Nematodes in Soil Ecosystems* (Ed. by D.W. Freckman), pp. 14–28. University of Texas Press, Austin.

Syers, J.K., Sharpley, A.N. & Keeney, D.R. (**1979**). Cycling of nitrogen by surface-casting earthworms in a pasture ecosystem. *Soil Biology and Biochemistry*, **11**, 181–185.

Telegdy-Kovats, L. de (**1932**). The growth and respiration of bacteria in sand cultures in the presence and absence of protozoa. *Annals of Applied Biology*, **19**, 65–86.

Tisdall, J.M. & Oades, J.M. (**1982**). Organic matter and water-stable aggregates in soil. *Journal of Soil Science*, **33**, 141–163.

Vereijken, P. (**1986**). From conventional to integrated agriculture. *Netherlands Journal of Agricultural Science*, **34**, 387–393.

Vereijken, P. (**1989**). From integrated control to integrated farming, an experimental approach. *Agriculture, Ecosystems and Environment*, **26**, 37–43.

Verhoef, H.A. & Brussaard, L. (**1990**). Decomposition and nitrogen mineralization in natural and agro-ecosystems; the contribution of soil animals. *Biogeochemistry*, **11**, 175–211.

Waksman, S.A. (**1916**). Studies on soil protozoa. *Soil Science*, **1**, 135–152.

Whitford, W.G., Freckman, D.W., Santos, P.F. Elkins, N.Z. & Parker, L.W. (**1982**). The role of nematodes in decomposition in desert ecosystems. *Nematodes in Soil Ecosystems* (Ed. by D.W. Freckman), pp. 98–166. University of Texas Press, Austin.

Woldendorp, J.W. (**1981**). Nutrients in the rhizosphere. *Proceedings of the 16th Colloqium of the International Potash Institute*, Bern, pp. 99–125.

Woods, L.E., Cole, C.V., Elliott, E.T., Anderson, R.V. & Coleman, D.C. (**1982**). Nitrogen transfers in soil as affected by bacterial–microfaunal interactions. *Soil Biology and Biochemistry*, **14**, 93–98.

PART 3
SPECIES INTERACTIONS

Session Chairman: T. LEWIS
Session Organizer: N. CARTER

It is now 147 years since Sir John Lawes began the winter wheat experiment on Broadbalk field, Rothamsted. Every year since then winter wheat has been sown and harvested in all or part of the field so it has become a classic reference point for study of a great range of interactions between weather, rotations, cultivations, cultivars, fertilizers, pesticides, the herbaceous flora and the invertebrate fauna. A brief reference to some of the organisms previously studied on Broadbalk will serve as a fitting introduction to the range of ecological interactions to be considered in this part of the volume.

It was on this field in 1935 that eyespot (*Pseudocercosporella herpotrichoides* (Fron.)) was first identified in the UK. Later, comparisons of yields and differences in amounts of take-all (*Gaeumannomyces graminis*, var. *tritici* Walker) between continuous wheat on Broadbalk and other fields in short sequences of cereals over a period of years culminated in the hypothesis of 'take-all decline'. For nearly 40 years fluctuations in the numbers of wheat blossom midges (*Contarinia tritici* (Kirby) and *Sitodiplosis mosellana* (Géhin)) were studied as well as the affects on yield of wheat bulb fly (*Delia coarctata* (Fallén)) on crops sown after fallow. More recently the field has served as one of the sites for observations on interactions between barley yellow dwarf virus, aphid vectors and yield. In another long-term study weeds and weed seeds were surveyed regularly. About fifty annual and ten perennial weed species occur in the field. Where weedkillers have never been applied each plot has its characteristic ten to twenty species and the ground is covered with weeds after harvest, except on the unmanured plot.

The papers in this part touch on and develop the type of work pioneered on Broadbalk, bringing a welcome broadening of the science base and a widening of the geographical spread of observations. Peter Blakeman explores in intriguing detail the microflora colonizing the cereal leaf surface and the many airborne or splashborne colonists from seed, leaf and soil. Once established, the source of nutrients for these organisms, arising from leaf exudates, pollen, aphid honeydew and metabolites from other colonists, encourage or discourage growth leading to varying degrees of proliferation. Nick Carter and Roger Plumb describe the epidemiology of the aphid-vectored barley yellow dwarf viruses and the fungally vectored cereal mosaic viruses. The switch to early autumn-sown cereals has benefitted both groups, indicating yet again that many of the problems associated with present-day cereal production arise from practices developed to push crop yield towards the maximum limits attainable with present cultivars in highly competitive farming systems.

Cereal weed ecology has suffered from financial cutbacks this last decade and a tendency to be overshadowed by herbicide research. Despite the remarkable advances and benefits conferred by the latter, there are now signs of problems similar to those that developed in insect pest control, particularly the occurrence of resistance to weedkillers. Les Firbank's paper on competition between weeds and crop and new approaches to modelling weed population dynamics is most welcome, especially as he recognizes the need to base models on data collected from realistically large-sized experimental plots.

The encouragement of natural enemies of pest aphids has long been an objective of integrated control in cereals. Wilf Powell and Steve Wratten bring together the different emphases of the Rothamsted and Southampton University approaches. The former has concentrated more on the interactions between aphid parasitoids and host aphids, and on devising novel ways of manipulating parasitoid density in cereal fields using, for example, semiochemicals derived from aphids or their food plants. By contrast, Southampton are trying to manipulate the habitat in and around cereal fields to encourage successful overwintering of polyphagous predators and encourage their early spread into fields in spring. The final paper by Miguel Altieri underlines some of the basic tenets of husbandry affecting pests and diseases with a warning that socio-economic and public concerns will increasingly influence the approach to farming.

The five papers form an intricate web of interactions which will broaden the experience and knowledge of many of the specialists studying particular aspects of farmland ecology.

9. MICROBIAL INTERACTIONS IN THE PHYLLOPLANE

J.P. BLAKEMAN

Plant Pathology Research Division, Department of Agriculture for Northern Ireland and Department of Mycology and Plant Pathology, The Queen's University of Belfast, Newforge Lane, Belfast BT9 5PX, Northern Ireland

INTRODUCTION

The shoots of cereals, like other plants, possess a microflora consisting of bacteria, yeasts and filamentous fungi. In this article the origin of this microflora is discussed together with a brief survey of the different groups of organisms and their ecological succession in this habitat.

Saprophytic micro-organisms colonizing cereal leaves may markedly influence the growth and well-being of the cereal plant. Certain of these activities, such as interactions with potential parasites leading to control of major foliar diseases, fixation of nitrogen in tropical regions and production of growth hormones, may benefit the plant. Others may impair the growth of the cereal plant and the yield and quality of its grain. The nature of these interactions and their effect on the plant will be examined.

THE PHYLLOPLANE MICROFLORA OF CEREALS

The phylloplane microflora of cereal leaves consists of bacteria, yeasts and filamentous fungi. Bacteria are often the initial colonists developing on newly formed leaves prior to the later arrival of yeasts and filamentous fungi. This underlying seasonal sequence of microbial development may be greatly modified by prevailing weather conditions. Micro-organisms on barley leaves are shown in scanning electron micrographs (Fig. 9.1).

Bacteria

Early in the growing season, nutrient levels on leaves are low and are derived almost entirely from leakage from internal tissues. Leaked products consist chiefly of amino acids and carbohydrates which are rapidly utilized by multiplying bacterial cells (Brodie & Blakeman 1976; Rodger & Blakeman 1984).

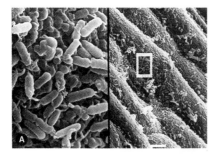
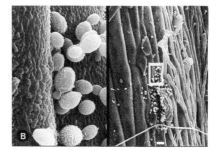
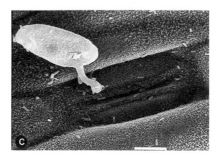

FIG. 9.1. (A) Underside of barley leaf showing rod-shaped bacteria. Magnified area shows details of bacteria. (B) Upperside of barley leaf showing filamentous fungi and yeasts. Details of yeasts are shown in magnified area. (C) Underside of barley leaf showing germinated powdery mildew conidium with appressorium and associated bacteria near stoma. Bar markers equal 10 μm.

There have been few comprehensive studies of the phylloplane bacterial flora of the Gramineae. Austin, Goodfellow & Dickinson (1978) studied the bacteria flora on the leaves of *Lolium perenne* L. analysing populations using numerical taxonomic methods. The authors concluded that the use of standard diagnostic schemes was of limited value because of the difficulty of identifying strains beyond genus level. They showed that xanthomonads and pseudomonads formed dominant populations in the middle of the growing season with less characteristic phylloplane organisms such as listeriae and staphylococci occurring at the end of the growing season. They found that leaves of *L. perenne* contained distinctive bacterial populations which differed from those of nearby litter and soil.

Bacterial populations on *L. perenne* leaves were low in spring and summer but rose substantially in September. Direct observations indicated that the area of leaf colonized by bacteria was around 0·0001% in May increasing to 0·1% in September (Dickinson, Austin & Goodfellow 1975).

Estimates of the abundance of bacteria on *Lolium* are smaller than for most other crop plants studied.

Bacterial pathogens frequently form epiphytic populations on the surfaces of healthy leaves (Hirano & Upper 1983). Principal genera of pathogens of the Gramineae are *Clavibacter, Erwinia, Pseudomonas* and *Xanthomonas* (Paul & Smith 1989). De Cleene (1989) studied the establishment of *X. campestris* (Pammel) Dowson pathovars (pvs.) on Italian ryegrass and maize. The three pathovars studied showed different distribution patterns, particularly with respect to attachment to trichomes. Pathovars *oryzae* (Ishiyama) Dye and *oryzicola* (Fang *et al.*) Dye became more strongly attached to trichomes of the hairier non-host maize leaves than to leaves of the ryegrass host. This suggests that maize, as well as possibly other non-host plants, may provide a habitat for survival of the pathogen. Pathovars *oryzae* and *oryzicola* could also be detected on trichomes of ryegrass host leaves but pv. *graminis* could not. *X. campestris* pvs. *secalis* (Reddy *et al.*) Dye and *translucens* (Jones *et al.*) Dowson are specific pathogens to barley and rye, respectively, forming watersoaked lesions on their hosts (El-Banoby & Rudolph 1989). Pathovar *translucens* has been shown to be ice-nucleation active (Kim *et al.* 1987) which suggests these bacteria may incite frost damage to cereals, thereby facilitating colonization of internal leaf tissues.

Yeasts

One of the earliest studies of yeasts on wheat and barley leaf surfaces was by Last (1955). The yeast population was determined using the spore fall technique, whereby ballistospores of the yeasts discharged from leaves placed above an agar surface, are incubated to form colonies. The predominant species was the pink yeast *Sporobolomyces roseus* Kluyver et Van Niel. This species became numerous only after the middle of the growing season in spring-sown crops. Thereafter, yeast numbers increased reaching a maximum after leaves had died. Distal ends of leaves supported more colonies than proximal ends. It was assumed that numbers of yeasts were primarily influenced by the presence of sugars on the leaf together with sufficiently high atmospheric humidity to support growth. Last (1955) showed that numbers of *Sporobolomyces roseus* colonies were greatly decreased in the presence of *Tilletiopsis minor* Nyland, another ballistospore former. *Bullera alba* (Hanna) Derx was also isolated from cereal leaves. Yeasts appeared to have no effect on the development of powdery mildews on cereal leaves.

Another abundant yeast genus on cereal leaves, which was not detected

by Last using the spore-fall method because it does not produce ballistospores, is the white yeast *Cryptococcus*. Fokkema *et al.* (1979) showed
that both *Cryptococcus* and *Sporobolomyces* responded by greatly increasing cell numbers in the presence of naturally-occurring enhanced
sugar levels derived from pollen or aphid honeydew. By contrast with the
biotrophic parasite powdery mildew (Last 1955), these yeasts decreased
development of necrotrophic parasites as discussed further in a later
section.

Yeasts formed a significant part of the mycoflora of flag leaves, bracts
and caryopses of wheat from ear emergence to harvest. *Cryptococcus
albidus* (Saito) Skinner, *Sporobolomyces roseus* and *Torulopsis ingeniosa*
di Menna were predominant (Flannigan & Campbell 1977).

Filamentous fungi

Studies on the fungal flora of wheat and barley have shown that comparatively few species are active on green leaves. Direct microscopic observation (Diem 1974) showed *Cladosporium* spp. to be predominant. Their
activity was greatly influenced by pollen deposits on the leaves and also by
extended wet periods. The area of leaf covered by colonies normally
ranged from 0·002 to 0·774 mm^2/cm^2. Microscopic observation was essential for determining the number, location and extent of fungal colonies
on leaf surfaces. Indirect plating indicated only the number of visible
propagules removable from a unit area of leaf but gave no information
about their activity.

In another study, Flannigan & Campbell (1977) used a dilution plating
technique to study the microflora of flag leaves, bracts and caryopses of
wheat. The first filamentous fungi to become established on these organs
were *Aureobasidium pullulans* (de Bary) Arnaud, a black mycelial yeast-
like fungus, *Cladosporium cladosporioides* (Fresen) De Vries and *C.
herbarum* (Pers.) Link ex Fr. Later, on the dying flag leaf and bracts and
the ripening caryopsis, other fungi appeared — chiefly *Alternaria alternata*
(Fr.) Kiessler, *Verticillium lecanii* (Zimm.) Viegas and *Epicoccum pur-
purascens* Ehrenb. ex Schlecht. In contrast, Magan & Lacey (1986) found
that *Alternaria alternata* could be isolated from wheat flag leaves earlier
in the season, from July onwards. Using direct microscopy, *Alternaria
alternata* was shown to be the dominant species on ripening barley grain
(Hill & Lacey 1983).

The diversity of species and the extent of colonization of leaves greatly
increased as leaves started to senesce. Fungal hyphae were able to enter
dying leaves and bring about the first stages of saprophytic breakdown
prior to further decomposition of the leaf as leaf litter (Dickinson 1981).

THE ORIGIN OF THE
PHYLLOPLANE MICROFLORA

Cereal crops originate from seed planted in autumn or spring in northern temperate climates. Microbial inocula, for colonization of foliar surfaces of cereals, originate from four main sources — seed, soil, air and over-wintering inocula on shoots in the case of wintersown cereals. Because there are differences in the nature and means of spread of propagules of the main groups of phylloplane colonists viz. bacteria, yeasts and filamentous fungi, each group will be examined separately.

Bacteria

There is more information on the origin of bacterial inocula for colonization of the phylloplane than for the other groups of micro-organisms, partly because a large number of species of pathogens have epiphytic phases which have been studied intensively.

Many plant pathogenic bacteria are seedborne, of which pathovars of *P. syringae* van Hall and *X. campestris* are known epiphytic colonists.

Foliar populations of *P. syringae* pathovar *syringae* van Hall, cause of bacterial leaf necrosis of wheat, can originate from seedborne inocula (Fryda & Otta 1978). Migration of the bacteria from seed to seedling leaves is enhanced by high relative humidity. Epiphytic bacteria on cucumber seedlings may be transferred similarly (Leben 1961). However, even in dry conditions, some *P. syringae* cells could still migrate from wheat seed to seedling and persist on lower leaves but movement to upper leaves was prevented. There were no differences in the movement of bacteria on wheat cultivars resistant or susceptible to bacterial leaf necrosis (Fryda & Otta 1978).

Foliar epiphytic bacteria do not readily survive in soil but plant debris previously colonized by pathogenic bacteria may provide an inoculum source (Hirano & Upper 1983). These and soil bacteria such as species of *Bacillus*, especially *B. cereus* var. *mycoides* (Flügge) Smith and some other Gram-positive bacteria, are splashed from soil onto lower leaves during heavy rainfall.

Large numbers of airborne bacteria are deposited on leaves throughout the growing season, supplementing populations derived from seeds or plant debris. Bacterial cells on leaves may be transported short distances onto neighbouring leaves by rainsplash or slightly greater distances in rain-generated aerosols. However, there is little likelihood of long distance transport of aerosols because of the strong downward movement of air during rain and the collection of aerosols by raindrops (Hirano & Upper

1983). Different serotypes of *P. syringae* pathovar *syringae* from those present on seed appeared on upper leaves of spring wheat following rainfall. These same leaves could not be colonized by seed-derived migrating *P. syringae* pathovar *syringae* under dry conditions which demonstrated the importance of rainborne inocula in contributing to epiphytic colonization (Fryda & Otta 1978). Aerosols can also be generated during pulverization of plant material, e.g. potato, at harvest (Perombelon, Fox & Lowe 1979).

Plants are important sources of airborne bacteria (Lindemann *et al.* 1982). About four times more bacteria were found above a field of alfalfa than over an adjacent field of bare, dry soil sampled at the same time but plant species differed in their contribution to the airborne bacterial flora. For example, a maturing field of winter wheat produced more airborne bacteria than an upwind field of young maize. It was concluded that an upward flux of bacteria over crop plants could be generated under dry conditions to provide intermediate or long distance transport of bacteria which, on deposition, could initiate epiphytic populations if still viable.

Because of the overlap of winter- and spring-sown cereal crops, bacterial cells overwintered on winter cereal crops can transfer from these to spring crops by the mechanisms discussed above. As many components of the phylloplane bacterial flora of cereals are not host specific, transfer of inocula from non-cereal crops to cereals will also readily occur.

Yeasts

Unlike leaf bacteria, there is no transfer of yeasts from seeds to shoot surfaces. New leaves are initially free of yeasts but become colonized subsequently by deposition of airborne inocula or transfer of vegetative cells from neighbouring leaves by water splash (Fokkema *et al.* 1979). The majority of phylloplane yeasts are basidiomycetous, producing ballistospores that are released into the air. Such propagules then initiate new colonies on newly formed leaves remote from the source of inoculum.

Filamentous fungi

Filamentous fungi are primarily transferred to newly formed leaves by airborne spores. Spores are deposited on leaves from the air throughout the growing season though the number deposited tends to increase as the season progresses, reflecting greater amounts of saprophytic colonization as leaves begin to senesce which leads to enhanced sporulation. Deposited spores will often remain dormant on leaves until favourable conditions for germination and colonization occur, usually when leaves commence to senesce which is often accompanied by higher relative humidities and greater nutrient release to promote spore germination (Dickinson 1967).

PROPERTIES OF THE PHYLLOPLANE MICROFLORA

Interactions with pathogens

The nature of pathogens and their relationship with their hosts will determine what characteristics potential antagonists need to give effective control. A range of antagonistic mechanisms is found in different constituents of the phylloplane microflora, the most widespread of these being nutrient competition, parasitism and antibiosis. However, few examples of antibiosis on cereal leaves have been identified and this aspect will not be considered further. The major phylloplane colonists of cereals, the yeasts, are largely not antibiotic producers.

Nutrient competition

Necrotrophic foliar cereal pathogens, e.g. *Cochliobolus sativus* (Ito & Kuribayashi) Drechsler ex Dastur and *Septoria nodorum* Berk., respond to saprophytes competing for nutrients because these pathogens require nutrients during their initial development on the leaf prior to penetration. If the prepenetration phase can be inhibited, penetration and subsequent infection of the leaf fail, resulting in control of the disease. Nutrients to support the prepenetration phase include leaf exudates, pollen and aphid honeydew (Blakeman 1985). The larger the amounts of such nutrient substances, the greater the stimulation of infection by this group of pathogens. If the level of nutrients on leaves can be decreased, through utilization by saprophytes, lower disease levels will result (Fokkema 1981).

The scavenging activity of the saprophytic microflora of cereal leaves undoubtedly constitutes a form of natural biological control of necrotrophic pathogens. This natural system may be developed into a form of biological control by applying cells of appropriate organisms to enhance the natural population of competing saprophytes. Fokkema *et al.* (1979) showed that, for effective antagonism against *C. sativus* and *S. nodorum* on wheat leaves, natural yeast populations must be at a level of approximately 10^4 cells/cm^2 of leaf surface. Often the population of yeasts on wheat leaves was less than this and insufficient to interfere significantly with prepenetration development by these pathogens. The most successful method of increasing yeast populations to inhibitory levels was inoculation, on successive occasions, with additional cells in a 2% sucrose and 0·1% yeast extract solution. The treatment decreased infection of flag leaves by both pathogens by 50% during the first 3 weeks after treatment compared to control treatments sprayed only with water. However, at later stages in crop growth, disease development on the yeast-treated plants did not differ from the control plants, probably because the natural yeast population on the latter had also reached inhibitory levels. Conversely, increased infection by *C. sativus* was demonstrated on rye plants treated with benomyl (Fokkema *et al.* 1975). The fungicide decreased growth of the saprophytic mycoflora so allowing increased infection by the pathogen which was relatively insensitive to the fungicide.

At certain times of the year, additional nutrient sources, such as pollen grains and aphid honeydew, are deposited on leaves giving a nutritional stimulus to necrotrophic pathogens. In rye, pollen grains are shed onto leaves in large numbers over a short time period stimulating infection by *C. sativus* and *S. nodorum* relative to control plants kept free of pollen (Fokkema 1971). Similarly, pollen stimulates infection of maize leaves by *Fusarium graminearum* Schwabe (Naik & Busch 1978). On glasshouse rye plants, added cells of *Cladosporium* spp., *Aureobasidium pullulans*, *Cryptococcus laurentii* (Kuff.) Skinner (a white yeast) and *Sporobolomyces roseus* (a pink yeast) decreased this stimulatory effect (Fokkema 1973). Likewise saprophytes applied to leaves of wheat, to which aphid honey-dew had been added, prevented an increase in infection by *S. nodorum* (Riphagen *et al.* 1979). When honeydew was added to wheat leaves in advance (e.g. 5 days) of inoculation by *S. nodorum*, its stimulatory effect on infection by the pathogen was decreased because the natural saprophytic mycoflora could consume the additional nutrients during the time interval thus neutralizing the potential stimulus to the pathogen (Fokkema 1981).

Germination and appressorium formation by conidia of anthracnose fungi on leaves is normally not inhibited by the presence of nutrient-competing yeasts and bacteria (Blakeman & Parbery 1977). However, Williamson & Fokkema (1985) have shown that yeasts on the surfaces of maize plants could decrease lesion development by the anthracnose pathogen, *Colletotrichum graminicola* (Ces.) Wils., by 50%. The yeasts inhibited the normal functioning of the infection peg formed from the appressorium perhaps because the yeasts starved the fungus of nutrients at the penetration site, preventing production of cutinase enzymes and normal penetration of the leaf.

Bacteria effectively compete for nutrients to inhibit necrotrophic pathogens such as *Botrytis cinerea* (Pers.) ex Fr. and *Phoma betae* (Oudem.) Frank (Blakeman & Brodie 1977). Fungal development was limited by competition for amino acids which was of greater significance than competition for carbohydrates. However, on cereal leaves, bacterial populations are accompanied by a substantial rise in yeast cell numbers as the season advances. At the time when necrotrophic pathogens such as *C. sativus* and *S. nodorum* start to increase, yeasts become very significant competitors (Fokkema & Schippers 1986).

Parasitism

Both bacteria and filamentous fungi are capable of lysing germinating spores and fruiting structures of foliar pathogens of cereals. A number of parasitic associations have been reported in cereal rusts. For example, uredospores of cereal rusts have been shown to be lysed by isolates of *Bacillus*. One of these, *B. pumilis* Meyer & Gottheil, produced a heat stable lytic substance which destroyed germ tubes (Morgan 1963). This bacterium was able to control infection by *Puccinia recondita* (Rob. & Desm.) even when the bacterium was applied up to 9 days prior to inoculation with the rust. Parasitic bacteria have also been shown to lyse pycnia, aecidia and uredia pustules of cereal rusts. A *Bacillus* sp. was particularly effective (Levine, Bamburg & Atkinson 1936). The soilborne bacterium *Erwinia uredovora* (Pon *et al.*) Dye (= *X. campestris* pathovar *uredovorus* Pon *et al.*) can be splashed onto the surfaces of cereal leaves to infect erupting pustules causing their lysis (Pon *et al.* 1954). Lytic enzymes are probably involved since heat-killed bacterial cells had no effect (Hevesi & Mashaal 1975).

Hyperparasitic fungi can also cause the destruction of cereal rust fruiting structures. Perhaps the best known is *Eudarluca caricis* (Fr.) O. Eriks. which can attack a wide range of cereal rust species. Although spores of *E. caricis* may germinate on wheat leaves in the absence of rust spores, germination is stimulated by the presence of rust uredospores on the leaf (Stahle & Kranz 1984). The uredospore wall is penetrated by both mechanical and enzymatic means (Carling, Brown & Millikan 1976). *E. caricis* conidia appear to have only a limited ability to remain viable on cereal leaves in the absence of the rust host (Swendsrud & Calpouzos 1972). To achieve significant disease control, pycnidia of the hyperparasite must cover about half of the total surface of the rust pustule (Hau & Kranz 1978).

Verticillium lecanii can attack several cereal rusts including *Puccinia graminis* f.sp. *tritici* (McKenzie & Hudson 1976). *Aphanocladium album* (Preuss) W. Gams is also a hyperparasite of cereal rusts but induces telia formation. A purified cell-free extract from the hyperparasite can cause precocious telia formation in several rusts including *P. graminis* f.sp. *tritici* (Pers.). After penetrating uredospores, the hyperparasite causes a dissolution of fungal structures, probably as a result of chitinase production (Srivastava, Defago & Boller 1985) then covers the fungal fruiting structure with a fine white mycelium (Koc, Forrer & Defago 1983).

Bacterial population and frost damage

Bacteria on plant surfaces can be major sources of ice nuclei, inciting frost damage at temperatures from around -2 to $-8\,°C$ (Lindow 1986). There have been no studies on ice-nucleation-active bacteria on cereals in the UK but in the US the occurrence and control of such bacteria on maize and barley has recently been investigated.

Arny, Lindow & Upper (1976) first showed that frost sensitivity of maize was increased by application of an ice-nucleation-active strain of *Pseudomonas syringae*, the most abundant species of epiphytic colonists of plant surfaces. A strain of *Erwinia herbicola* (Löhnis) Dye, a yellow pigmented bacterium, also a widespread epiphytic colonist on many plants including cereals, can be ice-nucleation-active on maize at temperatures around $-5\,°C$ (Lindow, Arny & Upper 1978).

Numbers of ice nuclei and, consequently, degree of incidence of frost damage depend on the numbers of bacterial cells forming an ice nucleus. The number differs between strains. For example, on maize, suspensions used for inoculation containing about 10^4 *P. syringae* and 6×10^4 *E. herbicola* cells/ml were required to induce frost damage (Lindow, Arny & Upper 1978). Consequently a longer time interval was required for cell multiplication after applying *E. herbicola* cells to maize leaves before sufficient ice nuclei could be formed to induce frost damage. Because of the waxy nature of seedling maize leaves only a small proportion of bacteria applied to leaves in droplets are retained on the leaf surface and therefore able to provide a source of inoculum for succeeding generations of cells.

It is known that maize plants in the field do not supercool to temperatures colder than about $-4\,°C$ and Lindow, Arny & Upper (1978) believe that at this temperature natural populations of epiphytic bacteria contribute sufficient ice nuclei to be active in inciting frost damage.

An isolate of *E. herbicola* lacking ice-nucleation activity decreased populations of ice-nucleation-active *E. herbicola* and frost damage on maize plants in both growth room and field experiments (Lindow, Arny & Upper 1983). Thus biological control through the application of bacteria may be a practical method of limiting frost damage in the field.

As mentioned earlier the pathogenic bacterium *Xanthomonas campestris* pathovar *translucens* on barley could incite frost injury at temperatures ranging from −2 to −8 °C (Kim *et al.* 1987). Most strains produced ice nuclei that were active at warmer subfreezing temperatures (> −5 °C). It is likely that this property is related to increased incidence of bacterial leaf streak on barley plants during periods of mild frosts and may therefore be an important factor influencing pathogenicity. Similarly incidence of black chaff disease on wheat was increased when plants were exposed to temperatures below freezing after inoculation with *X. campestris* pathovar *translucens* (Azad & Schaad 1986).

All strains of *X. campestris* pathovar *translucens* were ice-nucleation-active while strains of *X. campestris* pathovar *campestris* do not appear to have this property. The latter are widespread on ryegrass (Austin, Goodfellow & Dickinson 1978; De Cleene 1989) and on maize (De Cleene 1989) but their presence would not be expected to increase frost damage.

Production of metabolites

Phylloplane micro-organisms may interact with each other or with the plant through the action of metabolites. Interactions between microbes are largely antagonistic, involving lytic enzymes (discussed in the section on interactions with pathogens) or antibiotics. Although antibiotics have been found in a small number of phylloplane bacteria and fungi, evidence of their active involvement in this habitat is largely lacking. The size and extent of microbial populations and of the nutrients to support them on healthy leaves in temperate regions is restricted and probably suggests that significant metabolite production is unlikely.

A number of microbes on aerial plant surfaces has been shown to produce growth substances belonging to the indole group of compounds. For instance epiphytic bacteria on shoots of seedlings produce indole-3-acetic acid (IAA) (Libbert, Kaiser & Kunert 1969). Using ^{14}C-labelled tryptophan, applied to the tips of maize coleoptiles, as a precursor both with and without epiphytic bacteria, it was shown that significantly more IAA could be detected at the base of the coleoptiles of seedlings with bacteria (Libbert & Silhengst 1970), indicating a transfer of the growth hormone from the bacterium to the plant.

Relatively large amounts of IAA were produced from tryptophan in cultures of two yeasts, *Sporobolomyces roseus* and *Candida muscorum* di Menna, isolated from the surfaces of barley leaves (Diem 1971). Species of filamentous fungi common on cereal leaves, such as *Aureobasidium pullulans*, *Epicoccum nigrum* Link and *Cladosporium herbarum*, also produce IAA and indole-3-acetonitrile (IAN) from tryptophan in culture (Valodon & Lodge 1970; Buckley & Pugh 1971).

Nitrogenous substances, resulting from the fixation of atmospheric nitrogen, are produced by *Beijerinckia*, a bacterium widely distributed on leaves of tropical plants (Ruinen 1956), and by mixed populations of bacteria on a tropical grass (Ruinen 1971). It is unlikely that significant amounts of nitrogen are fixed by bacteria on cereal leaves in temperate regions because the bacteria can only grow intermittently due to limitations in water availability.

INFLUENCE OF THE PHYLLOPLANE MICROFLORA ON YIELDS

Numerous reports of yield increases (Table 9.1) after fungicide sprays on crops in which there was little or no visible disease suggest that the phylloplane microflora may influence cereal yields. Such yield increases have ranged from around 3% when using either broad spectrum MBC fungicides or dithiocarbamates (Cook 1980) up to 21% following a single application of a mixture of carbendazim and zineb to winter wheat shortly before or after anthesis (Dickinson & Walpole 1975). Five applications of zineb, and to a lesser extent benomyl, increased yields in winter wheat while tridemorph had no effect. Zineb had the greatest influence in inhibiting development of the leaf mycoflora while benomyl had a lesser effect. The mycoflora was unaffected by tridemorph which has specific activity against powdery mildews (Dickinson & Walpole 1975). Similarly yield increases of 10–16% have been achieved with two to five applications of captafol, triadimefon, propiconazole or prochloraz to barley plants substantially free of disease. The greatest effect was achieved with captafol which almost eliminated populations of *Cladosporium* (Smedegaard-Petersen & Tolstrup 1986). The possibility of the saprophytic microflora being negatively associated with yield increases following fungicide treatment is supported by the fact that such increases are less apparent in dry summers when the microflora is less active (Cook 1981; Smedegaard-Petersen & Tolstrup 1986).

A novel approach to demonstrate the effect of fungi, such as *Cladosporium*, on barley leaf surfaces on plant growth was to repeatedly inoculate plants in a growth room, almost free of a normal field microflora, with a mixture of *C. macrocarpum* Preuss and *C. herbarum* spores and to compare their performance with similar uninoculated plants. Although inoculated plants showed no visible symptoms, there was a 9% reduction in grain yield (Smedegaard-Petersen & Tolstrup 1986).

TABLE 9.1. Yield increase in response to fungicide application where no diseases were detected

Cereal	Growth stage	Fungicide	No. of fungicide applications	Mean yield increase (%)	Reference
Winter wheat	71	Carbendazim/Zineb mixture	1	21·0	Dickinson & Walpole (1975)
Winter wheat	59–85	Zineb Benomyl	5 5	17·7 5·4	Dickinson & Walpole (1975)
Winter wheat	30–31	Benzimadozole/Captafol mixture	1	2–3	Fehrmann et al. (1978)
Winter wheat	37–75	Carbendazim fungicides	1	3·3	Cook (1980)
Winter wheat	50 & 60	Benomyl/Mancozeb/Maneb	2	1–12	Magan & Lacey (1986)
Spring barley	—	Zineb	—	3·0	Yarham (unpubl.) quoted in Cook (1981)
Spring barley	— — — —	Captafol Triadimefon Propiconazole Prochloraz	5 2 2 2	16·0 6·0 10·0 10·0	Smedegaard-Petersen & Tolstrup (1986)
Spring barley	75, 85, 91	Benomyl	3	2–4	Hill & Lacey (1983)

A number of hypotheses have been put forward to try to explain how partial or total destruction of the phylloplane microflora might affect cereal yields. The microflora on leaves may control pathogens (discussed in an earlier section on interactions with pathogens) by direct antagonism or competition but certain filamentous fungi, normally thought of as saprophytes, can also act as minor pathogens and thus cause yield losses. The saprophytic microflora may also increase the rate of leaf senescence by accelerating chlorophyll breakdown. The latter two possibilities will be discussed in more detail.

Minor pathogenic activities of saprophytes

Dickinson (1981) identified three organisms, *Alternaria alternata, Cladosporium herbarum* and *C. cladosporioides*, which may regularly attack plant tissues prior to senescence. Each organism produces many wind-dispersed conidia throughout the summer which are deposited on leaves of a wide variety of plants including cereal crops (Dickinson & Wallace 1976). The majority of such conidia do not germinate immediately but intermittently when conditions are favourable, especially towards the end of the growing season when relative humidities and nutrient availability within the crop canopy are higher. These fungi have developed mechanisms that enable survival during unfavourable conditions. The large dark pigmented conidia of *Alternaria* can survive for long periods under dry conditions and are able to resist the effects of sunlight. Even germinated conidia of *Cladosporium* spp. were not necessarily destroyed if dry conditions ensued. Although growth stopped this could recommence on the return of moisture to the leaf surface (Diem 1971). *Cladosporium* can also survive desiccation by producing thick-walled darkly pigmented microsclerotia (Pugh & Buckley 1971). Conidia of *Alternaria* form long germ tubes enabling epiphytic colonization without an initial source of nutrients from the leaf (Dickinson & Bottomley 1980). *Cladosporium herbarum*, with smaller conidia grows more slowly and covers shorter distances than *C. cladosporioides* which produces microcyclic conidia on short germ tubes to allow further colonization of foliar parts at a later stage.

Alternaria can penetrate green wheat leaves on the point of senescence via stomata. Internal hyphae developed from such sites to form extensive intercellular colonies without visible symptoms apart from a few cells showing necrosis around the penetration site (Dickinson 1981). By contrast *Cladosporium* spp. entering via stomata, remain within the substomatal cavity (Dickinson & Bottomley 1980) although Ebrahim-Nesbat &

Fehrmann (1983) showed that penetration of wheat leaves by *C. herbarum* caused cell necrosis and accelerated loss of chlorophyll. Both genera of fungi usually made many attempts to penetrate wheat leaves but only a few were successful.

Dickinson's (1981) suggestion that *Cladosporium* species and *A. alternata* may behave as weak pathogens has been challenged by Smedegaard-Peterson & Tolstrup (1986). They examined large numbers of attempted penetration sites on barley leaves and established that these fungi fail to penetrate the outer layers of the leaf and enter internal tissues. The attempted penetrations always initiated defence reactions by the plant which took the form of papillae and haloes in the epidermal cell walls (Ride & Pearce 1979). Further strong indication of the involvement of these structures in defence was provided by Sherwood & Vance (1980) who found that an inhibitor of protein synthesis (cycloheximide) prevented formation of papillae in treated leaves allowing non-pathogenic fungi to penetrate. The formation of defence reactions in cereal leaves in response to attempted penetration by non-pathogens was associated with increases in the rate of respiration which depleted energy reserves of the plant reducing grain yields (Smedegaard-Petersen 1982; Smedegaard-Petersen & Tolstrup 1985).

A parallel situation can be seen in barley leaves where varieties highly resistant to the pathogens *Erysiphe graminis* DC. and *Pyrenophora teres* Drechsler showed a temporarily increased rate of respiration due to additional energy being expended in defence reactions (Smedegaard-Petersen 1980). Leaves continuously exposed to inoculum of these pathogens prematurely senesced although no other symptoms were apparent. The decrease in available energy for growth resulted in decreased grain yield and lower grain quality (Smedegaard-Petersen & Stolen 1981).

Influence of grain microflora

Both *Cladosporium* and *Alternaria* are largely responsible for the head blackening syndrome which can develop after anthesis. Control of such fungi leads to higher quality of grain for flour milling (Dickinson 1981). Sooty mould development is encouraged by the presence of aphid honey-dew (Fokkema 1981). Late fungicide treatment, especially with captafol, resulted in yield increases of up to 4% and, in some cases, an increase in subsequent grain germination (Hill & Lacey 1983). Captafol was most effective in decreasing the grain microflora but was less effective against *Cladosporium* than benomyl while the latter failed to control *Alternaria*.

Effect on senescence

A detailed *in vitro* study of the effect of phylloplane fungi on barley leaf senescence has been carried out by Skidmore & Dickinson (1973). Pathogenic fungi such as *Septoria nodorum* and *Drechslera sativus* (Sacc.) Pamm. King & Bakke, as might be expected, caused most breakdown of chlorophylls a and b. With *S. nodorum* this was clearly visible as large areas of yellowing surrounding the elongated brown lesions. *Stemphylium botryosum* Wallroth and *Alternaria alternata*, both common leaf saprophytes, also enhanced the rate of leaf senescence and accelerated chlorophyll breakdown. A metabolite that inhibits chlorophyll production is produced by a species of *Alternaria* (Fulton, Bollenbacher & Templeton 1965). Other common leaf saprophytes, such as *Cladosporium* and *Aureobasidium*, did not accelerate leaf senescence or induce chlorophyll breakdown (Skidmore & Dickinson 1973). Fokkema, Kastelein & Post (1979) reached a similar conclusion in experiments to compare the chlorophyll contents of yeast and nutrient-sprayed wheat leaves with increased microbial populations with those of captafol-treated leaves with a much smaller microflora. The chlorophyll content did not differ between the two treatments over periods of a month following the treatments. However, Smedegaard-Petersen & Tolstrup (1986) found larger chlorophyll contents in captafol-treated barley leaves with smaller *Cladosporium* populations than in unsprayed leaves with larger populations. Plants repeatedly sprayed with *Cladosporium* spores, but not fungicide, contained less chlorophyll in the leaves (especially flag leaves) than uninoculated plants.

Although treatment with benomyl affected the total microbial population of wheat leaves less than captafol treatment, the chlorophyll content of benomyl treated leaves was substantially greater (Fokkema, Kastelein & Post 1979). This suggested that fungicides with cytokinin-like properties may have a direct effect in enhancing chlorophyll retention. This phenomenon has been investigated by other workers, e.g. Staskawicz *et al.* (1978), who demonstrated significant chlorophyll retention in oat leaves even with very low levels of the benzimadozole fungicide, carbendazim.

CONCLUSION

A knowledge of the micro-organisms associated with the aerial parts of major crops, such as the temperate cereals, is important for understanding the different ways in which the microflora interacts with the plant and with potential pathogens. This is not only of academic interest but also of economic significance because the microflora can affect the yield and quality of the grain.

A knowledge of the microflora of a crop is also important for other reasons which may, in the future, radically alter agricultural practices. Within two decades or so, it is likely that use of fungicides, insecticides and nitrogenous fertilizers will be greatly decreased. The phylloplane microflora of cereals will almost certainly provide a source of potentially useful biocontrol and growth-enhancing organisms for use in the control of pests and diseases and for the stimulation of crop growth. Such organisms may be used in their natural 'wild type' state or more probably after suitable gene insertions or deletions to enhance their desirable characteristics and remove those which may be harmful.

A knowledge and understanding of the phylloplane microflora of the cereal crop as a whole will also form an essential baseline for assessing the risk from deliberate release of genetically engineered micro-organisms in the field. This would require monitoring and assessment of interactions and competitiveness of the introduced organism with the normal cereal microflora.

ACKNOWLEDGMENT

Assistance from Mr R.D. McCall, who took the scanning electron micrographs, is gratefully acknowledged.

REFERENCES

Arny, D.C., Lindow, S.E. & Upper, C.D. (1976). Frost sensitivity of *Zea mays* increased by application of *Pseudomonas syringae*. *Nature*, **262**, 282–284.

Austin, B., Goodfellow, M. & Dickinson, C.H. (1978). Numerical taxonomy of phylloplane bacteria isolated from *Lolium perenne*. *Journal of General Microbiology*, **104**, 139–155.

Azad, H. & Schaad, N.W. (1986). The effect of frost on black chaff development in wheat. *Phytopathology*, **76**, 1134 (Abstr.).

Blakeman, J.P. (1985). Ecological succession of leaf surface micro-organisms in relation to biological control. *Biological Control on the Phylloplane* (Ed. by C.E. Windels & S.E. Lindow), pp. 6–30. The American Phytopathological Society, St. Paul, Minnesota.

Blakeman, J.P. & Brodie, I.D.S. (1977). Competition for nutrients between epiphytic micro-organisms and germination of spores of plant pathogens on beetroot leaves. *Physiological Plant Pathology*, **10**, 29–42.

Blakeman, J.P. & Parbery, D.G. (1977). Stimulation of appressorium formation in *Colletotrichum acutatum* by phylloplane bacteria. *Physiological Plant Pathology*, **11**, 313–325.

Brodie, I.D.S. & Blakeman, J.P. (1976). Competition for exogenous substrates *in vitro* by leaf surface micro-organisms and germination of conidia of *Botrytis cinerea*. *Physiological Plant Pathology*, **9**, 227–239.

Buckley, N.G. & Pugh, G.J.F. (1971). Auxin production by phylloplane fungi. *Nature*, **231**, 332.

Carling, D.E., Brown, M.F. & Millikan, D.F. (1976). Ultrastructural examination of the *Puccinia graminis–Darluca filum* host–parasite relationship. *Phytopathology*, **66**, 419–422.

Cook, R.J. (1980). Effects of late-season fungicide sprays on yield of winter wheat. *Plant Pathology*, **29**, 21–27.

Cook, R.J. (1981). Unexpected effects of fungicides on cereal yields. *EPPO Bulletin*, **11**, 277–285.

De Cleene, M. (1989). Scanning electron microscopy of the establishment of compatible and incompatible *Xanthomonas campestris* pathovars on the leaf surface of Italian ryegrass and maize. *EPPO Bulletin*, **19**, 81–88.

Dickinson, C.H. (1967). Fungal colonization of *Pisum* leaves. *Canadian Journal of Botany*, **45**, 915–927.

Dickinson, C.H. (1981). Biology of *Alternaria alternata, Cladosporium cladosporioides* and *C. herbarum* in respect of their activity on green plants. *Microbial Ecology of the Phylloplane* (Ed. by J.P. Blakeman), pp. 169–184. Academic Press, London.

Dickinson, C.H. & Walpole, P.R. (1975). The effect of late application of fungicides on the yield of winter wheat. *Experimental Husbandry*, **29**, 23–28.

Dickinson, C.H. & Wallace, B. (1976). Effects of late applications of foliar fungicides on activity of micro-organisms on winter wheat flag leaves. *Transactions of the British Mycological Society*, **67**, 103–112.

Dickinson, C.H. & Bottomley, D. (1980). Germination and growth of *Alternaria* and *Cladosporium* in relation to their activity in the phylloplane. *Transactions of the British Mycological Society*, **74**, 309–319.

Dickinson, C.H., Austin, B. & Goodfellow, M. (1975). Quantitative and qualitative studies of phylloplane bacteria from *Lolium perenne*. *Journal of General Microbiology*, **91**, 157–166.

Diem, H.G. (1971). Effect of low humidity on the survival of germinated spores commonly found in the phyllosphere. *Ecology of Leaf Surface Micro-organisms* (Ed. by T.F. Preece & C.H. Dickinson), pp. 211–219. Academic Press, London.

Diem, H.G. (1974). Micro-organisms of the leaf surface: estimation of the mycoflora of the barley phyllosphere. *Journal of General Microbiology*, **80**, 77–83.

Ebrahim-Nesbat, F. & Fehrmann, H. (1983). Electron microscopical evidence of pathogenicity of *Cladosporium herbarum* in wheat leaves. *Proceedings of the 4th International Congress of Plant Pathology, Melbourne*, p. 179.

El-Banoby, F.R. & Rudolph, K.W.E. (1989). Multiplication of *Xanthomonas campestris* pvs *secalis* and *translucens* in host and non-host plants (rye and barley) and development of water soaking. *EPPO Bulletin*, **19**, 105–111.

Fehrmann, H., Reinecke, P. & Weihofen, U. (1978). Yield increase in winter wheat by unknown effects of MBC-fungicides and captafol. *Phytopathologische Zeitschrift*, **93**, 359–362.

Flannigan, B. & Campbell, I. (1977). Pre-harvest mould and yeast floras on the flag leaf, bracts and caryopsis of wheat. *Transactions of the British Mycological Society*, **69**, 485–494.

Fokkema, N.J. (1971). The effect of pollen in the phyllosphere of rye on colonisation by saprophytic fungi and on infection by *Helminthosporium sativum* and other leaf pathogens. *Netherlands Journal of Plant Pathology*, **77**, (1), 1–60.

Fokkema, N.J. (1973). The role of saprophytic fungi in antagonism against *Drechslera sorokiniana* (*Helminthosporium sativus*) on agar plates and on rye leaves with pollen. *Physiological Plant Pathology*, **3**, 195–205.

Fokkema, N.J. (1981). Fungal leaf saprophytes, beneficial or detrimental. *Microbial Ecology of the Phylloplane* (Ed. by J.P. Blakeman), pp. 433–454. Academic Press, London.

Fokkema, N.J. & Schippers, B. (1986). Phyllosphere versus rhizosphere as environments for saprophytic colonization. *Microbiology of the Phyllosphere* (Ed. by N.J. Fokkema & J. van den Heuvel), pp. 135–159. Cambridge University Press, Cambridge.

Fokkema, N.J., den Houter, J.G., Kosterman, Y.J.C. & Nelis, A.L. (1979). Manipulation of yeasts on field-grown wheat leaves and their antagonistic effect on *Cochliobolus sativus* and *Septoria nodorum*. *Transactions of the British Mycological Society*, 72, 19–29.

Fokkema, N.J., Kastelein, P. & Post, B.J. (1979). No evidence for acceleration of leaf senescence by phyllosphere saprophytes of wheat. *Transactions of the British Mycological Society*, 72, 312–315.

Fokkema, N.J., van de Laar, J.A.J., Nelis-Blomberg, A.L. & Schippers, B. (1975). The buffering capacity of the natural mycoflora of rye leaves to infection by *Cochliobolus sativus* and its susceptibility to benomyl. *Netherlands Journal of Plant Pathology*, 81, 176–186.

Fryda, S.J. & Otta, J.D. (1978). Epiphytic movement and survival of *Pseudomonas syringae* on spring wheat. *Phytopathology*, 68, 1064–1067.

Fulton, N.D., Bollenbacher, K. & Templeton, G.E. (1965). A metabolite from *Alternaria tenuis* that inhibits chlorophyll production. *Phytopathology*, 55, 49–51.

Hau, B. & Kranz, J. (1978). Modelrechnungen zur Wirkung des Hyperparasiten *Eudarluca caricis* auf Rostepidemien. *Zeitschrift fur Pflanzenkrankheiten und Pflanzenschutz*, 85, 131–141.

Hevesi, M. & Mashaal, S.F. (1975). Contributions to the mechanism of infection of *Erwinia uredovora*, a parasite of rust fungi. *Acta Phytopathologica Academiae Scientarum Hungaricae*, 10, 275–280.

Hill, R.A. & Lacey, J. (1983). The microflora of ripening barley grain and the effects of preharvest fungicide application. *Annals of Applied Biology*, 102, 455–465.

Hirano, S.S. & Upper, C.D. (1983). Ecology and epidemiology of foliar bacterial plant pathogens. *Annual Review of Phytopathology*, 21, 243–269.

Kim, H.K., Orser, C., Lindow, S.E. & Sands, D.C. (1987). *Xanthomonas campestris* pv. *translucens* strains active in ice nucleation. *Plant Disease*, 71, 994–997.

Koc, N.K., Forrer, H.R. & Defago, G. (1983). Hyperparasitism of *Aphanocladium album* on aecidiospores and teliospores of *Puccinis graminia* f.sp. *tritici*. *Phytopathologische Zeitschrift*, 107, 219–223.

Last, F.T. (1955). Seasonal incidence of *Sporobolomyces* on cereal leaves. *Transactions of the British Mycological Society*, 38, 221–239.

Leben, C. (1961). Micro-organisms on cucumber seedlings. *Phytopathology*, 51, 553–557.

Levine, M.N., Bamburg, R.H. & Atkinson, R.E. (1936). Micro-organisms antibiotic or pathogenic to cereal rusts. *Phytopathology*, 26, 99–100.

Libbert, E. & Silhengst, P. (1970). Interactions between plants and epiphytic bacteria regarding their auxin metabolism. VIII. Transfer of ^{14}C-indoleacetic acid from epiphytic bacteria to corn coleoptiles. *Physiologia Plantarum*, 23, 480–487.

Libbert, E., Kaiser, W. & Kunert, R. (1969). Interactions between plants and epiphytic bacteria regarding their auxin metabolism. VI. The influence of the epiphytic bacteria on the content of extractable auxin in the plant. *Physiologia Plantarum*, 22, 432–439.

Lindemann, J., Constantinidou, H.A., Barchett, W.R. & Upper, C.D. (1982). Plants as sources of airborne bacteria, including ice nucleation-active bacteria. *Applied and Environmental Microbiology*, 44, 1059–1063.

Lindow, S.E. (1986). Strategies and practice of biological control of ice nucleation-active bacteria on plants. *Microbiology of the Phyllosphere* (Ed. by N.J. Fokkema & J. van den Heuvel), pp. 293–311. Cambridge University Press, Cambridge.

Lindow, S.E., Arny, D.C. & Upper, C.D. (1978). *Erwinia herbicola*: a bacterial ice nucleus active in increasing frost injury to corn. *Phytopathology*, 68, 523–527.

Lindow, S.E., Arny, D.C. & Upper, C.D. (1983). Biological control of frost injury: establishment and effects of an isolate of *Erwinia herbicola* antagonistic to ice nucleation active bacteria on corn in the field. *Phytopathology*, 73, 1102–1106.

Magan, N. & Lacey, J. (1986). The phylloplane microflora of ripening wheat and effect of late fungicide applications. *Annals of Applied Biology*, **109**, 117–128.

McKenzie, E.H.C. & Hudson, H.J. (1976). Mycoflora of rust-infected and non-infected plant material during decay. *Transactions of the British Mycological Society*, **66**, 223–238.

Morgan, F.L. (1963). Infection inhibition and germ-tube lysis of three cereal rusts by *Bacillus pumilus*. *Phytopathology*, **53**, 1346–1348.

Naik, D.M. & Busch, L.V. (1978). Stimulation of *Fusarium graminearum* by maize pollen. *Canadian Journal of Botany*, **56**, 1113–1117.

Paul, V.H. & Smith, I.M. (1989). Bacterial pathogens of Gramineae: systematic review and assessment of quarantine status for the EPPO region. *EPPO Bulletin*, **19**, 33–42.

Perombelon, M.C.M., Fox, R.A. & Lowe, R. (1979). Dispersion of *Erwinia carotovora* in aerosols produced by the pulverisation of potato haulm prior to harvest. *Phytopathologische Zeitschrift*, **94**, 167–173.

Pon, D.S., Townsend, C.E., Wessman, G.E., Schmitt, C.G. & Kingsolver, C.H. (1954). A *Xanthomonas* parasitic on uredia of cereal rusts. *Phytopathology*, **44**, 707–710.

Pugh, G.J.F. & Buckley, N.G. (1971). The leaf surface as a substrate for colonisation by fungi. *Ecology of Leaf Surface Micro-organisms* (Ed. by T.F. Preece & C.H. Dickinson), pp. 431–445. Academic Press, London.

Ride, J.P. & Pearce, R.B. (1979). Lignification and papilla formation at sites of attempted penetration of wheat leaves by non-pathogenic fungi. *Physiological Plant Pathology*, **15**, 79–92.

Riphagen, I., Fokkema, N.J., Kastelein, W.J. & Vereijken, P.H. (1979). Effect of aphid honeydew on saprophytic and pathogenic fungi of wheat leaves under controlled environmental conditions. *Acta Botanica Neerlandica*, **28**, 240–241.

Rodger, G. & Blakeman, J.P. (1984). Microbial colonization and uptake of ^{14}C label on leaves of sycamore. *Transactions of the British Mycological Society*, **82**, 45–51.

Ruinen, J. (1956). Occurrence of *Beijerinckia* species in the 'Phyllosphere'. *Nature*, **177**, 220–221.

Ruinen, J. (1971). The grass sheath as a site for nitrogen fixation. *Ecology of Leaf Surface Micro-organisms* (Ed. by T.F. Preece & C.H. Dickinson), pp. 567–579. Academic Press, London.

Sherwood, R.T. & Vance, C.P. (1980). Resistance to fungal penetration in Gramineae. *Phytopathology*, **70**, 273–279.

Skidmore, A.M. & Dickinson, C.H. (1973). Effect of phylloplane fungi on the senescence of excised barley leaves. *Transactions of the British Mycological Society*, **60**, 107–116.

Smedegaard-Petersen, V. (1980). Increased demand for respiratory energy of barley leaves reacting hypersensitively against *Erysiphe graminis*, *Pyrenophora teres* and *Pyrenophora graminea*. *Phytopathologische Zeitschrift*, **99**, 54–62.

Smedegaard-Petersen, V. (1982). The effect of defence reactions on the energy balance and yield of resistant plants. *Active Defense Mechanisms in Plants* (Ed. by R.K.S. Wood), pp. 299–315. Plenum Press, New York.

Smedegaard-Petersen, V. & Stolen, O. (1981). Effect of energy-requiring defense reactions on yield and grain quality in a powdery mildew-resistant barley cultivar. *Phytopathology*, **71**, 396–399.

Smedegaard-Petersen, V. & Tolstrup, K. (1985). The limiting effect of disease resistance on yield. *Annual Review of Phytopathology*, **23**, 475–490.

Smedegaard-Petersen, V. & Tolstrup, K. (1986). Yield-reducing effect of saprophytic leaf fungi in barley crops. *Microbiology of the Phyllosphere* (Ed. by N.J. Fokkema & J. van den Heuvel), pp. 160–171. Cambridge University Press, Cambridge.

Srivastava, A.K., Defago, G. & Boller, T. (1985). Secretion of chitinase by *Aphanocladium album*, a hyperparasite of wheat rust. *Experientia*, **41**, 1612–1613.

Stahle, U. & Kranz, J. (1984). Interactions between *Puccinia recondita* and *Eudarluca caricis* during germination. *Transactions of the British Mycological Society*, **82**, 562–563.

Staskawicz, B., Khar-Sawhney, R., Slaybough, R., Adams, W. & Galston, A.W. (1978). The cytokinin-like action of methyl-2-benzimidazolecarbamate on oat leaves and protoplasts. *Pesticide Biochemistry and Physiology*, **8**, 106–110.

Swendsrud, D.P. & Calpouzos, L. (1972). Effect of inoculation sequence and humidity on infection of *Puccinia recondita* by the mycoparasite *Darluca filum*. *Phytopathology*, **62**, 931–932.

Valodon, L.R.G. & Lodge, E. (1970). Auxins and other compounds of *Cladosporium herbarum*. *Transactions of the British Mycological Society*, **55**, 9–15.

Williamson, M.A. & Fokkema, N.J. (1985). Phyllosphere yeasts antagonize penetration from appressoria and subsequent infection of maize leaves by *Colletotrichum graminicola*. *Netherlands Journal of Plant Pathology*, **91**, 265–276.

10. THE EPIDEMIOLOGY OF CEREAL VIRUSES

N. CARTER AND R.T. PLUMB

AFRC Institute of Arable Crops Research, Rothamsted Experimental Station, Harpenden, Hertfordshire AL5 2JQ, UK

INTRODUCTION

Infection by pathogens is the rule rather than the exception for small grain cereal crops; there can be no crops whose growth is not affected in some way by viruses, fungi and bacteria. The ecology of fungi and bacteria on leaves is discussed by Blakeman (this volume, pp. 171–191) and on the roots by Jagnow (this volume, pp. 113–137) (see also Krupa & Dommergues 1979). This chapter compares the epidemiologies of commonly occurring viruses. These viruses are spread only through an intermediary agent or vector.

The most widespread virus of cereals is barley yellow dwarf virus (BYDV). This is the name used to describe a range of isolates which are more appropriately described as two viruses (Waterhouse, Gildow & Johnstone 1988). They are spread by aphids and infect all types of cereal and many other species in the Gramineae. The next most frequently encountered viruses are those transmitted by root-infecting fungi. In the UK these viruses infect barley or oats. Each species is infected by two different viruses which cause mosaics or stripes on infected plants. Other viruses with similar properties and the same vector affect wheat in many other parts of the world but have not been reported in Britain.

The biology, especially the dispersal, of the two types of vector differs markedly and as a consequence the epidemiologies of the viruses they transmit are very different. In addition, over twenty aphid species are known vectors of the various isolates of BYDV (A'Brook 1981; Jedlinski 1981) although not all occur in any one place, and each species has a different life cycle, host preference and population dynamics. For all the fungally transmitted viruses of cereals only one vector species is cited, *Polymyxa graminis* (Ledingham). However, while there is much circumstantial evidence for its role as a vector, definitive evidence is lacking for most viruses (Adams 1988). In addition, evidence is accumulating that the wide host range reported for this fungus may be due to biotypes or variants specialized to particular host species.

This paper compares the epidemiology of these two groups of viruses, especially as it is influenced by the ecology of their vectors and host plants.

THE VIRUSES

Barley yellow dwarf virus

Yellowing and reddening of cereals were referred to in scientific literature long before the cause was attributed to an aphid-transmitted virus. In 1951, there was a very widespread epidemic in California and adjacent States which drew attention to the problem and resulted in the identification of the causal agent and vectors (Oswald & Houston 1951). The delay in identification was because earlier techniques to separate pathogenic from physiological or nutritional disorders were inadequate. Even now, diagnosis of BYDV by symptoms is not easy. Those familiar with the disease may be able to identify symptoms of infection with 90% confidence but symptoms alone provide no information on isolate type or vector species. There is, thus, a great need for diagnostic methods that distinguish isolates. Those currently used are based on immunological methods, using monoclonal or polyclonal antibodies in an enzyme-linked immunosorbent assay (ELISA). These methods have confirmed the separation of isolates determined by vector specificities (Rochow 1969). Transmission tests using different aphid vector species, while time consuming, continue to be a valuable way of assessing virus presence.

A combination of biological, cytopathological, immunological and nucleic-acid based evidence has accumulated to justify the separation of the various isolates, all previously known as BYDV, into two distinct viruses (Waterhouse, Gildow & Johnstone 1988). The different isolates of these two viruses are identified by the initial letters of their principal vectors. Under the proposed new classification, the MAV (*Macrosiphum (Sitobion) avenae* (F.)), PAV (*Rhopalosiphum padi* (L.) and *S. avenae*), and SGV (*Schizaphis graminum* (Rondani)) isolates are grouped as BYDV-I. The RPV (*R. padi*) and RMV (*R. maidis* (Fitch)) isolates are identified as BYDV-II. In Britain, isolates of RMV and SGV have not been reported. While *R. maidis* does occur, usually only on barley, it is uncommon, and *S. graminum* does not occur in the UK. Therefore, the absence of these isolates may be a direct consequence of the scarcity of their vectors. However, it is noteworthy that in parts of continental Europe, where these two species are more common, RMV and SGV are also unknown (Plumb 1990).

As well as the vectors identified in the isolate names, other aphids can also transmit. Indeed absolute specificity of transmission of an isolate by a single aphid species is, in our experience, unknown. The efficiency of transmission of PAV, MAV and RPV by different aphid species is given in Table 10.1 (Herrera 1989).

For convenience, in this paper we refer to BYDV when the isolate is not known or is irrelevant to the context, and use the appropriate identifying initials when the isolate is known. However, it should be noted that the most accurate description would be PAV-, MAV- and RPV-like isolates as there is no evidence that UK isolates are identical to the type isolates originally identified in North America.

Despite its name, BYDV infects a very wide host range within the family Gramineae. Thus, in most temperate regions there is no shortage of perennial grass species, either wild or cultivated, that are susceptible to infection. Usually infection with one, or often several, of the BYDV isolates is common. Consequently there is a large reservoir to act as a source for infection of the annual cereal crops. Work in Australia (Guy, Johnstone & Morris 1987) has indicated an association between virus isolates and different subfamilies of the Poaceae. In the UK no similar work has been done and it is assumed, from the evidence of surveys (Doodson 1967; Plumb 1977), that perennial grasses throughout Britain provide a relatively uniform source of inoculum. However, isolates of BYDV found in cereal crops differed between regions of the country (Plumb 1974) and alate aphids caught over a wide area transmitted the MAV and PAV isolates to test plants much more frequently than RPV (Torrance *et al.* 1986; Herrera 1989). The earlier results (Plumb 1974) were based on spring and summer infections because autumn-sown crops, usually wheat, were sown in October and avoided infection in autumn when *R. padi* was common (Taylor *et al.* 1981). Consequently, the prevalence of BYDV isolates was determined by the relative ability of *R. padi*

TABLE 10.1. Single aphid transmission efficiencies (%) of three isolates of barley yellow dwarf virus

Aphid species	Isolates		
	PAV	MAV	RPV
Rhopalosiphum padi	73	9	68
Sitobion avenae	20	54	0
Metopolophium dirhodum	26	38	9
Rhopalosiphum maidis	16	2	13
Metopolophium festucae	2	20	21

and *S. avenae* to survive winter in different areas and then introduce virus to crops. Isolates transmitted by *S. avenae* were predominant in the colder eastern regions because this aphid species survived, either holocyclically (as eggs after sexual reproduction) or anholocyclically (as aphids through-out the winter) on Gramineae. It could therefore introduce virus to the crop as soon as migrant forms were produced, whereas *R. padi* only survived holocyclically on a rosaceous host, *Prunus padus* L., that is immune to BYDV. Thus *R. padi* would not have introduced BYDV to crops except via infected grass hosts, although it may have spread virus isolates introduced by other species.

The area of autumn-sown crops has increased and sowing in September is now common. Crops are thus exposed to infection by the predominant autumn migrant, *R. padi*, as well as infection by the much smaller number of *S. avenae* (Tatchell, Plumb & Carter 1988). There are regional dif-ferences in the incidence of BYDV isolates (McGrath & Bale 1989) and seasonal effects can also influence which aphid species, and thus virus isolates, survive best. In the Rothamsted area the mild winters of 1987/88 and 1988/89 both favoured survival and dispersal of *S. avenae* and, as there was little infection by *R. padi*-transmitted isolates, almost all infection in these years was by the MAV isolates.

Mixtures of the virus isolates are found and are usually more frequent when the two isolates are from BYDV I and BYDV II. Dual infection by isolates of the same group is minimized by cross-protection, when prior infection with one virus precludes subsequent infection by a second, related virus. This has been reported in plants (Jedlinski & Brown 1965) and aphids (Rochow, Muller & Gildow 1983). A consequence of dual infection is that as the two isolates are multiplying together in the plant the nucleic acid of one can become enclosed in the coat protein of the other. As it is the virus coat protein that determines vector specificity, this can result in transmission of some virus isolates by aphids that are not normally vectors or are very inefficient. The isolate B identified in Britain by Watson & Mulligan (1960), and maintained in the glasshouse ever since, is a relatively stable mixture of RPV and PAV isolates (Herrera 1989) but generally the association between isolates is transient.

Fungally transmitted viruses

The two viruses transmitted to barley by *P. graminis* are barley yellow mosaic virus (BaYMV) and barley mild mosaic virus (BaMMV). Previously thought to be the same virus they were distinguished by the relative susceptibility of different cultivars (Table 10.2) and subsequently by immunology and studies of their nucleic acids (Batista *et al.* 1989).

TABLE 10.2. Susceptibility of three cultivars of winter barley to barley yellow mosaic and barley mild mosaic virus by inoculation with zoospores of *Polymyxa graminis*

	Percentage plants (angles)	
	BaYMV	BaMMV
Maris Otter	4	34
Igri	40	11
Torrent	0	0
SED	8·3	

There has also now been a report (M.J. Adams, personal communication) of a virus that infects cultivar Torrent but so far it has not been possible to separate it, except by host range, from BaYMV. In Japan, several other variants of BaYMV have been reported based on their ability to overcome host resistance genes (Kashiwazaki *et al.* 1989).

The two other fungally transmitted viruses present in the UK that infect cereals are oat mosaic (OMV) and oat golden stripe viruses (OGSV). In crops the presence of OMV is recorded more often than OGSV. While OGSV is usually reported each year it is often present together with OMV but the symptoms do not indicate the presence of two viruses. OGSV is also found in the roots of apparently healthy plants and in leaves of plants with few or no symptoms (Plumb *et al.* 1977).

VIRUS–VECTOR RELATIONSHIPS

BYDV is restricted to the phloem of infected plants and is only acquired by aphids during feeding. There are reports of transmission by the parasitic plant dodder (*Cuscuta campestris* Yuncker) (Timian 1964), through seeds (Mills, Mercer & McGimpsey 1986) and by frit fly *Oscinella frit* L. (Jess & Mowat 1986). These observations need to be confirmed but there is little evidence that these methods of transmission are important in the dissemination of the disease. The virus is not transmitted manually.

Aphids can acquire virus as soon as their stylets reach the phloem of an infected plant although aphids given acquisition feeds of less than 24 h rarely subsequently transmit BYDV. The efficiency of transmission increases with the length of feeding period and reaches a maximum after about 48 h (Toko & Bruehl 1959). Once acquired by an aphid the virus passes into the hindgut and from there into the haemocoel. The site of specificity appears to be at the accessory salivary gland membrane. A compatible interaction between virus and vector tissue is required to allow the virus to pass into the salivary duct and be injected into a plant with saliva (Gildow 1987). Infection has a similar time-scale to acquisition and reaches a maximum efficiency with a feeding period of 48 h. This process of acquisition and infection is known as 'persistent' transmission because the virus once acquired is usually retained for the rest of the aphid's life and, if acquired by an immature stage, is retained through moults. However, virus is not passed to progeny whether produced viviparously or through eggs and does not multiply in the aphid. In nature, there is rarely the opportunity to test the persistence of virus, as an aphid, once it has acquired virus, probably feeds on infected plants for the rest of its life. However, when retention of infectivity, that is the ability to transmit the virus that the vector has acquired, is tested by moving the aphid to fresh plants, the frequency of transmission often decreases (Watson 1972). There is no evidence that the transmission process or the retention of virus differs among vectors or virus isolates.

It is very rare to find that any aphid–BYDV-isolate combination results in 100% transmission by single aphids. In many ways the reasons for the failure to transmit are as interesting as the reasons for success. For example, virus is not uniformly distributed throughout the phloem of infected plants and some cells remain free. Also, some aphids may acquire an insufficient amount of virus because of little virus in the host or poor feeding. However, when virus content in aphids has been tested by immunological methods, those that contain most are usually vectors, but there is no threshold concentration that separates vectors from non-vectors. A further variable is the vector: some biotypes of the same species are more efficient than others (Rochow & Eastop 1966).

It is difficult to separate any direct effects that acquiring virus may have on an aphid from the indirect effects through virus infection of its host plant. These latter effects are described under virus–host-vector interactions.

All the fungally transmitted cereal viruses are apparently carried within the fungus rather than externally on the spore surface (Teakle 1983). Thus, they are less exposed to environmental conditions than other fungally-transmitted viruses that adhere to the spore. This is especially so when a fungus produces resting spores or cystosori (Fig. 10.1). The infectivity of soil can be retained for at least 10 years when stored in polythene bags at 4 °C (Plumb & Macfarlane 1977). Various soil sterilants, such as methyl bromide, can kill the fungus but this is not an economic treatment and once soil is infested, with current methods of control, infection remains indefinitely. The motile stage of the fungus, the zoospore, is the most vulnerable stage (Fig. 10.1) and requires a film of water in which to swim.

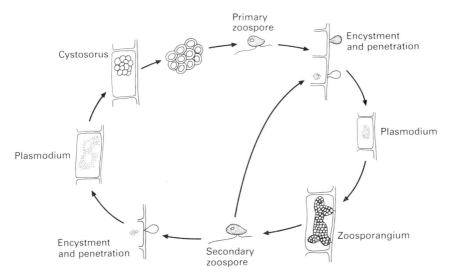

FIG. 10.1. Life cycle of *Polymyxa graminis*.

Such spores can travel only short distances but in very wet conditions, when there is free water, the spores, and the virus they carry, can apparently move downhill as the water drains. In contrast, in very dry and windy conditions resting spores can probably be blown large distances with the soil.

Little is known about how virus is acquired or transmitted by fungi, but the penetration of plants cells, from zoospores encysted on the surface, by the fungus protoplasm may release virus into the host. Even less is known about the specificity of transmission and whether it is mediated through interactions between the virus and the fungus, between the virus and the host, or between the fungus and the host. There is an interaction between virus and host as barley viruses will not infect wheat or oats. There is also some evidence that there is a host-fungus specificity but nothing is known of fungus–virus relationships although vector and non-vector strains of barley-infecting *P. graminis* are known (M.J. Adams, personal communication). There are no known effects on the fungus of virus infection either via the host plant or by the virus directly.

VECTOR LIFE CYCLES

Although many aphids are vectors of BYDV isolates, and in Britain more than ten vector aphid species have been reported, in few regions are more than two or three species of importance. In Britain, the principal vectors are *R. padi* and *S. avenae*. They have different life cycle strategies which influence the epidemiology of BYDV, as well as the prevalence of the different virus isolates.

In autumn in eastern England most migrant *R. padi* are seeking the primary woody host, bird cherry (*P. padus*) and survive the winter as eggs (Tatchell, Plumb & Carter 1988). So, even though alate *R. padi* constitute a large proportion of the total number of aphids caught in suction traps in the autumn (Taylor *et al.* 1981), when much primary infection of winter cereals with BYDV occurs, populations on crops do not mirror this and are much smaller than might be expected (Dewar & Carter 1984). The majority of the autumn migrants are gynoparae (winged females which produce egg-laying females or oviparae), and males which are flying to *P. padus*. While these aphids may carry BYDV and may settle in cereals they are unlikely to contribute significantly to virus spread as gynoparae do not normally reproduce on cereals (Dixon 1971) and males cannot.

The different aphid morphs migrating in the autumn have different flight behaviours. The gynoparae are caught more frequently in suction traps at 12·2 m than in traps at 1·5 m above ground level, while those that can colonize cereals are equally numerous at both heights (Tatchell, Plumb & Carter 1988). It was concluded that the gynoparae were flying further and that most virus spread was over only short distances of a few kilometres. In other parts of the world, such as in the USA, the establishment of jet streams can carry aphids many hundreds of kilometres from crops in the south, on which they are flourishing, to crops in more northern regions that are free of aphids and virus infection (Wallin & Loonan 1971).

In the UK, there may be annual and regional variations in the proportion of alates returning to cereals in the autumn. Although these variations are unquantified, anholocyclic overwintering on grasses and cereals is likely to be commoner in the milder south-west of England than in the east and north where winters are colder, and in mild rather than cold winters. Indeed, *R. padi* is one of the least cold hardy of the cereal aphids (Hand 1980) and does not survive on cereals in most winters at Rothamsted (Dewar & Carter 1984).

The spring migration of *R. padi* is usually small and initially consists largely of survivors on cereals and grasses rather than emigrants from *P. padus* (Carter *et al.* 1980). Migrants from Gramineae can carry BYDV but those from *P. padus* will be virus-free and will have to acquire virus from an infected Graminaceous host before they can contribute to virus spread. *R. padi* is usually uncommon on cereals during the summer in the UK and is generally considered to oversummer on grasses (Carter *et al.* 1980), although in Scandinavia it frequently reaches damaging levels on spring-sown cereals (Wiktelius & Ekbom 1985).

In contrast to *R. padi*, *S. avenae* is monoecious, spending its entire life cycle on members of the Gramineae. It is thought to be largely anholocyclic (Dewar & Carter 1984; Loxdale *et al.* 1985) although oviparae and males (Dewar & Carter 1984) and eggs (Hand 1980) have occasionally been found in cereals and grassland, respectively. Thus, although the numbers of *S. avenae* migrating in the autumn and caught in suction traps are much less than those of *R. padi* (Taylor *et al.* 1981), all are potential colonizers of cereals and thus the relative proportions of these two species in crops are more similar than in suction trap samples (Dewar & Carter 1984; Tatchell, Plumb & Carter 1988). All alate *S. avenae* are therefore potential virus vectors. *S. avenae* is also more cold-hardy than *R. padi* (Hand 1980) and is usually more common in crops, and often the only species present, by the end of the winter (Dewar & Carter 1984).

The spring migration of *S. avenae*, as with *R. padi*, is generally small (Taylor *et al.* 1981) although there is sufficient annual variation to give an indication of the size of populations on wheat in the summer (Dewar & Carter 1984). The time when migration starts, and its size, are important in determining the risk of infection in spring-sown crops. Early (Walters & Dewar 1986), large (Dewar & Carter 1984) migrations, following milder than average winters usually result in most infection. The emigration from ripening crops in the summer involves the largest number of alates (Taylor *et al.* 1981) but very little is known of their fate. In some regions where maize is grown this crop may be colonized by migrants from small grains and bridge the gap between the maturity of one crop and the emergence of the next, itself then acting as a source of aphids and BYDV (Hand & Carillo 1982).

The life cycle of *P. graminis* is no less complex than that described for the aphid virus vectors (Fig. 10.1) but it is much more difficult to observe and there is still uncertainty about the various stages (Barr 1988). If the roots of infected plants are washed free of soil and examined, the most obvious structures of *P. graminis* are the thick walled cystosori or resting spore clusters. They look rather like a mulberry fruit and are formed from a cluster, often 100 or more, of resting spores. These may remain in the soil for many years and are the principal method of survival for the fungus. When these germinate they produce biflagellate zoospores which swim in water films until they reach a root hair to which they attach and encyst. Encysted spores develop a dagger-like body (Barr 1988) which is injected through the host wall and the contents of the zoospore cyst follow it, although the mechanism is unclear. The injected material can move in the root tissue and forms a plasmodium which develops into a zoosporangium and produces zoospores for dispersal. The events that produce a cystosorus are also uncertain and may involve a fusion of zoospores. While the zoospores have a limited range, except if carried in flowing water, and a limited life, cystosori are resistant, long-lived and have the potential for dispersal over long distances. Blown soil, or soil adhering to machinery and the feet of migrating birds have been invoked as potential carriers of *P. graminis*. It is also not unusual for harvested seed to be contaminated with small amounts of soil and this could also carry the fungus. Indeed this may be one of the ways in which such apparently immobile vectors in their normal environment have been dispersed so widely throughout the world.

VIRUS–HOST–VECTOR INTERACTIONS

The effects of virus infection on the host plant differ with the isolate of the virus, the host, its growth stage and environmental conditions. Generally young plants are more damaged by the virus than more mature ones. Thus, for BYDV, autumn infection of autumn-sown crops causes more yield loss than spring infection and, for the spring-sown crops, the coincidence of infective aphids and young crop causes most damage (Plumb 1983). Thus, later sowing in autumn can avoid infection if emergence is after the aphid migration has finished, and early sowing in spring can minimize damage because plant growth is well advanced before infective aphids colonize the crop.

There are differences in effect among the isolates: PAV-like isolates cause most damage and can kill plants, MAV-like isolates cause less, but still unacceptable, damage whereas the RPV-like isolates can cause almost as much damage as PAV (Herrera 1989).

Of the hosts, barley and oats are more severely damaged than wheat. There is less information on the effects on grasses. However, yield losses of 20% have been reported (Catherall & Parry 1987). Infection by BYDV has obvious effects, such as discolouration and stunting, on the various hosts that are susceptible, but these are only the external manifestation of physiological and metabolic changes in the plant. Respiration is increased and soluble carbohydrate and starch is accumulated (Jensen 1972; Jensen & von Sambeek 1972) in infected plants. Indigestible crude protein (Jensen 1969) and free amino acids (Markkula & Laurema 1964) are also increased but chlorophyll content, leaf area and photosynthetic rate are decreased (Jensen 1972; Jensen & von Sambeek 1972).

As well as the consequences these changes have for the host, they also affect the aphids that transmit BYDV and other pathogens that infect the crop. Thus, populations of *Metopolophium dirhodum* (Wlk.) and *S. avenae* were significantly increased on BYDV-infected wheat, oats and barley, apparently because of greater colonization of infected plants (Ajayi &

Dewar 1982). This was presumably because of the greater nutritional value of the sap from infected than from healthy plants. Virus infection also increases the frequency of alate aphid production (Gildow 1983) and an additional effect of this could be the production of fitter offspring on favoured hosts (Leather 1989).

In field crops single diseases and their vectors cannot be considered in isolation. Infection by BYDV increases the damage done by some diseases such as take-all of cereal roots caused by *Gaeumannomyces graminis* (Sacc.) von Arx & Oliver var. *tritici* (Sward & Kollmorgen 1986), *Pyrenophora avenae* Ho & Kuribay (Sommerfeld, Frank & Gildow 1987), *P. teres* (Drechsl.) (Delserone, Cole & Frank 1987), *Septoria* spp. (Pelletier, Comeau & Couture 1974), *Erysiphe graminis* DC (Potter 1984) on leaves and *Alternaria* and *Cladosporium* spp. on ears (Smith 1963). In addition BYDV-infected plants are more likely to be killed during winter than healthy plants (Paliwal & Andrews 1979; Andrews & Paliwal 1983, 1986).

The significance of these interactions relative to the direct effects of BYDV differ. Clearly if virus infection alone causes more than 90% loss of yield there is little opportunity for interaction and even if it does occur it is of little consequence. However, changes that may increase aphid populations or their dispersal could have a marked effect on virus epidemiology.

There is less evidence for differences between the different fungally-transmitted viruses in their effects on their hosts and little evidence of virus infection modifying host response to *P. graminis* or to other pathogens. However in glasshouse tests, plants infected with BaMMV had larger concentrations of BYDV when subsequently infected with this virus than did plants free from BaMMV infection (S. Forde, personal communication). The frequent association of infection with oat golden stripe virus in plants also infected with oat mosaic virus suggests that some dependent relationship may exist between these viruses but there is no definitive evidence for this.

DISCUSSION

The aphid-transmitted BYDV and the fungally transmitted viruses of cereals offer interesting contrasts in epidemiology. Both are very successful, if measured by their frequency but, on cereals themselves, have been very strongly influenced by man's interference through the husbandry of the crop. This has been especially true in Britain where the shift from spring to winter sowing of barley and the greatly increased area of wheat, also sown in the autumn, has greatly favoured infection and dispersal of both aphid- and, on barley, fungally transmitted viruses. Both groups of viruses no doubt existed long before they were recognized in the UK but when most autumn-sown crops were wheat, sown from late October onwards, infection was slight and problems of BYDV only became apparent on late-sown spring crops. Similarly, while spring barley is very susceptible to infection by BaYMV and BaMMV, if sown in infected soil in autumn, naturally occurring infection of spring-sown crops has not been reported, presumably because either the warming and drying of soils in spring does not favour infection and/or the symptoms of infection are not expressed at the temperatures that occur soon after sowing. Therefore, it is possible that spring barley crops in the past, often grown consecutively, may have slowly built up infection that was not evidenced by symptoms. When barley was then sown on this land in autumn, conditions favoured infection and symptom expression. Since autumn-sown barley has been grown widely the fungally transmitted viruses have been recorded increasingly frequently and have certainly spread within and between fields. The presence of compatible hosts during the winter, when soil moisture is normally plentiful, encourages resting spore germination and facilitates spread. Thus, the fungally transmitted viruses of barley are now an important factor in crop husbandry and have led to the development of resistant cultivars, the recognition of virus isolates that can overcome this immunity, and in severe cases, farmers have decided not to grow winter barley again on infected sites.

Therefore, in the past decade a previously unknown problem has become of national concern despite the relative natural immobility of the virus' vector. A similar virus, beet necrotic yellow vein virus, also transmitted by a fungus, *Polymyxa betae* Keskin, is the cause of Rhizomania of sugar beet. It is subject to strict quarantine requirements and susceptible crops cannot be grown on infected or adjacent land. When it was realized that the soilborne barley viruses were present it was too late to do more than try to limit their spread to new areas. However, wheat yellow mosaic (wheat spindle streak) and soilborne wheat mosaic viruses transmitted by *P. graminis* occur in France but have not been reported in the UK. The evidence from the barley viruses is that if they did become established they would spread quickly.

In contrast to the fungally transmitted viruses, which appear to be largely restricted in the UK to one of the two genera, *Hordeum* or *Avena*, BYDV has a wide host range in the Gramineae. Thus, in regions such as Britain, where perennial grass cover is the dominant vegetation type, there is no shortage of sources of infection. Interactions of potential aphid vectors with hosts, natural enemies and environment are vital requirements in any study of the epidemiology and control of the disease.

Irrespective of future methods for control, be they based (i) on chemicals, as at present for aphids; (ii) on resistance, whether conventionally obtained or through genetic manipulation; or (iii) on avoidance by husbandry methods based on knowledge of the epidemiology of the pathogens and phenology of their vectors, viruses of temperate cereals and their vectors seem certain to remain an important element in any study of the ecology of cereal crops.

ACKNOWLEDGMENTS

We are grateful to Dr G. Herrera and Dr M.J. Adams for the data in Tables 10.1 and 10.2 respectively.

REFERENCES

A'Brook, J. (1981). Vectors of barley yellow dwarf virus. *Euraphid; Rothamsted 1980* (Ed. by L.R. Taylor), p. 21. Rothamsted Experimental Station, Harpenden.

Adams, M.J. (1988). Evidence for virus transmission by plasmodiophorid vectors. *Viruses with Fungal Vectors*. (Ed. by J.I. Cooper & M.J.C. Asher), pp. 203–211. Association of Applied Biologists, Wellesbourne.

Ajayi, O. & Dewar, A.M. (1982). The effect of barley yellow dwarf virus on honeydew production by the cereal aphids *Sitobion avenae* and *Metopolophium dirhodum*. *Annals of Applied Biology*, **100**, 203–212.

Andrews, C.J. & Paliwal, Y.C. (1983). The influence of pre-infection cold hardening on disease development period and the interaction between barley yellow dwarf virus and cold stress tolerance in wheat. *Canadian Journal of Botany*, **61**, 1935–1940.

Andrews, C.J. & Paliwal, Y.C. (1986). Effects of barley yellow dwarf virus infection and low temperature flooding on cold stress tolerance of winter cereals. *Canadian Journal of Plant Pathology*, **8**, 311–316.

Barr, D.J.S. (1988). Zoosporic plant parasites as fungal vectors of viruses: taxonomy and life cycles of species involved. *Viruses with Fungal Vectors* (Ed. by J.I. Cooper & M.J.C. Asher), pp. 123–137. Association of Applied Biologists, Wellesbourne.

Batista, M.F., Antoniw, J.F., Swaby, A.G., Jones, P. & Adams, M.J. (1989). RNA/cDNA hybridization studies of UK isolates of barley yellow mosaic virus. *Plant Pathology*, **38**, 226–229.

Carter, N., McLean, I.F.G., Watt, A.D. & Dixon, A.F.G. (1980). Cereal aphids: a case study and review. *Applied Biology* (Ed. by T.H. Coaker), Vol. 5, pp. 272–348. Academic Press, London.

Catherall, P.L. & Parry, A.L. (1987). Effects of barley yellow dwarf virus on some varieties of Italian, hybrid and perennial ryegrasses and implications for grass breeders. *Plant Pathology*, **36**, 148–151.

Delserone, L.M., Cole, H. & Frank, J.A. (1987). The effects of infection by *Pyrenophora teres* and barley yellow dwarf virus on the freezing hardiness of winter barley. *Phytopathology*, **77**, 1435–1437.

Dewar, A.M. & Carter, N. (1984). Decision trees to assess the risk of cereal aphid (Hemiptera: Aphididae) outbreaks in summer in England. *Bulletin of Entomological Research*, **74**, 387–398.

Dixon, A.F.G. (1971). The life cycle and host preferences of the bird cherry-oat aphid, *Rhopalosiphum padi* L., and their bearing on the theories of host alternation in aphids. *Annals of Applied Biology,* **68,** 135–147.

Doodson, J.K. (1967). A survey of BYDV in S.24 perennial ryegrass in England and Wales, 1966. *Plant Pathology,* **16,** 42–45.

Gildow, F.E. (1983). Influence of BYDV-infected oats and barley on morphology of aphid vectors. *Phytopathology,* **73,** 1196–1199.

Gildow, F.E. (1987). Virus membrane interaction involved in circulative transmission of luteoviruses by aphids. *Current Topics in Vector Research,* **4,** 95–120.

Guy, P.L., Johnstone, G.R. & Morris, D.J. (1987). Barley yellow dwarf virus in, and aphids on, grasses (including cereals) in Tasmania. *Australian Journal of Agricultural Research,* **38,** 139–152.

Hand, S.C. (1980). Overwintering of cereal aphids. *IOBC, WPRS Bulletin,* **3,** 59–61.

Hand, S.C. & Carillo, J.R. (1982). Cereal aphids on maize in southern England. *Annals of Applied Biology,* **100,** 39–47.

Herrera, G.F. (1989). *Interactions between host plants and British isolates of barley yellow dwarf virus.* PhD thesis, University of London, Imperial College.

Jedlinski, H. (1981). Rice root aphid, *Rhopalosiphum rufiabdominalis,* a vector of barley yellow dwarf virus in Illinois, and the disease complex. *Plant Disease,* **65,** 975–978.

Jedlinski, H. & Brown, C.M. (1965). Cross protection and mutual exclusion by three strains of barley yellow dwarf virus in *Avena sativa* L. *Virology,* **26,** 613–621.

Jensen, S.G. (1969). Composition and metabolism of barley leaves infected with barley yellow dwarf virus. *Phytopathology,* **59,** 1694–1698.

Jensen, S.G. (1972). Metabolism and carbohydrate composition in barley yellow dwarf-infected wheat. *Phytopathology,* **62,** 587–592.

Jensen, S.G. & van Sambeek, J.W. (1972). Differential effects of barley yellow dwarf virus on the physiology of tissues of hard red winter wheat. *Phytopathology,* **62,** 290–293.

Jess, S. & Mowat, J. (1986). Transmission of BYDV by larvae of frit fly, *Oscinella frit* (L.) and effects of sward-killing herbicide on transmission. *Record of Agricultural Research, Northern Ireland,* **35,** 821–830.

Kashiwazaki, S., Ogawa, K., Usugi, T., Omura, T. & Tsuchizaki, T. (1989). Characterization of several strains of barley yellow mosaic virus. *Annals of the Phytopathological Society of Japan,* **55,** 16–25.

Krupa, S.V. & Dommergues, Y.R. (Eds) **(1979).** *Ecology of Root Pathogens.* Elsevier, Amsterdam.

Leather, S.R. (1989). Do alate aphids produce fitter offspring? The influence of maternal rearing history and morph on life-history parameters of *Rhopalosiphum padi* (L). *Functional Ecology,* **3,** 237–244.

Loxdale, H.D., Tarr, I.J., Weber, C.P., Brookes, C.P., Digby, P.G.N. & Castanera, P. (1985). Electrophoretic study of enzymes from cereal aphid populations. III. Spatial and temporal genetic variation of populations of *Sitobion avenae* (F) (Hemiptera: Aphididae). *Bulletin of Entomological Research,* **75,** 121–141.

Markkula, M. & Laurema, S. (1964). Changes in the concentration of free amino acids in plants induced by virus disease and the reproduction of aphids. *Annales Agriculturae Fenniae,* **3,** 265–271.

McGrath, P.F. & Bale, J.S. (1989). Cereal aphids and the Infectivity Index for barley yellow dwarf virus (BYDV) in Northern England. *Annals of Applied Biology,* **114,** 429–442.

Mills, P.R., Mercer, R.C. & McGimpsey, H.C. (1986). Barley yellow dwarf virus. *Annual Report on Research and Development, Department of Agriculture for Northern Ireland.*

Oswald, J.W. & Houston, B.R. (1951). A new virus disease of cereals transmitted by aphids. *Plant Disease Reporter,* **35,** 471–475.

Paliwal, Y.C. & Andrews, C.J. (1979). Effects of barley yellow dwarf virus and wheat spindle streak mosaic virus on cold hardiness of cereals. *Canadian Journal of Plant Pathology*, **1**, 71–75.

Pelletier, G., Comeau, A. & Couture, L. (1974). Interaction contre le virus de la juille rouge de l'avoine (BYDV) *Septoria avenae* et *Puccinia coronata* sur *Avena sativa*. *Phytoprotection*, **55**, 9–12.

Plumb, R.T. (1974). Properties and isolates of barley yellow dwarf virus. *Annals of Applied Biology*, **77**, 87–91.

Plumb, R.T. (1977). Grass as a reservoir of cereal viruses. *Annales de Phytopathologie*, **9**, 361–364.

Plumb, R.T. (1983). Barley yellow dwarf virus — a global problem. *Plant Virus Epidemiology — The Spread and Control of Insect-borne Viruses* (Ed. by R.T. Plumb & J.M. Thresh), pp. 185–198. Blackwell Scientific Publications, Oxford.

Plumb, R.T. (1990). The epidemiology of barley yellow dwarf in Europe. *CIMMYT Workshop on Barley Yellow Dwarf* (Ed. by P.A. Burnett), pp. 215–227. CIMMYT, Mexico.

Plumb, R.T. & Macfarlane, I. (1977). A 'new' virus of oats. *Annual Report of Rothamsted Experimental Station for 1976*, pp. 256–257.

Plumb, R.T., Catherall, P.L., Chamberlain, J.A. & Macfarlane, I. (1977). A new virus of oats in England and Wales. *Annales de Phytopathologie*, **9**, 365–370.

Potter, L. (1984). Interaction between barley yellow dwarf virus and mildew in oats. *Welsh Plant Breeding Station Report for 1983*, pp. 79–80.

Rochow, W.F. (1969). Biological properties of four isolates of barley yellow dwarf virus. *Phytopathology*, **69**, 1580–1589.

Rochow, W.F. & Eastop, V.F. (1966). Variation within *Rhopalosiphum padi* and transmission of barley yellow dwarf virus by clones of four species. *Virology*, **30**, 286–296.

Rochow, W.F., Muller, I. & Gildow, F.E. (1983). Interference between two luteoviruses in an aphid: lack of reciprocal competition. *Phytopathology*, **73**, 919–922.

Smith, H.C. (1963). Control of barley yellow dwarf virus in cereals. *New Zealand Journal of Agricultural Research*, **10**, 445–466.

Sommerfield, J.A., Frank, J.A. & Gildow, F.E. (1987). Effect of barley yellow dwarf virus and *Pyrenophora avenae* infections, singly and in combination, on yield components of oats. *Phytopathology*, **77**, 989.

Sward, R.J. & Kollmorgen, J.F. (1986). The separate and combined effects of BYDV and take-all fungus (*Gaeumannomyces graminis* var. *tritici*) on the growth and yield of wheat. *Australian Journal of Agricultural Research*, **37**, 11.

Tatchell, G.M., Plumb, R.T. & Carter, N. (1988). Migration of alate morphs of the bird cherry aphid (*Rhopalosiphum padi*) and implications for the epidemiology of barley yellow dwarf virus. *Annals of Applied Biology*, **112**, 1–11.

Taylor, L.R., French, R.A., Woiwod, I.P., Dupuch, M. & Nicklen, J. (1981). Synoptic monitoring for migrant insect pests in Great Britain and Western Europe. I. Establishing expected values for species content, population stability and phenology of aphids and moths. *Rothamsted Experimental Station Report 1980*, Part 2, 41–104.

Teakle, D.S. (1983). Zoosporic fungi and viruses. Double trouble. *Zoosporic Plant Pathogens* (Ed. by S.T. Buczaki), pp. 233–248. Academic Press, London.

Timian, L.G. (1964). Dodder transmission of barley yellow dwarf virus. *Phytopathology*, **54**, 910.

Toko, H.V. & Bruehl, G.W. (1959). Some host and vector relations of a strain of barley yellow dwarf virus. *Phytopathology*, **49**, 343–347.

Torrance, L., Pead, M.T., Larkins, A.P. & Butcher, G.W. (1986). Characterization of monoclonal antibodies to a UK isolate of barley yellow dwarf virus. *Journal of General Virology*, **67**, 549–556.

Wallin, J.R. & Loonan, D.V. (1971). Low level jet winds, aphid vectors, local weather and barley yellow dwarf virus outbreaks. *Phytopathology*, **61**, 1068–1070.

Walters, K.F.A. & Dewar, A.M. (1986). Overwintering strategy and the timing of the spring migration of the cereal aphids *Sitobion avenae* and *Sitobion fragariae*. *Journal of Applied Ecology*, **23**, 905–915.

Waterhouse, P.M., Gildow, F.E. & Johnstone, G.R. (1988). Luteovirus Group. AAB Descriptions of Plant Viruses No. 339, 9pp.

Watson, M.A. (1972). Transmission of plant viruses by aphids. *Principles and Techniques in Plant Virology* (Ed. by C.I. Kado & H.O. Agrawal), pp. 131–167. Van Nostrand Reinhold, New York.

Watson, M.A. & Mulligan, T.E. (1960). Comparison of two barley yellow dwarf viruses in glasshouse and field experiments. *Annals of Applied Biology*, **48**, 559–574.

Wiktelius, S. & Ekbom, B.S. (1985). Aphids in spring sown cereals in central Sweden. Abundance and distribution 1980–1983. *Zeitschrift für angewandte Entomologie*, **100**, 8–16.

11. INTERACTIONS BETWEEN WEEDS AND CROPS

L.G. FIRBANK

Department of Applied Sciences, Anglia Polytechnic,
Cambridge CB1 1PT, UK

INTRODUCTION

The cereal field habitat is typified by moderate to high levels of nutrients and water, and high levels of disturbance designed in part to reduce the populations of species other than the crop. Within this habitat, weeds and crop plants interact in a variety of ways. They compete for resources, so that a successful weed infestation reduces the yield of the crop, and the crop in turn may strongly suppress the growth of weeds. Their distributions may be interdependent, as the weed seeds may be dispersed with the crop and its residues, or perhaps the weed thrives only on land where crops are grown. Weeds and crops may evolve in similar ways; some weed species can come to mimic the life history and even growth form of the crop plants. There are also indirect interactions, as weed populations may affect the levels of herbivores and pathogens of the crop.

The distinction between weeds and crops is far less clear-cut than many people imagine. Both weeds and crops are able to endure the highly disturbed habitat of the cereal field. Seeds of sugar beet or oil-seed rape can germinate a season (or more) too late and act as weeds in a subsequent cereal crop. Cereal grain spilled during harvest can germinate to act as volunteer weeds in a later crop. On the other hand weed species have in the past become domesticated to become crop plants — oats is an example of such a secondary crop (e.g. Zohary & Hopf 1988). The only characteristic difference between a weed and a crop is the attitude and actions of the farmer.

It is clear that weeds and crops interact over a variety of scales of space and time. Competition operates over a very local scale and within a single season, although the *effects* of competition may have implications for future seasons, by influencing the population dynamics of the weeds and possibly the management of the crops. The effect of management on weed–crop systems is seen at larger spatial scales: the level of the field, the

This chapter is dedicated to the memory of Kees Spitters, who died suddenly in April 1990. That I found his clear vision of agricultural systems a source of inspiration is evident from what I have written. His warm, engaging personality will be much missed.

farm and beyond. The influences of soils and climate may be very local and short term, for example, a sudden heavy rainstorm may cause lodging of the crop and encourage weed growth. They may also operate over higher scales, by affecting the geographic range of weeds and crops and the processes of selection. The synoptic diagram in Fig. 11.1 helps to show, in a very general way, the scales at which different ecological processes operate on weed–crop systems. In this paper, I will begin with the most local interaction, competition, before considering larger scale changes in weed populations and communities.

COMPETITION

There are many definitions of plant competition (Grace & Tilman 1990), but it is here defined as 'an interaction between individuals brought about by a shared requirement for a resource in limited supply, and leading to a reduction in the survivorship, growth and/or reproduction of the individuals concerned' (Begon, Harper & Townsend 1986). Competition is very important in practice, as it affects the yield of the crop and the sizes of the weed populations. It operates over local scales and within single seasons, and so is amenable to study using experimental manipulation.

The mechanism of competition between plants is thought to be well understood (e.g. Harper 1977; Grace & Tilman 1990). Any area of land

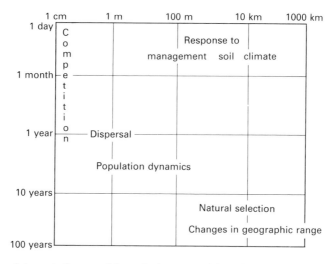

FIG. 11.1. Schematic diagram of the scales in space and time of the ecological processes which affect weed–crop interactions. These scales are clearly indicative only.

contains a limited amount of resources — water, minerals and light flux. Each individual plant requires a certain amount of these if it is to grow to its maximum size. At very low densities of plants, there are enough resources to go round, but at higher densities the resources can only support a certain amount of plant growth, so yield per unit area levels off and yield per plant is reduced (see Biscoe & Costigan, this volume, pp. 59–61). Crops are sown at quite high densities to maximize yield per unit area. Any weeds present will reduce the pool of resources available to the crop, thereby reducing crop yield. The amount of the reduction depends on the number and size of the weed plants and on their demand for resources needed by the crop.

This process has been simulated most effectively by Spitters (1991; Spitters & Aerts 1984). He simulated competition for light by dividing the canopy into a series of horizontal layers. The leaves of the topmost layer intercept a certain proportion of light, and the remainder is available to leaves in the next layer, and so on. How much light is intercepted by the weed and how much by the crop depends upon the vertical distribution of leaves of the two species. The model allocates light energy between the weed and crop and calculates the growth up to the next increment of time (expressed in degree days, to take into account temperature) according to species-specific relationships. The light capture algorithm is then repeated. Competition for water and minerals is represented by a reduction of growth when they are in short supply. This modelling approach is an extension of crop production models (see Biscoe & Costigan, this volume, p. 58). It would not apply readily to parasitic or climbing weeds, where growth is also dependent upon plants of other species.

Such a model can provide good fits to data (Spitters & Aerts 1984), but is too imprecise to be of widespread practical use (Spitters, in press). However, it does provide valuable insights into the circumstances which affect weed–crop competition. In particular it demonstrates the importance of starting position. Competition for light is intrinsically asymmetrical. Plants with a small initial advantage grow more quickly than their shaded neighbours, and so the advantage to the early emerged plants can be great. Spitters & Aerts (1984) claim that a 12-h difference in emergence times can give the early emerging species a clear competitive advantage. The importance of emergence time is supported by repeated experiments showing that crop yield loss is greatest when early emerged weeds are left alive (e.g. Weaver 1984). Differences between crop varieties are to be expected; those which achieve high levels of ground cover quickly can suppress weed infestations better than others (Richards 1989). Detailed demographic studies of the weed *Bromus sterilis* L. growing in winter wheat show that, as expected from Spitters's analysis, the early emerging plants grow to be larger and set more seeds than the late emerging cohorts (Fig. 11.2; Firbank & Watkinson 1990; see also Peters 1984 and included references). This may not be the case when the early emerging plants endure harsher conditions than later emerging individuals, whether due to management or adverse weather.

Most competition experiments in agriculture deal with a single species of weed (if the numerous experiments testing herbicides on mixed or even unspecified weed floras are excluded), and the weed density and sometimes the crop density are manipulated. The results are usually presented in terms of crop yield loss (Radosevich & Roush 1990), and the data necessary for detailed simulation are available only very rarely. It is far more usual to use experimental designs and models looking simply at the relationships between yield at harvest and density. In the replacement series design (de Wit 1960; de Wit & van den Bergh 1965), the total weed and crop density is kept constant and the proportions are varied. The relative yields in mixture are assumed to reflect relative competitive

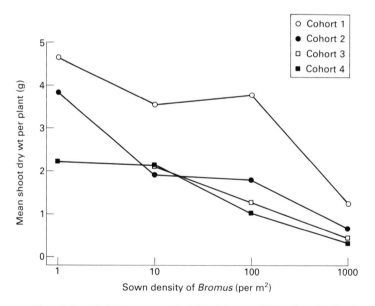

FIG. 11.2. The relationship between mean individual shoot weight and sowing density and age for plants of *Bromus sterilis* growing among winter wheat. Four *Bromus* cohorts are shown, where the first cohort emerged within 3 weeks of the crop, the second within the next 3 weeks and so on for the third and fourth cohorts. Unpublished data from Firbank from an experiment reported by Firbank *et al.* (1984).

abilities (de Wit & van den Bergh 1965) which in turn depend upon environmental conditions (Fleming, Young & Ogg 1988) and relative emergence times (Firbank & Watkinson 1985). The design can also indicate any niche differentiation between the species (de Wit & van den Bergh 1965; Hall 1974a,b).

An alternative experimental design, the additive series, keeps the density of one species — the crop — constant, while varying that of the weeds (Harper 1977). This reveals the yield loss of the crop to be a function of weed density (Zimdahl 1980). The widely used model of Cousens (1985) assumes an hyperbolic function:

$$Y = Y_{wf}(1 - [IN/(1 + IN/A)]), \tag{1}$$

where Y is the crop yield per unit area, Y_{wf} is the crop yield in the absence of weeds, A is the crop yield at an asymptotically high weed density, expressed as a proportion of Y_{wf}, I is the yield loss resulting from a single weed at low density, and N is the density of the weeds (Fig. 11.3). This model can be adapted to include a term to account for differences in emergence times explicitly (Cousens *et al.* 1987a).

A third experimental design, termed the addition series (Spitters 1983), requires the densities of both weed and crop to vary. Such an experiment is larger than the replacement and the addition series, and so is much less common. However, it allows a much improved comparison of inter- and

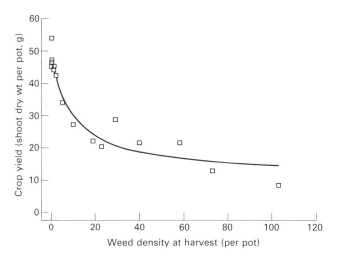

FIG. 11.3. The relationship between crop yield and weed density for *Bromus sterilis* and winter wheat in a pot experiment on sandy soil, showing the fitted curve using eqn (1). Data from Firbank *et al.* (1990).

intraspecific competition between the two species. In particular, if we look
at the yield per unit area Y of one species in terms of its own density N and
the density of the other species N' by using a model such as

$$Y = Nw_m(1 + a[N + \alpha N'])^{-b} \qquad (2)$$

it is possible to estimate competition coefficients α for both the weed and
crop. These are parameters which quantify the mean effect of an individual
of one species on the yield of individuals of the other; the other parameters
are determined by the yield and effects of competition within the species
when grown in monoculture (Firbank & Watkinson 1985, 1990).

To summarize, the eventual yields of the weeds and crops are functions
of density (which in part determines the abundance of leaf surface down
the profile) and emergence time. These functions may be derived empirically (such as eqns (1) and (2)) or as results of a more mechanistic
approach such as Spitters's. The parameters of these functions depend in
turn upon environmental variables and on the growth rates, physiology and
morphology of the plants.

In principle, crop yield loss–weed density functions can be used to help
determine threshold levels for herbicide application or other control measures (Auld, Menz & Tisdell 1987). Unfortunately, the functions are usually
determined *post hoc*, whereas to be of practical use, they must be known in
advance. Gill, Poole & Holmes (1987) looked at data for *Bromus diandrus*
Roth taken over a range of sites and years in Australia. They concluded
that the crop yield function was robust to differences between sites and
seasons, and so could be used as management tools. This finding relates
to the magnitude of differences deemed to be important (no statistical
comparisons were given) and to a low level of variation from place to place
and year to year.

It should not be assumed that the functions are always so reliable. One
common problem is that the yield loss due to weeds depends upon relative
emergence times and early growth rates. Yet they in turn depend upon the
weather (e.g. Roberts 1984), and the data may be time-consuming to
collect in practice. Kropff (1988) has suggested that the best compromise
between accuracy and ease of collecting data is to compare weed and crop
ground cover early in the season — thus taking into account emergence
time, growth and density of the more competitive early emerging weeds.
This may come too late to decide upon the use of some herbicides.

More seriously, all of the models discussed so far are based on the
analysis of small-scale experiments and assume that the environment is
constant through space. This is incorrect even in a small field. The weed
infestations themselves are commonly clumped, due to ingress from field

margins, patchy herbicide application, differences in soils and drainage within fields or the local dispersal and growth of many weeds. Estimates of weed density based on a small number of samples can be highly inaccurate for both shoots (Marshall 1988) and for the seed-bank (Benoit, Kenkel & Cavers 1989). Also, competition among weeds is increased if they are clumped, often reducing the yield loss of the crop compared with an even weed distribution at the same overall density (Auld & Tisdell 1988). While this situation can be described by extending existing mean yield models (Brain & Cousens 1990), the results are unlikely to be used widely unless farmers or their advisers are prepared to adopt intensive weed density sampling strategies.

Extending the spatial scale, differences between fields are, of course, going to be even greater. Nutrient availability, pH, drainage, climate and management history all have the capacity to affect the outcome of competition (e.g. Blackman & Templeman 1938; Harper 1977; Zimdahl 1980) by affecting the growth rates of weed and crop plants. For example, Carlson & Hill (1985) varied the nitrogen supply to weed-infested plots to obtain a multiple regression describing crop yield as a function of weed density and quantity of nitrogen. Studies of natural vegetation show that it is possible to relate relative competitive ability of different species to the balance between nutrient uptake and loss (Berendse & Elberse 1990). This dynamic approach may need to be followed if the effects of timing of application of fertilizer on weed–crop competition are to be modelled, especially as the response to pulses of nutrients can be difficult to predict otherwise (Benner & Bazzaz 1987). The effects of another factor, waterlogging, may be seen in the differences in wheat yield — *Bromus sterilis* density relationships observed by Cousens *et al.* (1988) and Firbank *et al.* (1990); the competitiveness of the weed was reduced on waterlogged sites as its emergence was delayed.

Of course, few fields have only one major weed species present. One approach to describe the effects of several weed species on crop yield is to assume that the competitive effects of each weed species on the crop can simply be added together. The concept of weeds having crop equivalent values is based on this idea. Each weed species is given a competitive value based upon the ratio of its individual plant dry weight to that of the crop. The crop yield loss is estimated by summing the crop equivalences of all weeds present (Cussans, Cousens & Wilson 1987). This model is naïve for a number of reasons; it assumes that total yield of the field is constant from year to year and regardless of species composition and densities, and it ignores variation in weed size at all scales (see Wilson & Wright 1990). It also ignores the possible effect of interspecific competition between weeds, although in plant communities, interspecific competition can be reduced if the different species are found in different patches (e.g. Czaran & Bartha 1989). It may well be that interspecific competition between weed species with clumped distributions is often at low levels in practice.

Other density-dependent factors

One problem about the analysis of the effects of competition is that other factors also operate in a density-dependent manner. For example, claims have been made that there are negative chemical-induced (allelopathic) interactions between crop and weed species (Rice 1974). Unequivocal evidence is hard to obtain (Williamson 1990), except where the effect results from the release of chemicals during the breakdown of crop residues (e.g. Putnam 1985). Increases in mean plant yield with density are possible in such cases, as the toxins are shared out among more plants (Weidenhamer, Hartnett & Romeo 1989).

Indirect interactions between weeds and crops are also possible. Weeds can act as reservoirs to herbivores and disease vectors, thereby acting to reduce crop yield, or can support predators and parasitoids of crop herbivores, therefore increasing crop yield (Altieri, this volume, p. 262, Table 13.1)

These and other factors affect the parameters of the competition models outlined above, possibly even to the extent that competition for resources is inferred when it is not actually happening (Holt 1977; Connell 1990).

WEED POPULATION DYNAMICS

The propagule production of weeds depends upon their size at maturity. Therefore weed densities in future seasons, and future crop yield losses, are also functions of density, emergence time and growth rates in the current season. Weather patterns and management practices in 1 year have the potential to affect weed levels several seasons into the future.

Attempts have been made to incorporate such future losses into proposed threshold levels for weed control (Cousens et al. 1986; Auld, Menz & Tisdell 1987). The basic approach is simple. For many weeds, especially annuals, seed production is highly correlated with shoot biomass at harvest which can be estimated in turn from yield-density models. Knowing seed mortality and dormancy rates, it is often a simple matter to model weed density at the start of the following season. Submodels may be required to describe the behaviour of a persistent seed-bank. In some cases it is helpful to model the behaviour of specific age-categories of the weed (see reviews by Mortimer 1983; Cousens *et al.* 1987b). The weed population model can then be incorporated into an economic decision-making system (e.g. Cousens *et al.* 1986).

A rare example of using a population model to forecast future infestations deals with the annual grass weed *Bromus sterilis*. Firbank, Mortimer & Putwain (1985) planted this species in winter wheat and in monoculture over a range of densities and recorded survivorship and fecundity as functions of density. After harvesting the crop, the plots were subject to four control practices, and seed survivorship functions were derived using seedling densities in the following season. As *Bromus* is believed to have no persistent seed-bank, it was assumed that seeds which had not germinated had died. The densities of seedlings a year ahead were then forecast by assuming that the competition, fecundity and seed survival functions remained constant. The forecasts were too low by the unacceptable margin of approximately 20%, apparently because the *Bromus* plants were more competitive with the crop than predicted.

There are many reasons why forecasts of weed densities for future seasons are unlikely to be accurate (Firbank 1989). Estimates of the various parameters can never be obtained with perfect precision and the effects of errors can become magnified through time (Firbank 1989). Competitive relationships may differ from year to year at the same site in an unpredictable way (Reader 1986), even when models are used which explicitly account for emergence time (Cousens *et al.* 1987a). The efficiency of control practices also may vary from year to year and from place to place (e.g. Tottman *et al.* 1989).

Population models are still important, however, as they show the potential for rapid changes in weed numbers by highlighting the way that density-dependent factors and density-independent factors jointly determine population size. Most weeds have the capacity to increase rapidly from low densities (e.g. Grime 1979), but are prevented from doing so by the control practices of the farmer. *Bromus sterilis*, for example, can multiply 100-fold between seasons from low densities, and so requires an annual kill rate of 99% to control it (Firbank *et al.* 1984). At these levels of control, slight fluctuations in mortality can have large effects on the weed dynamics; a small increase in mortality can cause rapid extinction, a small decrease can allow a rapid expansion in numbers.

Firbank & Watkinson (1986) used a population model to help explain the sudden decline of the formerly common annual weed, the corncockle

Agrostemma githago L. This species is unusual, as it depends upon being harvested and resown with the crop. The degree of control depends upon the efficiency of seed cleaning (Firbank 1988). Using a combination of modern experimental data and herbarium specimens, they suggested that a small increase in seed cleaning efficiency was all that was needed to cause the population to crash.

The high mortality rates usually experienced by cereal weeds encourages natural selection of those genotypes which can survive human control. In the corncockle, this natural selection resulted in crop mimicry (Thompson 1973). The seeds least likely to be separated from the grain were those of the same size as the grain. Different varieties of corncockle evolved in different parts of Europe; one matched in seed size to wheat, and one to flax. Furthermore, the individuals producing the most seeds are those which emerge early, encouraging selection for rapid germination and against delayed emergence from a seed-bank. Because spilled seeds would germinate and be controlled during cultivations, the corncockle also developed retention of the seeds on the plants (Firbank 1988). Such crop mimicry reaches its apogee in secondary crops, such as oats, which are former weed species which have now become fully domesticated (Zohary & Hopf 1988).

The greater the selection pressures, the more rapid the possible evolutionary response of the weed. It is not surprising then that the reliance on herbicides for weed control is being threatened by the evolution of resistance in certain weeds; already fifty-five species have evolved resistance to triazine herbicides and resistance to other herbicides is increasing — even to paraquat which is non-residual (Putwain & Mortimer 1989). While resistant weeds tend to be smaller and have lower rates of increase than susceptible conspecifics, population studies show that they are capable of rapid expansion when the appropriate herbicide is being used (Putwain & Mortimer 1989, and included references). For example, control failures of several weed species including *Lactuca serriola* L. and *Stellaria media* (L.) Vill. were observed after only five years of consecutive use of sulphonyl urea herbicides (Reed *et al.* 1989).

Changes in mortality rates do not depend solely upon man. They also vary across the geographic range of the weed; in extreme habitats, such as at the edge of the range, natural mortality is likely to be high and the rate of increase low, leaving the species more susceptible to the effects of human control. Even where individual plants may be able to survive, they may be unable to form viable populations (Prince & Carter 1985). We should expect a rapidly changing weed flora if our climate changes appreciably.

WEED COMMUNITIES

We have seen that studies of weed populations almost always assume that there is only one species of weed present. In the short term this may be justified as often only one species is seen to be economicaly important in a field. However, this attitude fosters the idea that the weed community is a static system, whereas it is dynamic. In practice, if you alter the management to control one species, other species may be promoted. This effect can be seen on a large scale by the increases in grass weeds and reductions in dicotyledonous weeds in the UK during the current century (Fryer & Chancellor 1970).

One way of evaluating the dynamics of weed communities is to produce population models of each species in turn. However, the resulting models retain all the difficulties of single species models and may be further complicated because of possible interactions between weed species.

Firbank (1989) developed a simple stochastic simulation of a three-species weed community growing in a continuous cereal crop. In the model, even when all three species have identical characteristics, the populations can fluctuate markedly due to random changes in control levels. In particular, one species often becomes dominant for a long period of time before random fluctuations allow another species to increase (Fig. 11.4). The model implies that competitive interactions between weed species are important in determining community structure, although this importance may well have been exaggerated by the spatial uniformity and the similarity of the species in the model. The model also suggests that precise, numerical forecasts of future weed densities may be unobtainable and that analyses based on equilibrium densities of weeds are of limited value (see Mortimer, Sutton & Gould (1989) for a more optimisitic view).

It may be more practicable to discover how weed community structure as a whole is determined and how it changes in response to management. There is currently much interest in the structure of natural communities of plants (e.g. Grabherr *et al.* 1990; Grace & Tilman 1990) and studies concerning the relationship between the abundance of species and the characteristics of the plants on the one hand (e.g. Grime 1979) and resource levels on the other (e.g. Tilman 1982, 1988; Goldberg 1990) are clearly relevant to weed–crop systems (e.g. Roush & Radosevich 1985). Yet the special features of weed communities — in particular the effects of repeated disturbance, crop rotation and the pulsed applications of nutrients — have received little consideration.

There have been few studies looking at changes in weed communities at a local level. Chancellor (1985) recorded the weed flora throughout

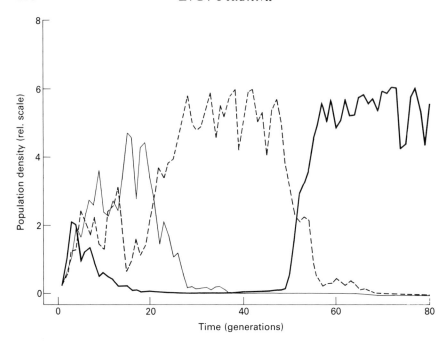

FIG. 11.4. A simulation of three identical weed species growing in a continuous cereal crop. All three species are subject to random changes in control from year to year. From Firbank (1989).

a 20-year period of cropping of one field, and showed the importance of sowing date of the crop on the weed flora. Spring sowing favoured spring-germinating weeds which avoided the period of maximum disturbance, while autumn sowing favoured autumn-germinating weeds which emerge early relative to the crop (see also Bazzaz 1990, and included references).

In a more carefully designed experiment, Post (1988) showed that weed communities (as identified by cluster analysis) corresponded well with different cultivation dates which had been in force for 3 years. These effects were not seen in the seed-bank, however, presumably because the time scale was too short. In this volume, Aebischer (pp. 311–313) and El Titi (p. 405) give accounts of the responses of weed populations to management; both authors show that changes in weed communities can be detected at the level of the individual field.

Long-term designed experiments looking at weed communities are most unusual. Hume (1987) applied high, low and no application of the herbicide 2, 4-D consistently to plots for 12 years. Each plot was under a wheat–wheat–fallow rotation which started in 1947. He classified the more abundant weed species into susceptible, moderately tolerant and highly tolerant to the herbicide. He then found that while the densities of susceptible species were reduced on the plots with the herbicides, these species did not become extinct. The tolerant species became more abundant on these plots, very possibly in response to reduced competition from other weeds. If so, it may be unwise to assume that weed–weed interactions can be safely ignored in community studies (see below).

The most important long-term experiment is of course the Broadbalk classical experiment at Rothamsted, looking at the effects of different management practices on crop production (Woiwod, this volume, pp. 283–287). The plots have had continuous wheat sown on them since 1843, and the abundance of weed species was recorded regularly until the late 1960s. As in Hume's (1987) data, while relative abundance of species has varied between plots, species did not become extinct in any treatment (Thurston 1968). This may, however, simply reflect migration between plots.

A full analysis of these data has never been attempted. Their potential is illustrated by the behaviour of three species during the 1950s and 1960s on plots which had been switched to a herbicide treatment. Thurston's (1968) data clearly show the steady decline of *Cirsium arvense* (L.) Scop. and the increase of the less herbicide-susceptible *Alopecurus myosuroides* Hudson (Fig. 11.5). However, the phase diagrams suggest that the magnitude of year to year variations differ between the species; the perennial *Cirsium* occupies a narrower and hence more predictable region of phase space than the annuals *Ranunculus arvensis* L. and *Alopecurus*. The *Alopecurus* phase diagram hints at cycles around two equilibria which may correspond to population levels without and with herbicide; the cycles may correspond to population oscillations resulting from overcompensating density dependence (Symonides, Silvertown & Andreasen 1986).

The weed community depends upon the pool of species which can invade the area as well as their success at maintaining their populations. The species pool depends in turn upon human activities and the dispersal characteristics of the weeds in space and time.

It is often stated that weeds are capable of wide dispersion (e.g. Grime 1979), but while some species such as *Papaver rhoeas* L. exhibit a long-lived seed-bank to allow dispersal in time and small wind-blown seeds to allow long-range dispersal through space, many others are dispersed much less effectively. Marshall's (1989) study of weed distribution at cereal field boundaries revealed that several species were found no further than 2.5 m into the crop, implying in some cases poor dispersal from hedgerows. Recruitment from the seed-bank reflects the reproductive output of at least

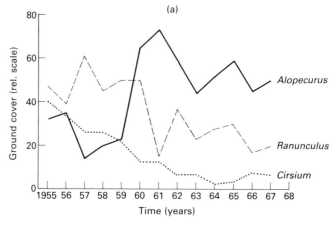

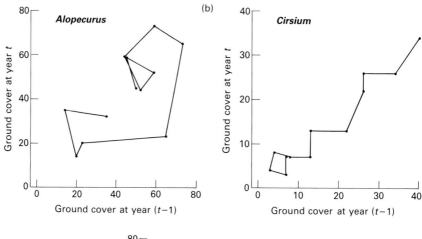

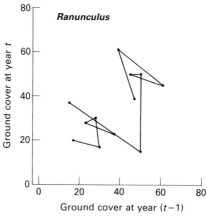

some species from years past, adding an historical dimension to weed community dynamics. Finally, weeds may be introduced into the field directly by man, his machinery, or via the seed grain. For example, Chancellor (1985) observed an invasion of several species of *Polygonum* from contaminated seed grain. The corncockle *Agrostemma githago* depended upon such accidental introductions for its very survival in Britain (Firbank 1988).

Over a broader spatial scale, the abundance of different weed species is determined by the interaction between the biology of individual weed species, farming methods and climatic and edaphic factors. Such relationships can be monitored using large-scale repeated surveys (e.g. Chancellor 1977; Chancellor & Froud-Williams 1984; Froud-Williams & Chancellor 1981), as long as they are truly comparable in method (Whitehead & Wright 1989).

The recent increases in density of many grass weeds in British cereal crops, and decline of many dicotyledonous weeds, have been linked to changing management practices, especially the use of herbicides (Fryer & Chancellor 1970). Considering the increased use of autumn sowing over the last 20 years (Carter & Plumb, this volume, pp. 203–204), and its effects upon the weed flora (Brenchley & Warrington 1933), it is nevertheless surprising how little the frequencies of the most common dicotyledonous weeds in the UK have changed (Table 11.1).

While more experiments are needed to look at the development of weed communities, we also need some way of organizing and using the results to help forecast future changes. Detailed population models are too restricted in space and time, do not cope with variation from year to year and rarely deal with weed–weed and other interactions. A better approach might be to construct simple models, conceptual or mathematical, to indicate which ecological strategies are compatible with a given form of management.

For example, Chancellor & Froud-Williams (1986) anticipate future weed problems in Britain by anticipating trends in farming practices. They suggest that direct drilling favours perennials and those annual species (mostly grasses) which do not require seed burial, and continued late summer and autumn sowing will favour weeds emerging at these times.

FIG. 11.5. Population changes in the arable weeds, *Alopecurus myosuroides*, *Cirsium arvense* and *Ranunculus arvensis* (dashed line) on eighteen plots on the Broadbalk experiment following the introduction of herbicides. Ground cover estimates each year are expressed as (a) time series and (b) phase diagrams. Data from Thurston (1968).

TABLE 11.1. Abundance in percentages (and ranking) of the ten most widely dispersed broad-leaved weeds in winter cereals compared with earlier surveys. The data refer to presence/absence in fields sampled in Great Britain. In the 1972 and 1967 surveys, not all species were included (—) and the *Veronica* species were not distinguished (*). From Whitehead & White (1989)

Fields infested	1989	1972	1967
Stellaria media	94 (1)	89 (1)	77 (1)
Veronica persica	72 (2)	55 (2)*	52 (2=)*
Matricaria spp.	67 (3)	53 (3)	52 (2=)
Galium aparine	58 (4)	52 (4)	49 (4)
Lamium purpureum	47 (5)	—	—
Viola arvensis	36 (7)	33 (6)	39 (6)
Sinapsis arvensis	36 (7)	33 (6)	39 (6)
Veronica hederifolia	30 (8)	*	*
Capsella bursa-pastoris	23 (9=)	11 (6)	—
Volunteer rape	23 (9=)	—	—

The inherent assumption is that the species which will thrive under a given management system are those best able to survive it, and that the competitive abilities of the weeds *within* a season need not be good indicators of success *between* seasons.

Perhaps it is sufficient to understand the life history of each species with respect to the different disturbances it must endure. I suggest that it may be that the dynamics of some weed species are inherently more predictable than others because of differences in life history. Annuals may experience greater fluctuations than many perennials. The presence of seed-banks would be expected to dampen population changes. The analysis of phase diagrams from more examples of long-term weed studies would test such ideas.

The Broadbalk results suggest that that the dynamics of each individual species can be interpreted without reference to the other species present: the effects of weed–weed and weed–crop interactions may be regarded as noise when looking at the general trends of a single population over several seasons (Law & Watkinson 1989). Chancellor & Froud-Williams (1986) implicitly assume that a such a species-by-species analysis is adequate, and that detailed consideration of the behaviour of the weed community as a whole is not essential. A similar approach has been adopted by Svensson & Wigren (1986) for their analysis of the decline of weeds which are now rare.

The assumptions of these essentially autecological studies have not been tested. For example, we do not know how important the decline of dicotyledonous weeds has been in the increases in grass weeds witnessed by

Fryer & Chancellor (1970) and Hume (1987). If the autecological approach is acceptable, however, then general forecasts of the likely long-term effects of management on the weed flora are entirely achievable.

THE SCALE OF WEED–CROP INTERACTIONS

Some aspects of weed–crop ecology, such as competition, are highly localized in time and space, whereas other interactions, such as selection and evolution among weeds and crops, reveal themselves over much greater scales. Many studies have concentrated on competition and response of individual weeds to control because these can be studied using small-scale experiments.

Yet the information most easily obtained by the weed scientist may not be at the scale most useful for the farmer, adviser or policy maker. Mean weed thresholds assuming uniform weed distribution are of little value to the risk-averse farmer. Deterministic, site-specific population models will not help in effective planning for the future — the accuracy required will often prove to be unattainable. The key scale of desired information needed is probably at the field level over 1–5 years. Yet there are very few long-term weed experiments (Cousens *et al.* 1987b), and while there are more surveys over large areas, field-scale effects such as patchiness have received little attention.

As a result, there is a mismatch between the type of information required and that which is most often obtained. This can be illustrated using synoptic diagrams of the scales of ecological processes, investigative techniques and the information desired in practice (Figs 11.1 and 11.6). These should not be taken too literally, but I suggest that much of our knowledge about weed–crop ecology concerns small spatial and/or temporal scales, where the information desired by farmers concerns more larger scale studies (compare Figs 6a and b). These diagrams illustrate the importance of whole-farm studies such as at Boxworth (Greig-Smith, this volume, pp. 333–371) and at Lautenbach (El Titi, this volume, pp. 399–411). Even if there were more studies, it may prove impossible to produce the detailed and accurate forecasts of weed populations and crop yields the farmer might like; weed population processes are particularly variable at this scale of space and time. I do not argue that small-scale studies are worthless, but I do argue that their value is much clearer when interpreted alongside large-scale studies (see also Woiwod, this volume, pp. 275–304; Weins 1989).

Effort is needed where there may be true benefits. Assessments of risk of increase of weeds (resulting from natural selection, genetic engineering

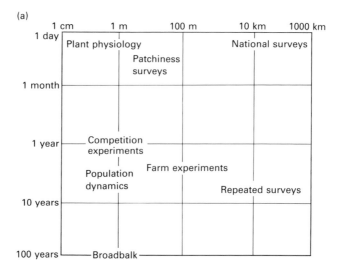

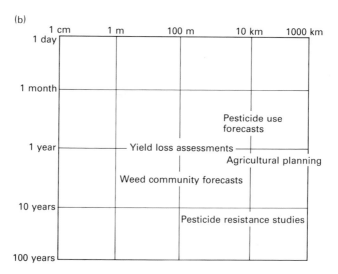

FIG. 11.6. Schematic diagrams of the scales in space and time of (a) typical investigations and (b) the information required by farmers and planners. These scales are clearly indicative only.

or climatic change); impacts of different control practices on weeds of different life histories; efficient sampling techniques for yield loss assessments and the breeding of crops for competitiveness against weeds are examples of practical applications of weed ecology.

The way is open to more practical weed ecology. After all, if we cannot apply ecological knowledge to help manage a system as simple as a weed infestation in a cereal field, our chances of successfully applying ecology to the myriad other more complex situations demanding our attention must be depressingly remote.

ACKNOWLEDGMENTS

I would like to thank Jon Marshall and Moya McCloskey for their comments on drafts of this chapter, and also to Jo and Jonathan for putting up with this symposium.

REFERENCES

Auld, B.A. & Tisdell, C.A. (1988). Influence of spatial distribution of weeds on crop yield loss. *Plant Protection Quarterly*, **3**, 81.

Auld, B.A., Menz. K.M. & Tisdell, C.A. (1987). *Weed Control Economics*. Academic Press, London.

Bazzaz, F.A. (1990). Plant–plant interactions in successional environments. *Perspectives on Plant Competition* (Ed. by J.B. Grace & D. Tilman), pp. 239–263. Academic Press, San Diego.

Begon, M., Harper, J.L. & Townsend, C.R. (1986). *Ecology: Individuals, Populations and Communities*. Blackwell Scientific Publications, Oxford.

Benner, B.L. & Bazzaz, F.A. (1987). Effects of timing of nutrient addition on competition within and between two annual plant species. *Journal of Ecology*, **75**, 229–245.

Benoit, D.L., Kenkel, N.C. & Cavers, P.B. (1989). Factors influencing the precision of soil seed bank estimates. *Canadian Journal of Botany*, **67**, 2833–2840.

Berendse, F. & Elberse, W.T. (1990). Competition and nutrient availability in heathland and grassland ecosystems. *Perspectives on Plant Competition* (Ed. by J.B. Grace & D. Tilman), pp. 93–116. Academic Press, San Diego.

Blackman, G.E. & Templeman, W.G. (1938). The nature of the competition between cereal crops and annual weeds. *Journal of Agricultural Science*, **28**, 247–271.

Brain, P. & Cousens, R. (1990). The effect of weed distribution on predictions of yield loss. *Journal of Applied Ecology*, **27**, 735–742.

Brenchley, W.E. & Warrington, K. (1933). The weed seed population of arable soil. II. Influence of crop, soil and methods of cultivation upon the relative abundance of viable seeds. *Journal of Ecology*, **21**, 103–127.

Carlson, H.L. & Hill, J.E. (1985). Wild oats (*Avena fatua*) competition with spring wheat: effects of nitrogen fertilization. *Weed Science*, **34**, 29–33.

Chancellor, R.J. (1977). A preliminary survey of arable weeds in Britain. *Weed Research*, **17**, 283–287.

Chancellor, R.J. (1985). Changes in the weed flora of an arable field cultivated for 20 years. *Journal of Applied Ecology*, **22**, 491–501.

Chancellor, R.J. & Froud-Williams, R.J. (1984). A second survey of cereal weeds in central southern England. *Weed Research*, **24**, 29–36.

Chancellor. R.J. & Froud-Williams, R.J. (1986). Weed problems of the next decade in Britain. *Crop Protection*, **5**, 66–72.

Connell, J.H. (1990). Apparent versus 'real' competition in plants. *Perspectives on Plant Competition* (Ed. by J.B. Grace & D. Tilman), pp. 9–26. Academic Press, San Diego.

Cousens, R. (1985). A simple model relating yield loss to weed density. *Annals of Applied Biology*, **107**, 239–252.

Cousens, R., Doyle, C.J., Wilson, B.J. & Cousens, G.W. (1986). Modelling the economics of controlling *Avena fatua* in winter wheat. *Pesticide Science*, **17**, 1–12.

Cousens, R., Brain, P., O'Donovan, J.T. & O'Sullivan, P.A. (1987a). The use of biologically realistic equations to describe the effects of weed density and relative time of emergence on crop yield. *Weed Science*, **35**, 720–725.

Cousens, R., Moss, S.R., Cussans, G.W. & Wilson, B.J. (1987b). Modeling weed populations in cereals. *Reviews of Weed Science*, **3**, 93–112.

Cousens, R., Firbank, L.G., Mortimer, A.M. & Smith, R.R. (1988). Variability in the relationship between crop yield and weed density for winter wheat and *Bromus sterilis*. *Journal of Applied Ecology*, **25**, 1033–1044.

Cussans, G.W., Cousens, R.D. & Wilson, R.J. (1987). Progress towards rational weed control strategies. *Rational Pesticide Use* (Ed. by J.K. Brent & R.K. Atkin), pp. 301–314. Cambridge University Press, Cambridge.

Czaran, T. & Bartha, S. (1989). The effect of spatial pattern on community dynamics; a comparison of simulated and field data. *Vegetatio*, **83**, 229–239.

Firbank, L.G. (1988). The biological flora of the British Isles. *Agrostemma githago* L. *Journal of Ecology*, **76**, 1232–1246.

Firbank, L.G. (1989). Forecasting weed infestations — the desirable and the possible. *Proceedings of the Brighton Crop Protection Conference — Weeds 1989*, pp. 567–572. British Crop Protection Council, Farnham.

Firbank, L.G. & Watkinson, A.R. (1985). On the analysis of competition within two-species mixtures of plants. *Journal of Applied Ecology*, **22**, 503–517.

Firbank, L.G. & Watkinson, A.R. (1986). Modelling the population dynamics of an arable weed and its effects upon crop yield. *Journal of Applied Ecology*, **23**, 147–159.

Firbank, L.G. & Watkinson, A.R. (1990). On the effects of competition: from monocultures to mixtures. *Perspectives on Plant Competition* (Ed. by J.B. Grace & D. Tilman), pp. 165–192. Academic Press, San Diego.

Firbank, L.G., Cousens, R., Mortimer, A.M. & Smith, R.G.R. (1990). Effects of soil type on crop yield–weed density relationships between winter wheat and *Bromus sterilis*. *Journal of Applied Ecology*, **27**, 308–318.

Firbank, L.G., Manlove, R.J., Mortimer, A.M. & Putwain, P.D. (1984). The management of grass weeds in cereal crops, a population biology approach. *Proceedings of the 7th International Symposium on Weed Biology, Ecology and Systematics*, pp. 375–384. Columa, Paris.

Firbank, L.G., Mortimer, A.M. & Putwain, P.D. (1985). *Bromus sterilis* in winter wheat: a test of a predictive population model. *Aspects of Applied Biology*, **9**, 59–66.

Fleming, G.W., Young, F.L. & Ogg, A.G. (1988). Competitive relationships among winter wheat (*Triticum aestivum*), Jointed Goatgrass (*Aegilops cylindrica*), and Downy Brome (*Bromus tectorum*). *Weed Research*, **36**, 479–486.

Froud-Williams, R.J. & Chancellor, R.J. (1981). A survey of grass weeds in cereals in central southern England. *Weed Research*, **22**, 163–171.

Fryer, J.D. & Chancellor, R.J. (1970). Herbicides and our changing weeds. *The Flora of a Changing Britain* (Ed. by F. Perring). Botanical Society of the British Isles, E.W. Classey.

Gill, G.S., Poole, M.L. & Holmes, J.E. (1987). Competition between wheat and brome grass in Western Australia. *Australian Journal of Experimental Agriculture*, **27**, 291–294.

Goldberg, D.E. (1990). Components of resource competition in plant communities. *Perspec-*

tives on Plant Competition (Ed. by J.B. Grace & D. Tilman), pp. 27–49, Academic Press, San Diego.

Grabherr, G., Mucina, L., Dale, M.B. & ter Braak, C.J.T. (Eds) **(1990).** *Progress in Theoretical Vegetation Science* (reprinted from *Vegetatio*, Vol. 83). Kluwer Academic, Dordrecht.

Grace, J.B. & Tilman, D. (Eds) **(1990).** *Perspectives on Plant Competition*. Academic Press, San Diego.

Grime, J.P. (1979). *Plant Strategies and Vegetation Processes*, Wiley, New York.

Hall, R.L. (1974a). Analysis of the nature of interference between plants of different species. I. Concepts and extension of the de Wit analysis to examine effects. *Australian Journal of Agricultural Research*, **25**, 739–747.

Hall, R.L. (1974b). Analysis of the nature of interference between plants of different species. II. Nutrient relations in Nandi Setaria and Greenleaf Desmodium association with particular reference to potassium. *Australian Journal of Agricultural Research*, **25**, 749–756.

Harper, J.L. (1977). *The Population Biology of Plants*. Academic Press, London.

Holt, R.D. (1977). Predation, apparent competition, and the structure of prey communities. *Theoretical Population Biology*, **12**, 197–229.

Humc, L. (1987). Long-term effects on the weed community in a wheat crop. *Canadian Journal of Botany*, **65**, 2530–2536.

Kropff, M. (1988). Modelling the effects of weeds on crop production. *Weed Research*, **28**, 465–471.

Law, R. & Watkinson, A.R. (1989). Competition. *Ecological Concepts* (Ed. by J.M. Cherrett), pp. 243–284. Blackwell Scientific Publications, Oxford.

Marshall, E.J.P. (1988). Field-scale estimates of grass weed populations in arable land. *Weed Research*, **28**, 191–198.

Marshall, E.J.P. (1989). Distribution patterns of plants associated with arable field edges. *Journal of Applied Ecology*, **26**, 247–257.

Mortimer, A.M. (1983). On weed demography. *Recent Advances in Weed Research* (Ed. by W.W. Fletcher), pp. 3–40. Commonwealth Agricultural Bureau, Farnham Royal.

Mortimer, A.M., Sutton, J.J. & Gould, P. (1989). On robust weed population models. *Weed Research*, **29**, 229–238.

Peters, N.C.B. (1984). Time of onset of competition and effects of various fractions of an *Avena fatua* L. population on spring barley. *Weed Research*, **24**, 305–315.

Post, B.J. (1988). Multivariate analysis in weed science. *Weed Reseaerch*, **28**, 425–430.

Prince, S.D. & Carter, R.N. (1985). The geographical distribution of prickly lettuce (*Lactuca serriola*) III. Its performance in transplant sites beyond its distribution limit in Britain. *Journal of Ecology*, **73**, 49–64.

Putnam, A.R. (1985). Allelopathy: a viable strategy for weed control? *Proceedings of the 1985 British Crop Protection Conference — Weeds*, pp. 583–589. British Crop Protection Council, Farnham.

Putwain, P.D. & Mortimer, A.M. (1989). The resistance of weeds to herbicides: a rational approach for containment of a growing problem. *Proceedings of the Brighton Crop Protection Conference — Weeds 1989*, pp. 285–294. British Crop Protection Council, Farnham.

Radosevich, S.R. & Roush, M.L. (1990). The role of competition in agriculture. *Perspectives on Plant Competition* (Ed. by J.B. Grace & D. Tilman), pp. 341–363. Academic Press, San Diego.

Reader, R.J. (1986). Temporal variation in recruitment and mortality for the pasture weed *Hieracium floribundum*: implications for a model of population dymanics. *Journal of Applied Ecology*, **22**, 175–183.

Reed, W.T., Saladini, J.L., Cotterman, J.C., Primiani, M.M. & Saari, L.L. (1989). Resistance in weeds to sulfonylurea herbicides. *Proceedings of the Brighton Crop Protection Conference — Weeds 1989*, pp. 295–300. British Crop Protection Council, Farnham.

Richards, M.C. (1989). Crop competitiveness as an aid to weed control. *Proceedings of the Brighton Crop Protection Conference — Weeds 1989*, pp. 573–578. British Crop Protection Council, Farnham.

Rice, E.L. (1974). *Allelopathy*. Academic Press, New York.

Roberts, H.A. (1984). Crop and weed emergence pattern in relation to time of cultivation and rainfall. *Annals of Applied Biology*, **105**, 263–275.

Roush, M.L. & Radosevich, S.R. (1985). Relationships between growth and competitiveness of four annual weeds. *Journal of Applied Ecology*, **22**, 895–905.

Spitters, C.J.T. (1983). An alternative approach to the analysis of mixed cropping experiments. 1. Estimation of competition effects. *Netherlands Journal of Agricultural Science*, **31**, 1–11.

Spitters, C.J.T. (1991). On descriptive and mechanistic models for inter-plant competition, with particular reference to crop – weed interaction. *Theoretical Production Ecology, Hindsight and Perspectives* (Ed. by R. Rabbinge *et al.*). Simulation Monographs, Pudoc, Wageningen (in press).

Spitters, C.J.T. & Aerts, R. (1984). Simulation of competition for light and water in crop–weed associations. *Aspects of Applied Biology*, **4**, 467–483.

Svensson, R. & Wigren, M. (1986). A survey of the history, biology and preservation of some retreating synanthropic plants. *Acta Universitatis Upsaliensis, Symbolae Botanicae Upsaliensis*, **25:4**.

Symonides, E., Silvertown, J. & Andreasen, V. (1986). Population cycles caused by overcompensating density dependence in an annual plant. *Oecologia (Berlin)*, **71**, 156–158.

Thompson, P.A. (1973). The effects of geographical dispersion by man on the evolution of physiological races of the corncockle *(Agrostemma githago L.)*. *Annals of Botany*, **37**, 413–421.

Thurston, J.M. (1968). Weed studies on Broadbalk. *Rothamsted Report for 1968*, pp. 186–208. HMSO, London.

Tilman, D. (1982). *Resource Competition and Community Structure*. Princeton University Press, Princeton.

Tilman, D. (1988). *Plant Strategies and the Dynamics of and Structure of Plant Communities*. Princeton University Press, Princeton.

Tottman, D.R., Steer, P.M., Orson, J.H. & Green, M.C.E. (1989). The effect of weather conditions in three seasons on the control of *Galium aparine* (cleavers) in winter wheat with fluroxypyr ester and mecoprop salt. *Proceedings of the Brighton Crop Protection Conference — Weeds 1989*, pp. 125–130. British Crop Protection Council, Farnham.

Weaver, S. (1984). Critical period of weed competition in three vegetable crops in relation to management practices. *Weed Research*, **24**, 317–325.

Weidenhamer, J.D., Hartnett, D.C. & Romeo, J.T. (1989). Density-dependent phytotoxicity: distinguishing resource and allelopathic interference in plants. *Journal of Applied Ecology*, **26**, 613–624.

Weins, J.A. (1989). Spatial scaling in ecology. *Functional Ecology*, **3**, 385–397.

Whitehead, R. & Wright, H.C. (1989). The incidence of weeds in winter cereals in Great Britain. *Proceedings of the Brighton Crop Protection Conference — Weeds 1989*, pp. 107–112. British Crop Protection Council, Farnham.

Williamson, G.B. (1990). Allelopathy, Koch's Postulates, and the Neck Riddle. *Perspectives on Plant Competition* (Ed. by J.B. Grace & D. Tilman), pp. 143–162. Academic Press, San Diego.

Wilson, B.J. & Wright, K.J. (1990). Predicting the growth and competitive effects of annual weeds in wheat. *Weed Research*, **30**, 201–211.

de Wit, C.T. (1960). On competition. *Verslagen van Landbouwkundige Onderzoekingen*, **660**, 1–82.

de Wit, C.T. & van den Bergh, J.P., (1965). Competition between herbage plants. *Netherlands Journal of Agricultural Science*, **13**, 212–221.

Zimdahl (1980). *Weed Crop Competition, A Review*. International Plant Protection Centre, Corvallis, Oregon.

Zohary, D. & Hopf, M. (1988). *Domestication of Plants in the Old World*. Clarendon Press, Oxford.

12. CEREAL APHIDS AND THEIR NATURAL ENEMIES

S.D. WRATTEN* AND W. POWELL[†]

*School of Biological Sciences, Bassett Crescent East,
University of Southampton, Southampton SO9 3TU, UK;
†Department of Entomology and Nematology, AFRC Institute of Arable
Crops Research, Rothamsted Experimental Station, Harpenden,
Hertfordshire AL5 2JQ, UK

INTRODUCTION

It is only since the late 1970s that substantial data have accrued in Western Europe on the natural enemies of cereal aphids. Much of this work was stimulated by Potts & Vickerman (1974) whose studies on the cereal ecosystem demonstrated that cereals are not an ecologically sterile environment but that dynamic ecological processes (including predation and parasitism) are operating within and between a wide range of arthropod guilds. These authors also outlined the probable role of beneficial arthropods in preventing cereal aphid outbreaks in some years, a role which was likely to be affected by changes in cereal agronomy. Nearly two decades later, major changes have indeed taken place in the cereal ecosystem in Europe, many of which are likely to have been detrimental to the populations of beneficial arthropods. The most important changes in this context include continuing hedgerow removal, a doubling of the area treated with herbicides, a ten-fold increase in fungicide use and a trend away from undersowing and from spring cereals. Grain surpluses leading to 'set-aside' and 'extensification' are more recent important factors. Against this changing ecological templet (*sensu* Southwood 1977), work on beneficial arthropods has developed along the following progression.

1   The demonstration that guilds of predators or parasitioids have the potential to limit the population growth of cereal aphids. This work has often taken the form of manipulative studies in which natural enemies are excluded from barriered or caged areas and aphid populations compared. Simulation modelling has begun to be used more recently to address this question also.

2   The ranking of individual species of predator in terms of their likely biological control. Criteria used have included density and diet information, voracity, aggregative numerical responses and phenology. Simulation modelling has helped in this area also.

3 The frequency and role of soil-surface predation compared with feeding at the level of the leaves or ears. Many of the predators identified as important rarely climb plants and until recently it was not clear how these groups obtained their aphid prey.

4 The enhancement of predator and parasitoid communities on farmland via habitat creation/modification schemes and behavioural manipulation as part of developing integrated pest management strategies.

A thorough understanding of natural enemy ecology and behaviour is essential for successful manipulation. This review will cover the most noteworthy of these developments, but for the polyphagous groups in particular, a more detailed review has recently been written (Wratten 1987). For the parasitoid groups, methods of evaluation have been rather different compared with those for predatory groups, so parasitoids will be dealt with separately.

PARASITOIDS

The cereal aphid parasitoid community

Cereal aphids are attacked directly by primary parasitoids and indirectly by hyperparasitoids which use the primary parasitoids as hosts. The primary parasitoids belong to two hymenopteran groups, the Aphidiinae (Braconidae) and the Aphelinidae, whilst the hyperparasitoids include species from five hymenopteran families, Pteromalidae, Encyrtidae, Eulophidae, Megaspilidae and Cynipidae. All are solitary endoparasitoids which means that a single parasitoid develops internally within each host individual. In Britain, seven primary parasitoid and seven hyperparasitoid species regularly attack aphids on cereal crops (Table 12.1), although other

TABLE 12.1. The principal primary parasitoids and hyperparasitoid species which attack cereal aphids in Britain

Primary parasitoids	Hyperparasitoids
Braconidae (Aphidiinae)	Megaspilidae
Aphidius rhopalosiphi (de Stefani-Perez)	*Dendrocerus carpenteri* (Curtis)
Aphidius picipes (Nees)	
Aphidius ervi (Haliday)	Cynipidae
Praon volucre (Haliday)	*Phaenoglyphis villosa* (Hartig)
Ephedrus plagiator (Nees)	*Alloxysta victrix* (Westwood)
Toxares deltiger (Haliday)	*Alloxysta macrophadna* (Hartig)
Aphelinidae	Pteromalidae
Aphelinus abdominalis (Dalman)	*Asaphes vulgaris* (Walker)
	Asaphes suspensus (Nees)
	Coruna clavata (Walker)

species are occasionally recorded (Powell 1982). Most of these occur simultaneously in the same fields, forming a complex parasitoid community associated with cereal aphid populations. For example, during an outbreak of grain aphid *Sitobion avenae* F. in 1980, seven primary parasitoid and five hyperparasitoid species were reared from samples of this host collected in a single winter wheat field at Rothamsted Experimental Station in southern England (Fig. 12.1). Starỳ (1981) lists the cereal aphid–primary parasitoid associations recorded in the western Palaearctic region.

Aphidius rhopalosiphi De Stefani Perez (= *A. uzbekistanicus* Luzhetski (Pungerl 1983)) is most often the dominant species in the primary parasitoid community attacking cereal aphids in Europe. However, the relative abundance of the different species varies from year to year and from place to place (Table 12.2). This is also true for the hyperparasitoids, the dominant species usually being one of three: *Dendrocerus carpenteri* (Curtis) (Megaspilidae), *Phaenoglyphis villosa* (Hartig) (Cynipidae) or *Asaphes vulgaris* Walker (Pteromalidae). Relative abundance also appears to vary depending upon the aphid host species (Starỳ 1976; Dean, Jones & Powell 1981). However, it is important to realize that relative abundance data can be strongly influenced by the sampling methods employed. Several methods were used to investigate the parasitoid community attacking populations of *Metopolophium dirhodum* (Walker) and *S. avenae* on winter wheat at Rothamsted in 1979 and 1980, respectively. In both years, the importance of the primary parasitoid *Toxares deltiger* (Haliday) would have been completely overlooked if sampling had been restricted to the collection of mummified aphids, the commonest method used in aphid studies (Table 12.2). This is because aphids parasitized by *T. deltiger* almost always leave their host plant prior to mummification (Powell 1980). The rearing out of primary parasitoids from random samples of live aphids and the sampling of adult parasitoids in cereal crops using a 'D-vac' suction

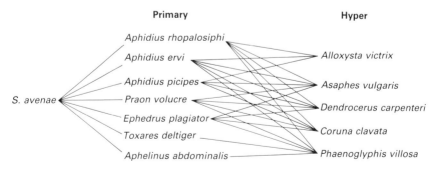

FIG. 12.1. Parasitoids reared from the aphid *Sitobion avenae* collected from a single winter wheat field in southern England in 1980.

TABLE 12.2. Relative abundance (%) of primary parasitoid species attacking cereal aphids in different years in different European countries

Year	Country	Sampling technique	Aphidius rhopalosiphi	Aphidius picipes	Aphidius ervi	Praon volucre	Toxares deltiger	Ephedrus plagiator	Aphelinus spp.
1979	England	Mummy collection	72	13	0	15	0	0	0
1979	England	Aphid rearing	35	1	0	5	59	0	0
1979	England	'D-vac' suction net	48	4	8	4	37	0	0
1980	England	Aphid rearing	33	6	15	7	38	0	0
1980	England	'D-vac' suction net	36	3	12	7	42	0	0
1981	England	Aphid rearing	80	7	10	1	1	0	0
1981	England	'D-vac' suction net	79	6	10	1	3	0	0
1982	England	Aphid rearing	56	2	19	13	2	6	1
1982	England	'D-vac' suction net	63	1	20	8	5	1	1
1956–75	Czechoslovakia[1]	Aphid rearing	60	7	19	5	0	8	0
1976	Czechoslovakia[2]	Mummy collection	35	1	8	6	0	50	0
1986	W Germany[3]	Mummy collection	11	43	23	19	0	0	3
1984	France[4]	Mummy collection	57	8	31	0	0	4	0
1980–83	Spain[5]	Mummy collection and Aphid rearing	55	0	44	1	0	0	0

[1] Starý (1976). [2] Starý (1978). [3] Kuo-Sell & Eggers (1987). [4] Fougeroux et al. (1988). [5] Castañera (1988).

net (Thornhill 1978) gave very similar data on the relative abundance of primary parasitoid species (Table 12.2). However, most hyperparasitoids, with the exception of the cynipids, attack their primary parasitoid hosts after aphid mummification and so are not detected by the sampling of live aphids.

The significance of early season activity

Parasitoids need to be active in the crop early in the year if they are to play a significant role in the control of summer cereal aphid populations (Powell 1983; Vorley 1986; Fougeroux *et al.* 1988). A high parasitoid : aphid ratio early in the season will slow down the initial growth rate of the aphid population. This reduces potential peak population levels and increases the likelihood that other natural enemies will be able to keep aphids below economic damage levels later in the season. Field surveys done in northern France in 1983 and 1984 demonstrated an inverse relationship between early levels of parasitism in cereal fields and subsequent peak aphid densities (Fougeroux *et al.* 1988). This type of relationship has also been shown for the polyphagous predatory groups (see below). In an attempt to demonstrate the significance of high parasitoid : aphid ratios early in the season in England, parasitoid populations in winter wheat plots were augmented in spring. This was done by releasing populations of the aphid *Metopolophium festucae* (Theobald) which had been exposed to *A. rhopalosiphi* in the laboratory, and therefore contained parasitoid larvae. Field populations of *S. avenae* which subsequently colonized these plots reached significantly lower peak densities (6·0 aphids per shoot) than they did in nearby control plots (9·4 aphids per shoot) (W. Powell, unpublished data).

Parasitoid abundance in cereal fields in early spring is influenced by a number of factors, including winter weather conditions and crop sowing date (Powell 1983; Vorley & Wratten 1987). Winter crops which are sown early (before mid-October) are likely to be colonized by cereal aphids in the autumn. If the winter is then mild without any prolonged periods with low temperatures, aphids, particularly *S. avenae* will survive on these crops and thereby provide an in-crop reservoir for parasitoids (Table 12.3). However, spring parasitoid densities can vary considerably between individual fields due to the influence of local environmental factors such as cultural practices, neighbouring crops and the proximity and local abundance of hedgerows and other semi-natural habitats (Starý 1978; Rabasse & Dedryver 1983; Fougeroux *et al.* 1988). *A. rhopalosiphi* is the most active parasitoid early in the year and this species is capable of

Table 12.3. Levels of parasitism in populations of *Sitobion avenae* in spring and subsequent adult parasitoid (*Aphidius rhopalosiphi*) densities in early summer in winter wheat fields following one cold and two mild winters in southern England

Year	Day-degrees below 0° C (Nov.–March)	% parasitism of *S. avenae* (early April)	No. of adult *A. rhopalosiphi* per m² (late May)
1978/79	112·3	No aphids found	0·9
1979/80	32·3	35·2	6·8
1980/81	30·6	29·4	18·7

attacking aphids throughout the winter if mild weather conditions prevail (Dedryver & Gellé 1982; Vorley 1986).

Late-sown winter crops and spring-sown crops must rely on parasitoid immigration in spring and early summer. Vorley & Wratten (1987) have shown that, in England, adult *Aphidius* species migrate into these crops from early-sown winter cereals and from grassland. It was calculated that one particularly early-sown winter wheat field amongst those sampled generated enough female *Aphidius* parasitoids per square metre to account for the observed immigration into approximately twenty-five nearby late-sown, winter wheat fields in the second half of May. The consequences of such immigration are that parasitoids can keep grain aphid populations at a level which is one seventh that which they would have been in the absence of parasitoids. This was demonstrated, using simulation modelling, by Vorley & Wratten (1985).

Host range and host preference

Host ranges vary amongst the primary parasitoids known to attack cereal aphids. For example, *A. rhopalosiphi* has been recorded only from aphids feeding on graminaceous plants whereas *Praon volucre* (Haliday) and *Ephedrus plagiator* (Nees) attack a wide range of hosts on a variety of food-plants. Several authors have suggested that alternative, non-pest host species could act as reservoirs for parasitoid populations at times when pest aphids are scarce or absent on nearby crops (Perrin 1975; Robert & Rabasse 1977; Powell 1986; Starỳ 1986a,b). The strategy is particularly pertinent to the control of aphids on cereals and other arable crops because of the temporary, unstable habitat that such crops provide. After monitoring the movement of cereal aphid parasitoids between grass fields and cereal crops, Vorley & Wratten (1987) concluded that, because of its early development on grasses, *M. festucae* was an important reservoir host in winter and early spring, particularly for *A. rhopalosiphi*. Similarly, Perrin (1975) concluded that the nettle aphid *Microlophium carnosum* (Buckt.) could act as a reservoir host for *Aphidius ervi* Haliday on patches of the stinging nettle *Urtica dioica* L. growing in field boundaries.

The concept of using alternative hosts to maintain and increase populations of parasitoids which attack pest species raises several interesting ecological questions. How readily do parasitoids transfer between different host species? Do some parasitoids occur as distinct biotypes associated with particular host species? How far do parasitoids disperse when searching for hosts? These questions, of course, are inter-related. Laboratory host transfer trials have shown that some cereal aphid parasitoids are capable of transferring between certain host species without any apparent behavioural problems or loss of fitness. However, others are unwilling to do so or successfully parasitize far fewer individuals in the first few generations following host transfer (Cameron, Powell & Loxdale 1984; Powell & Wright 1988). Inconsistent results in laboratory host-transfer trials with cereal aphid parasitoids led Powell & Wright (1988) to suggest that small founder numbers, genetic drift and genetic 'bottlenecks' in laboratory cultures result in the inadvertent selection of laboratory strains which vary in their biological and behavioural characteristics. Thus the use of laboratory strains in host transfer and other studies can give misleading results which may not accurately reflect the behaviour and ecology of more genetically diverse natural populations. Furthermore, it cannot be assumed from an apparent ability to transfer between different hosts in the laboratory that this actually occurs in the field. This problem is currently being addressed at Rothamsted in field studies of aphid parasitoid population dynamics and dispersal on a local scale.

The key question is — do parasitoid species occur locally as single metapopulations, attacking different hosts in an opportunistic manner depending upon their temporal and spatial availability, or as several discrete populations associated with particular hosts or particular habitats? Evidence from studies of host preference, olfactory responses to hosts and their food-plants, morphology and electrophoretic banding patterns indicate that the oligophagous species *A. ervi*, which attacks the nettle aphid *M. carnosum* and the pea aphid *Acyrthosiphon pisum* (Harris) in addition to cereal aphids, exists as distinct host biotypes (Powell & Zhang 1983; Němec & Starý 1983; Cameron, Powell & Loxdale 1984; Powell & Wright 1988). It has even been suggested that those attacking *M. carnosum* are a separate species (Pennacchio & Tremblay 1986). Electrophoretic techniques are currently being used at Rothamsted to study local field populations of *A. ervi*, following the detection of differences in esterase banding patterns between laboratory populations reared on different hosts (Powell & Wright 1988; Powell & Walton 1989). Detailed comparisons of allozyme frequencies in local field populations from different hosts and habitats should allow assessments of the extent of genetic exchange between those populations, and so clarify whether or not they really constitute a single homogeneous population.

Habitat is believed to be one of the principal factors determining the host range of aphid parasitoids (Starý 1970). It appears that many aphidiines are habitat dependent, normally only attacking hosts which occur in broad habitat types, such as deciduous forests or open steppes. For example, dioeceous aphid species are frequently attacked by different parasitoids on their primary host plants to those on their secondary host plants, if these are found in different habitats. However, within agricultural ecosystems parasitoids may undergo seasonal movements between different crops, and between crops and field boundaries (Starý 1981; Vorley & Wratten 1987). It is not known how far aphid parasitoids disperse but large numbers (predominantly females) are regularly caught in the Rothamsted Insect Survey suction traps which sample the insect fauna flying at a height of 12·2 m (Dean 1974).

Host location

After emergence, adult female parasitoids need to locate appropriate hosts to attack. Vinson (1984) defined five steps in the host location and selection process: host habitat location, host location, host acceptance, host suitability and host regulation (e.g. minimization of encapsulation of the parasitoid egg or larva by the host). During this process parasitoids respond to a series of environmental cues, many of which are chemical, olfactory cues originating from the host or host by-products such as honey-dew (kairomones) or from the food-plant of the host (synomones). These behaviour-controlling chemicals, collectively known as semiochemicals, are known to play key roles at all stages in the host finding and attack behaviour of parasitoids (Vinson 1984).

In olfactometer tests, both male and female, *A. rhopalosiphi* and *A. ervi* responded positively to volatile chemical cues from wheat leaves, but only females responded to volatiles from host aphids (Powell & Zhang 1983). This makes sense since females need to locate the correct host habitat and then the appropriate host in which to oviposit, whereas males need to locate female parasitoids with which to mate. The chances of a male finding a mate would be improved by efficient host habitat location. Male parasitoids also respond to sex pheromones produced by virgin females. This has been demonstrated in the laboratory for *A. rhopalosiphi*

and *A. ervi* (Powell & Zhang 1983) and in the field for *A. rhopalosiphi*
and *P. volucre* (Decker 1988). Decker (1988) caught large numbers of
conspecific males in delta traps placed in winter wheat fields and baited
with virgin females (Table 12.4).

Aphid parasitoids also use non-volatile, chemical cues in addition to
volatile substances during the host location and attack process. These are
detected by direct contact, usually via the antennae. Contact kairomones
in aphid honeydew are used by *A. rhopalosiphi* females as searching
stimuli, causing them to concentrate their searching activity on contaminated parts of the cereal plant (Gardner & Dixon 1985). Such a response
should increase the chances of the parasitoid encountering the aphid or
aphid colony which produced the honeydew. The primary parasitoids
A. rhopalosiphi, *A. ervi*, *Aphidius picipes* (Nees), *P. volucre* and *E.
plagiator* and the hyperparasitoids *D. carpenteri*, *P. villosa*, *Alloxysta
victrix* (Westwood) and *Alloxysta macrophadna* (Hartig) all spent longer
periods of time searching on filter paper discs contaminated with cereal
aphid honeydew than they did on clean discs (Budenberg & Powell 1988;
Budenberg 1990). This searching response was evident for both males and
females although females tended to show a stronger response than males.

Contact kairomones and/or short-range volatiles appear to be involved
in host recognition and the stimulation of oviposition following aphid
location. *A. rhopalosiphi* females attempted to lay eggs in model aphids in
the form of small pieces of vermiculite which had been treated with crude
aqueous extracts from cereal aphids or from wheat leaves before being
attached to wheat plants (Decker 1988). The percentage of encounters
which resulted in an oviposition attempt was significantly greater for
extract-treated vermiculite than for vermiculite treated with deionized
water but lower than for live aphids (Table 12.5).

The use of semiochemicals to manipulate parasitoids in the field offers

TABLE 12.4. Mean number of male parasitoids caught in pheromone traps in winter wheat in
July. Traps were baited with live virgin female parasitoids of either *Aphidius rhopalosiphi*
or *Praon volucre*, and placed either within the crop canopy (A) or just above it (B). Traps
ran for 24 h. From Decker (1988)

Species caught	Unbaited control traps		Traps baited with *P. volucre* females		Traps baited with *A. rhapolosiphi* females	
	A	B	A	B	A	B
P. volucre	0·0	0·0	141·5	148·5	1·0	1·0
A. rhopalosiphi	0·0	1·5	0·5	0·5	161·0	40·0
Other Aphidiinae	1·5	2·5	0·0	1·0	5·0	1·0

TABLE 12.5. Percentage of encounters by female *Aphidius rhopalosiphi* with vermiculite aphid models and with live *Sitobion avenae* on wheat seedlings which resulted in oviposition attempts. Vermiculite models were treated with deionized water (control) or aqueous *S. avenae* extract (treated) in experiment A and with deionized water or aqueous wheat leaf extract in experiment B. From Decker (1988)

	Live aphid	Control model	Treated model
Experiment A	77·4	5·4	49·2
Experiment B	74·0	2·7	32·7

exciting possibilities for increasing parasitoid impact on pest populations. However, in order to develop this approach, a sound understanding of the identity and role of semiochemicals in the interactions between organisms at the four trophic levels of plant, herbivorous pest, primary parasitoid and hyperparasitoid is essential.

PREDATORS

Evidence for the role of predator guilds in
cereal aphid population suppression — manipulative studies

Apart from parasitoid wasps, cereal aphids are attacked by a wide range of so-called aphid-specific predators (mainly Coccinellidae and Syrphidae) as well as by polyphagous groups dominated by the Carabidae, Staphylinidae and Araneae, together with the less well-studied predatory flies in the families Empidae and Dolichopodidae. In the south of England up to 400 species of predators and parasitoids may occur in winter wheat, but many of them have a limited role in the suppression of cereal aphid populations, for reasons associated with their density, phenology or polyphagy. Most work has been concerned with the role of natural enemies of the grain aphid, *S. avenae*, while the predator complex attacking the rose grain aphid *M. dirhodum* and the bird cherry aphid *Rhopalosiphum padi* L. are much less well known. This is partly because *M. dirhodum* is a much more sporadic pest than is *S. avenae*, while *R. padi* has a major role as a transmitter of barley yellow dwarf virus in the autumn, during which time the predator fauna in cereals has received little attention.

One approach to evaluate the role of epigeal guilds of predators in cereals has been to compare the numbers of the grain aphid inside and outside predator-exclusion barriers. This apparently simple technique is very difficult to use rigorously and some of the published results are seriously flawed because of the methodology used. Such problems are discussed in Putman & Wratten (1984), and in the worst published cases most potential problems did in fact appear to arise. Such difficulties include: barrier methods do not identify which predators are responsible for observed effects; higher prey levels brought about through predator population reduction could 'attract' other flying predators, etc. A decade of studies of this type had passed before unequivocal results appeared, the clearest examples coming from Chiverton (1986) and Winder (1990a,b). Even when experiments of this type appear to have coped with most of the methodological problems, the results still sometimes show reductions in already small numbers of grain aphids by polyphagous predators. In relation to damage and economic thresholds for this pest in Europe, few manipulative studies have demonstrated that polyphagous predators alone can keep cereal aphid populations below these thresholds. Recent manipulative work by Mauremootoo (1991) (Fig. 12.2) has shown that when polyphagous predators' densities are reduced, grain aphid numbers can exceed one of the accepted UK thresholds (that of five aphids per ear). In terms of the season of activity when polyphagous groups have the greatest effects, some manipulative studies had purported to show that these predators have greatest effects early in the season. However, this conclusion is usually not justified by the experimental protocol used (see Wratten 1987) and simulation modelling is more appropriate to answer this question (see below).

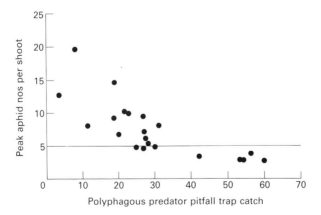

FIG. 12.2. The relationship ($r_s = 0.81$, $P<0.001$) between peak densities of cereal aphids (*S. avenae* and *M. dirhodum*) and the pitfall-trap catch of polyphagous predators during the week before the aphid peak. The horizontal line is one of the commonly-used spray thresholds in the UK but usually applies to flowering; peak aphid numbers usually occur after flowering. From Mauremootoo (1991).

Similar manipulative studies, this time involving the exclusion of aphid-specific predators from cages have also shown that the numbers of aphids are higher when numbers of predators are reduced. Numbers of aphids were about eight times higher in the experiments of Chambers *et al.* (1983) when cages were used to exclude Coccinellidae and Syrphidae in particular. Studies of this type usually deal with all or part of a particular predator guild, so even if effects are detected it is not usually possible to identify which predator species has the most important role. Rarely, single species manipulations are done in which each predator species in turn is *added* to enclosures, and the aphid populations developing compared with those in control plots. This was done for some species of the Staphylinidae by Dennis & Wratten (1991) and work showed that the most voracious of the Staphylinidae (*Philonthus* spp.) significantly decreased aphid populations as did *Tachyporus* spp. (Fig. 12.3).

These single-species studies are very labour intensive and rarely done, so other methods have been more frequently used to determine which members of a particular guild of predators have the greatest role to play in the suppression of aphid populations. These attempts at ranking of

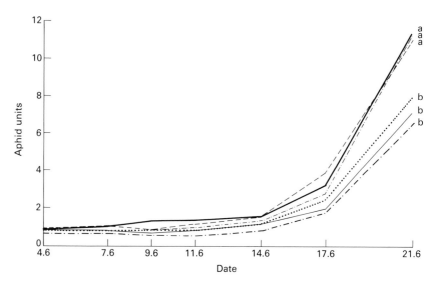

FIG. 12.3. Mean growth rates of *S. avenae* populations in exclusion cages where individual staphylinid beetle species were enclosed. Similar letters denote homogeneous groups (randomized block analysis of variance, followed by Tukey's test). Total exclusion (_ _ _ _); *Tachyporus hypnorum* (–·–··–); *T. chrysomelinus* (–·–··–); *T. obtusus* (———); *Tachyporus* spp. larvae (———); *Philonthus cognatus* (......). From Dennis & Wratten (1991).

individual predator species have included criteria such as the density and diet of the predators, their voracity, their ability to aggregate in 'patches' of their prey, and the extent to which their phenology matches that of the aphids. Methods for ranking predators, based on a combination of criteria, were outlined by Wratten *et al.* (1984) and developed in Wratten (1987). They were initiated by the very important early work of Sunderland & Vickerman (1980) who combined two criteria: predator density and the proportion of predator individuals containing aphid remains when dissected. This 'predation index' was robust enough to stimulate a range of autecological studies of predators in the 1980s but the ranking was flawed by its exclusion of species which take fluid food. The use of enzyme-linked immunosorbent assay (ELISA), (initially developed for virological studies) in predation ecology has permitted the inclusion of these guilds of predators in more recent analyses. Work by Sopp (1987) and Sunderland *et al.* (1987) has shown that such groups as Staphylinidae and Linyphiidae are more important than the Sunderland & Vickerman data suggested, and that a wide range of predatory groups contribute to aphid suppression in cereals. For the actively foraging species, unlike the Linyphiidae, the ability to locate and exploit areas of high aphid density should be another important factor for theoretical and practical reasons. Bryan & Wratten (1984) were able to rank polyphagous species on this criterion alone although, as they used pitfall traps to assess this response, the behavioural basis of the high capture rates of the traps which were set in aggregations of grain aphids remained unknown. However, a recent video analysis of predator foraging in the field and the laboratory has shown some Carabidae species to be efficient detectors of aphid patches (Halsall & Wratten 1988; Halsall 1990). Most work on the ranking of predators has been done in the southern counties of England. To improve the value of this information, regional comparisons of the relative abundances and phenologies of predator species are urgently needed, since these are likely to be affected by regional variations in factors such as climate and soil type.

Many of the Carabidae and Staphylinidae species invade cereals in the spring from field-boundary overwintering sites. Those species which do not fly, such as some of the carabids, often invade the field slowly so that at the time of aphid invasion many of them may still be closely associated with field boundaries (Coombes & Sotherton 1986). This phenological limitation can be partly overcome by using the information on the special overwintering requirements obtained by Sotherton (1985). The crop environment can be changed in a way which exploits knowledge of the predators' overwintering (see below).

Predation on the soil surface

In the Sunderland & Vickerman (1980) ranking of predators many of the species which were shown to have biological control potential, such as the carabid *Agonum dorsale* (Pont.), rarely climb the plant (Griffiths, Wratten & Vickerman 1985). Given that the grain aphid feeds on ears and flag leaves in spring and summer it is difficult to see how these predator species can attain their high ranking. Work in the late 1980s however, showed that up to 90% of the aphid population falls to the ground each day, while between 4 and 71% of aphid populations have been recorded on the ground at any one time (Sopp, Sunderland & Coombes 1987). Of these individuals, over 90% of the grain aphid were able to return to the crop in the absence of predation (Sopp 1987). These data implicate predation on the soil surface. Winder (1990a) used a novel trap and experimental design to establish a relationship between the mean number of aphids falling per shoot per day and the mean number per shoot. He then investigated how this relationship changed in the presence of low, intermediate and normal densities of non-climbing predators on the soil surface (Fig. 12.4). This work was the first good evidence that epigeal predators which do not climb the cereal plant can bring about significant reductions in grain aphid populations.

Modelling approaches

In cereals there are only a few examples of models used to assess the effects of predators and parasitoids. Chambers & Adams (1986) quantified the impact of hoverflies on cereal aphids in winter wheat by calculating two quantities: the estimated kill rate of aphids by hoverfly larvae and the kill required to provide the degree of control observed in the aphid population, i.e. to bring the aphid population increase down from potential to observed. These workers used a model based on that first used by Bombosch (1963) which incorporated, among other things, the density of predators on each day and the required kill per predator per day. Hoverfly larvae had the potential to halt the aphid population increase in four of the six field populations analysed. Despite this work, there is still a need for more modelling of predation of aphids in cereals, including that of the main coccinellid predator, *Coccinella septempunctata* L. For polyphagous predators, there were no useful models until recently. A preliminary model published by Winder, Carter & Wratten (1988) concerned the Carabidae only and revealed the importance of setting correct consumption rates for predators. Winder (1990b) incorporated predation by polyphagous groups

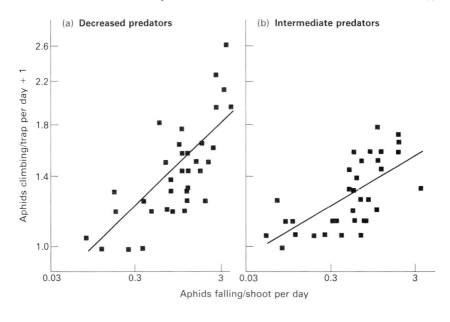

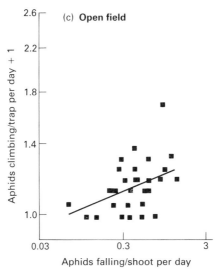

FIG. 12.4. The relationship between $\log_{10}$ aphid fall off/shoot per day and $\log_{10}$ climbs/trap per day for three levels of polyphagous predator density (from Winder 1990a).

into the grain aphid simulation model which was constructed by Carter, Dixon & Rabbinge in 1982. This new model used a 'required kill' approach (Fig. 12.5) similar to that of Chambers & Adams (1986) and demonstrated that polyphagous predators were most important during the early stage of aphid population development, accounting for between 40 and 100% of the 'required kill', depending on year, season and predatory group. Staphylinid beetles of the genus *Tachyporus* were the most important predators (Fig. 12.6). This conclusion supports the results from the manipulative work on staphylinids by Dennis & Wratten (1990) (see above). Winder (1990a) also concluded that the action of polyphagous predators in combination with that of other natural enemies such as parasitoids, aphid specific predators and fungal pathogens was necessary to prevent aphid outbreaks from occurring.

Habitat manipulation in farmland to increase
the predator : prey ratio

Given the conclusion above that a range of predators and parasitoid groups contribute to pest suppression in cereals, the logical next step is to enhance natural enemy densities. This practice in cereals has a fairly recent pedigree but at a number of sites in Europe the following range of approaches is being attempted.

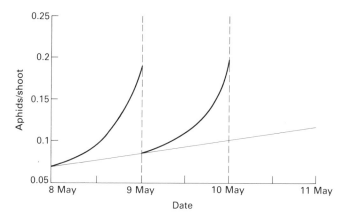

FIG. 12.5. Method of obtaining observed grain aphid population increase during simulation of polyphagous predators' effects. On each day, aphid development was simulated with no mortality (——). A proportional mortality (proportion of aphids killed/shoot per hour) was then applied so the simulated population increase matched that observed (——). The 'required kill' was the total number of aphids killed/shoot during the day when the above proportioned mortality was applied. From Winder (1990a).

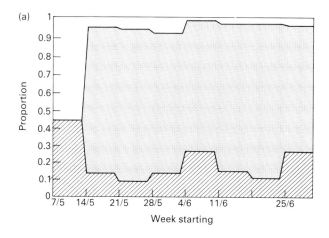

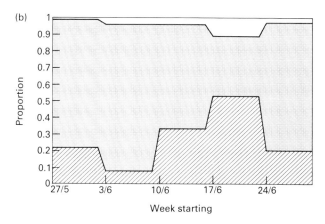

FIG. 12.6. Contribution of polyphagous predator groups to 'required kill' (see text and Fig. 12.5) of the grain aphid. Open area = Linyphiidae, shaded area = Staphylinidae, hatched area = Carabidae. From Winder (1990b).

Nentwig (1988, 1989) alternated 1 m wide strips of naturally occurring successional vegetation with 12 m wide strips of meadowland or winter wheat. The strips developed a larger number of species of beneficial arthropods and a greater diversity of these groups, although their role in the dynamics of aphid populations in the crop itself was not demonstrated. In the UK a more goal-orientated approach has been used by Thomas & Wratten (1988), Wratten & Thomas (1990) and Thomas, Wratten & Sotherton (1991), whereby simulated hedge banks, of the type identified by Sotherton (1985) as being important as overwintering refuges for polyphagous predators in 'natural' field boundaries, were created across winter wheat fields. These banks were sown with a range of grass species. *Dactylis glomerata* L. and *Holcus lanatus* L. in particular, which are tussock-forming species, harboured up to 1600 predatory individuals per m². This is an order of magnitude higher than the densities occurring in 'good' field boundaries identified by Sotherton (1985). As these banks traverse large cereal fields they also have the potential to permit colonization of field centres in early spring by the more poorly dispersing species (see above; Fig. 12.7). For aphid-specific predators such as hoverflies and ladybirds some work has identified the native species of flowers in field boundaries which provide useful sources of pollen and nectar for the adults. This has been done by Ruppert & Klingauf (1988) in Germany and by Cowgill (1989) in the UK. Cowgill showed flower species were not visited according to their abundance, particular species of hoverfly showing a 'preference' for certain weeds. A 'forage ratio' was used to quantify this preference which varied with the species of hoverfly and habitat composition (Fig. 12.8). It would be potentially useful to increase the densities of the 'preferred' wild flowers on farmland but some are perennials or biennials and the agronomic aspects of increasing their densities in farmland are unknown. Some non-native species have the combined advantages of high availability of pollen and/or nectar for hoverflies with readily available seed supplies and an established agronomy. Two such species are *Phacelia tanacetifolia* (Benth.) (Hydrophyllaceae) and *Fagopyrum esculentum* (Moench) (Polygonaceae). The former is grown in west Germany

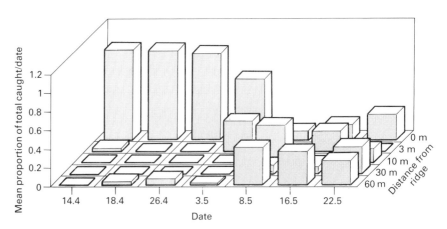

FIG. 12.7. Pitfall-trap captures of predatory carabid beetles during their spring dispersal from overwintering refuges in cereals. From Wratten & Thomas (1990).

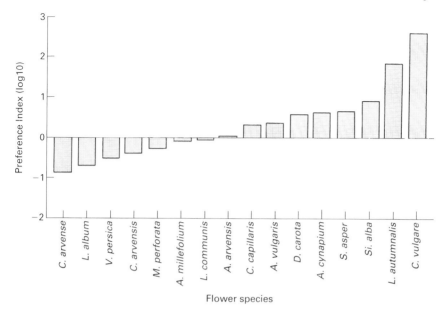

FIG. 12.8. Forage preferences of the hoverfly *Episyrphus balteatus* in field margins. A ratio greater than one indicates 'preference' for that plant species. Plant genera (from left to right) are: *Cirsium, Lamium, Veronica, Convolvulus, Matricaria, Achillea, Lapsana, Anagallis, Crepis, Artemisia, Daucus, Aethusa, Sonchus, Silene, Leontodon, Cirsium*. From Cowgill, unpublished.

as a nectar source for honeybees (*Apis mellifera* L.) and its role as a pollen source for hoverflies has been investigated in a preliminary way by Klinger (1987). In the UK *Phacelia* is planted as a cover crop in nitrogen-sensitive areas but in this case is not permitted to persist to flowering. Current work at Southampton in the UK is drilling metre-wide field-margin strips around winter wheat fields with both these species and the consequences for hoverfly foraging, oviposition and predation rate are being assessed.

Other crop manipulation studies initiated in the UK include The Game Conservancy's Conservation Headlands (Sotherton, this volume, pp. 373–397). Although not created with biological control of pests in mind, these strips do harbour larger numbers of aphid-specific predators than does the open field; the biological control consequences of this difference remain to be investigated. Van Emden (1988) reviewed in a wider context the potential for managing indigenous natural enemies of aphids in field crops including cereals, including the use of pesticides in a rational way in relation to biological control. Van Emden identified selective pesticides, selective placement, selective timing and dosage reduction as potential ways in which pesticides could be accommodated with maximized biological control. Reduced doses in particular have been demonstrated to raise

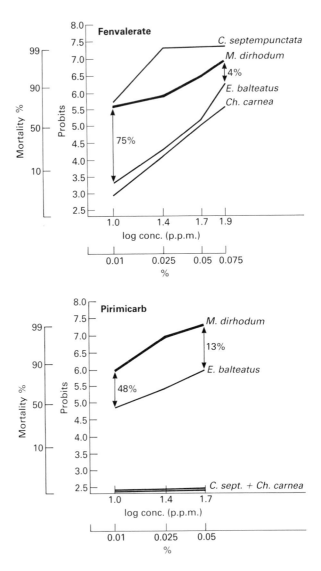

FIG. 12.9. The relationship between different dosages of pirimicarb and fenvalerate and the effects (mortality after 24 h) on *Metopolophium dirhodum* and the larvae of *Coccinella septempunctata*, *Episyrphus balteatus* and *Chrysoperla carnea*. (Semi-field test, winter wheat seedlings.) From Poehling (1989).

the natural enemy aphid ratio (Stern *et al.* 1959) and the attractive prospect of combining reduced-rate pesticide application with partially resistant host plants was also investigated in a theoretical way by van Emden (1987). In west Germany, Mann *et al.* (1991) showed that reduced rates of the insecticides pirimicarb and fenvalerate could give increased profits to the farmer under some circumstances while having the potential to leave greater numbers of predators surviving. Poehling (1989) investigated in more detail the prospects for enchanced levels of natural enemies in reduced-rate regimes, and the results in Fig. 12.9 illustrate the method's potential. Van Emden & Wratten (1991) reviewed the applications of tri-trophic level interactions between crop, pest and natural enemy and outlined a large number of theoretical and practical possibilities for maximizing the predator : prey ratio by exploiting these tritrophic links.

CONCLUSION

In the late 1980s and early 1990s two main developments concerning the natural enemies of cereal aphids have taken place. The first is that the body of knowledge concerning the effects of most predatory and parasitic groups has now reached the stage when it can be applied in attempts at integrated pest management (IPM) schemes. However, an important realization is that successful control depends upon the combined action of a range of natural enemies. Secondly, changes in the social and economic climate in western Europe, in particular driven by the combination of grain surpluses and an increasing concern about over-use of pesticides (Potts, this volume, pp. 15–18), make the application of the ideas outlined in this chapter a real prospect for the first time.

REFERENCES

Bombosch, S. (1963). Untersuchungen zur Vermehrung von *Aphis fabae* Scop. in Samenrübenbeständen unter besonderer Berücksichtigung der Schwebfliegen (Diptera. Syrphidae). *Zeitschrift für Angewandte Entomologie*, **32**, 105–141.

Bryan, K.M. & Wratten, S.D. (1984). The responses of polyphagous predators to prey spatial heterogeneity. Aggregation by carabid and staphylinid beetles to their cereal aphid prey. *Ecological Entomology*, **9**, 251–259.

Budenberg, W.J. (1990). Honeydew as a contact kairomone for aphid parasitoids. *Entomologia Experimentalis et Applicata*, **55**, 139–148.

Budenberg, W.J. & Powell, W. (1988). Honeydew as a kairomone for aphid parasitoids and hyperparasitoids. *Proceedings of the Brighton Crop Protection Conference — Pests and Diseases — 1988*, **3**, 1149–1154.

Cameron, P.J., Powell, W. & Loxdale, H.D. (1984). Reservoirs for *Aphidius ervi* Haliday (Hymenoptera, Aphidiidae), a polyphagous parasitoid of cereal aphids (Hemiptera: Aphididae). *Bulletin of Entomological Research*, **74**, 647–656.

Carter, N., Dixon, A.F.G. & Rabbinge, R. (1982). *Cereal Aphid Populations: Biology, Simulation and Prediction.* Pudoc, Wageningen.

Castañera, P. (1988). Present status of cereal pests in Spain with special reference to cereal aphids. *Integrated Crop Protection in Cereals* (Ed. by R. Cavalloro & K.D. Sunderland), pp. 13–24. *Proceedings of a Meeting of the EC Experts' Group*, Littlehampton, England, 1986. A.A. Balkema, Rotterdam.

Chambers, R.J. & Adams, T.H.L. (1986). Quantification of the impact of hoverflies (Diptera: Syrphidae) on cereal aphids in winter wheat: an analysis of field populations. *Journal of Applied Ecology*, **23**, 895–904.

Chambers, R.J., Sunderland, K.D., Wyatt, I.J. & Vickerman, G.P. (1983). The effects of predator exclusion and caging on cereal aphids in winter wheat. *Journal of Applied Ecology*, **20**, 209–224.

Chiverton, P. (1986). Predator density manipulation and its effects on populations of *Rhopalosiphum padi* (Hom: Aphididae) in spring barley. *Annals of Applied Biology*, **109**, 49–60.

Coombes, D.S. & Sotherton, N.W. (1986). The dispersal and distribution of polyphagous predatory Coleoptera in cereal fields. *Annals of Applied Biology*, **108**, 461–474.

Cowgill, S. (1989). The role of non-crop habitats on hoverfly (Diptera: Syrphidae) foraging on arable land. *Proceedings of the Brighton Crop Protection Conference — Weeds —* 1989, **3**, 1103–1108.

Dean, G.J. (1974). Effects of parasites and predators on the cereal aphids *Metopolophium dirhodum* (Wlk.) and *Macrosiphum avenae* (F.). (Hem., Aphididae). *Bulletin of Entomological Research*, **63**, 411–422.

Dean, G.J., Jones, M.G. & Powell, W. (1981). The relative abundance of the hymenopterous parasites attacking *Metopolophium dirhodum* (Walker) and *Sitobion avenae* (F) (Hemiptera Aphididae) on cereals during 1973–79 in southern England. *Bulletin of Entomological Research*, **71**, 307–315.

Decker, U.M. (1988). *Evidence for semiochemicals affecting the reproductive behaviour of the aphid parasitoids* Aphidius rhopalosiphi *De Stefani-Perez and* Praon volucre *Haliday (Hymenoptera: Aphidiinae) — A contribution towards integrated pest management in cereals.* PhD thesis, University of Hohenheim.

Dedryver, C.A. & Gellé, A. (1982). Biologie des pucerons des céréales dans L'Ouest de la France. IV. Étude de l'hivernation de populations anholocycliques de *Rhopalosiphum padi* L., *Metopolophium dirhodum* Wlk. et *Sitobion avanae* F. sur repousses de céréales, dans trois stations de Bretagne et du Bassin parisien. *Acta Oecologica Applicata*, **3**, 321–342.

Dennis, P. & Wratten, S.D. (1991). Field manipulation of populations of individual staphylinid species in cereals and their impact on aphid populations. *Ecological Entomology*, **16**, 17–24.

van Emden, H.F. (1987). Cultural methods: the plant. *Integrated Pest Management* (Ed. by A.J. Burn, T.H. Coaker & P.C. Jepson), pp. 27–68. Academic Press, London.

van Emden, H.F. (1988). The potential for managing indigenous natural enemies of aphids on field crops. *Philosophical Transactions of the Royal Society of London*, **318**, 183–201.

van Emden, H.F. & Wratten, S.D. (1991). Tri-trophic interactions involving plants in the biological control of aphids. *Aphid–Plant Interactions: Populations to Molecules* (Ed. by J.A. Webster & D.C. Peters). Oklahoma (in press).

Fougeroux, A., Bouchet, C., Reboulet, J.N. & Tisseur, M. (1988). Importance of Microhymenoptera for aphid population regulation in French cereal crops. *Integrated Crop Protection in Cereals* (Ed. by R. Cavalloro & K.D. Sunderland), pp. 61–68. *Proceedings of a Meeting of the EC Experts' Group*, Littlehampton, England, 1986. A.A. Balkema, Rotterdam.

Gardner, S.M. & Dixon, A.F.G. (1985). Plant structure and the foraging success of *Aphidius rhopalosiphi* (Hymenoptera: Aphidiidae). *Ecological Entomology*, **10**, 171–179.

Griffiths, E., Wratten, S.D. & Vickerman, G.P. (1985). Foraging by the carabid *Agonum dorsale* in the field. *Ecological Entomology*, **10**, 181–189.

Halsall, N.B. (1990). *Video analysis of aggregative responses of invertebrate predators of cereal aphids.* PhD thesis, University of Southampton.

Halsall, N.B. & Wratten, S.D. (1988). Video recording of aphid predation in a wheat crop. *Proceedings of Brighton Crop Protection Conference — Pests and Diseases*, **3**, 1047–1052.

Klinger, V.K. (1987). Auswirkungen eingesäter Randstreifen an einem Winterweizen-Feld die Raubarthropoden-fauna und den Getreideblattlausbefall. *Journal of Applied Entomology*, **104**, 47–58.

Kuo-Sell, H.L. & Eggers, G. (1987). Evaluierung der Wirkung von Parasitoiden auf die Populationsentwicklung von Getreideblattlausen durch Vergleich zwischen Mumifizierungs- und Parasitierungstrate in Winterweizen. *Zeitschrift für Pflanzenkrankheiten und Pflanzenschutz*, **94**, 178–189.

Mann, B.P., Wratten, S.D., Poehling, M. & Borgemeister, C. (1991). The economics of reduced-rate insecticide applications to control aphids in winter wheat. *Annals of Applied Biology* (in press).

Mauremootoo, J. (1990). *Manipulative studies of the polyphagous predators of cereal aphids.* PhD thesis, University of Southampton.

Němec, V. & Starý, P. (1983). Elpho-morph differentiation in *Aphidius ervi* Hal. biotype on *Microlophium carnosum* (Bckt.) related to parasitization on *Acyrthosiphon pisum* (Harr.) (Hym. Aphidiidae). *Zeitschrift für angewandte Entomologie*, **95**, 524–530.

Nentwig, W. (1988). Augmentation of beneficial arthropods by strip management I. Succession of predacious arthropods and long-term change in the ratio of phytophagous and predacious arthropods in a meadow. *Oecologia (Berlin)*, **76**, 597–606.

Nentwig, W. (1989). Augmentation of beneficial arthropods by strip management II. Successional strips in a winter wheat field. *Journal of Plant Diseases and Protection*, **96**, 89–99.

Pennacchio, F. & Tremblay, E. (1986). Biosystematic and morphological study of two *Aphidius ervi* Haliday (Hymenoptera, Braconidae) 'biotypes' with the description of a new species. *Bulletino del Laboratorio di Entomologia Agraria 'Filippo Silvestri'*, **43**, 105–117.

Perrin, R.M. (1975). The role of the perennial stinging nettle, *Urtica dioica*, as a reservoir of beneficial natural enemies. *Annals of Applied Biology*, **81**, 289–297.

Poehling, H.M. (1989). Selective application strategies for insecticides in agricultural crops. *Pesticides and Non-target Invertebrates* (Ed. by P.C. Jepson), pp. 151–175. Intercept, Wimborne, Dorset.

Potts, G.R. & Vickerman, G.P. (1974). Studies on the cereal ecosystem. *Advances in Ecological Research*, **8**, 108–197.

Powell, W. (1980). *Toxares deltiger* (Haliday) (Hymenoptera: Aphidiidae) parasitising the cereal aphid, *Metopolophium dirhodum* (Walker) (Hemiptera: Aphididae), in southern England: a new host–parasitoid record. *Bulletin of Entomological Research*, **70**, 407–409.

Powell, W. (1982). The identification of hymenopterous parasitoids attacking cereal aphids in Britain. *Systematic Entomology*, **7**, 465–473.

Powell, W. (1983). The role of parasitoids in limiting cereal aphid populations. *Aphid Antagonists* (Ed. by R. Cavalloro), pp. 50–56. *Proceedings of a Meeting of the EC Experts' Group, Portici, Italy, 1982.* A.A. Balkema, Rotterdam.

Powell, W. (1986). Enhancing parasitoid activity in crops. *Insect Parasitoids* (Ed. by J. Waage & D. Greathead), pp. 319–340. *Proceedings of the 13th Symposium of the Royal Entomological Society of London.* Academic Press, London.

Powell, W. & Walton, M.P. (1989). The use of electrophoresis in the study of hymenopteran parasitoids of agricultural pests. *Electrophoretic Studies on Agricultural Pests.* (Ed. by H.D. Loxdale & J. den Hollander), Systematics Association Special Vol. 39, pp. 443–465. Clarendon Press, Oxford.

Powell, W. & Wright, A.F. (1988). The abilities of the aphid parasitoids *Aphidius ervi*, Haliday and *A. rhopalosiphi* De Stefani Perez (Hymenoptera: Braconidae) to transfer between different known host species and the implications for the use of alternative hosts in pest control strategies. *Bulletin of Entomological Research*, **78**, 683–693.

Powell, W. & Zhang, Z.L. (1983). The reactions of two cereal aphid parasitoids *Aphidius uzbekistanicus* and *A. ervi* to host aphids and their food-plants. *Physiological Entomology*, **8**, 439–443.

Pungerl, N.B. (1983). Variability in characters commonly used to distinguish *Aphidius* species (Hymenoptera: Aphidiidae). *Systematic Entomology*, **8**, 425–430.

Putman, R.J. & Wratten, S.D. (1984). *Principals of Ecology.* Croom Helm, London.

Rabasse, J.M. & Dedryver, C.A. (1983). Biologie des pucerons des céréales dans L'Ouest de la France. III Action des hymenoptères parasites sur les populations de *Sitobion avenae* F., *Metopolophium dirhodum* Wlk. et *Rhopalosiphum padi* L. *Agronomie*, **3**, 779–790.

Robert, Y. & Rabasse, J.M. (1977). Role ecologique de *Digitalis purpurea* dans la limitation naturelle des populations de puceron strié de la pomme de terre *Aulacorthum solani* par *Aphidius urticae* dans L'Ouest de la France. *Entomophaga*, **22**, 373–382.

Ruppert, V. & Klingauf, F. (1988). Attraktivität Ausgewählter Blütenpflanzen für Nutzinsekten am Beispiel der Syrphinae (Diptera, Syrphidae) *Mitteilungen der Deutschen Gesellschaft für Allgemeine und Angewandte Entomologie*, 255–261.

Sopp, P.I. (1987). *The use of ELISA in assessing relative efficiencies of polyphagous predators in cereals.* PhD thesis, University of Southampton.

Sopp, P.I., Sunderland, K.D. & Coombes, D.S. (1987). Observations on the number of cereal aphids on the soil in relation to aphid density in winter wheat. *Annals of Applied Biology*, **111**, 53–57.

Sotherton, N.W. (1985). The distribution and abundance of predatory Coleoptera overwintering in field boundaries. *Annals of Applied Biology*, **106**, 17–21.

Southwood, T.R.E. (1977). Habitat, the templet for ecological strategies. *Journal of Animal Ecology*, **46**, 337–365.

Starý, P. (1970). *Biology of Aphid Parasites with Respect to Integrated Control.* Dr W. Junk, The Hague.

Starý, P. (1976). Parasite spectrum and relative abundance of parasites of cereal aphids in Czechoslovakia (Hymenoptera, Aphidiidae; Homoptera Aphidoidea). *Acta Entomologica Bohemoslovaca*, **73**, 216–233.

Starý, P. (1978). Seasonal relations between lucerne, red clover, wheat and barley agroecosystems through the aphids and parasitoids (Homoptera, Aphididae; Hymenoptera, Aphidiidae) *Acta Entomologica Bohemoslovaca*, **75**, 296–311.

Starý, P. (1981). Biosystematic synopsis of parasitoids on cereal aphids in the western Palaearctic (Hymenoptera, Aphidiidae; Homoptera, Aphidoidea). *Acta Entomologica Bohemoslovaca*, **78**, 382–396.

Starý, P. (1986a). Common elder, *Sambucus nigra*, as a reservoir of aphids and parasitoids (Hymenoptera, Aphidiidae). *Acta Entomologica Bohemoslovaca*, **83**, 271–278.

Starý, P. (1986b). Creeping thistle *Cirsium arvense*, as a reservoir of aphid parasitoids (Hymenoptera, Aphidiidae) in agroecosystems. *Acta Entomologica Bohemoslovaca*, **83**, 425–431.

Stern, V.M., Smith, R.S., van den Bosch, R. & Hagen, K. (1959). The interaction of chemical and biological control of the spotted alfalfa aphid. *Hilgardia*, **29**, 81–101.

Sunderland, K.D., & Vickerman, G.P. (1980). Aphid feeding by some polyphagous predators in relation to aphid density in cereal fields. *Journal of Applied Ecology*, **17**, 389–396.

Sunderland, K.D., Crook, N.E., Stacey, D.L. & Fuller, B.J. (1987). A study of feeding by polyphagous predators on cereal aphids using ELISA and gut dissection. *Journal of Applied Ecology*, **24**, 907–933.

Thomas, M.B. & Wratten, S.D. (1988). Manipulating the arable crop environment to enhance the activity of predatory insects. *Aspects of Applied Biology*, **17**, *Environmental Aspects of Applied Biology*, 57–66.

Thomas, M.B., Wratten, S.D & Sotherton, N.W. (1991). The creation of 'island' habitats in farmland to manipulate populations of beneficial arthropods: predator densities and emigration. *Journal of Applied Ecology*, **28** (in press).

Thornhill, E.W. (1978). A motorised insect sampler. *PANS*, **24**, 205–207.

Vinson, S.B. (1984). Parasitoid host relationships. *Chemical Ecology of Insects* (Ed. by W.J. Bell & R.T. Cardé), pp. 205–233. Chapman & Hall, New York.

Vorley, W.T. (1986). The activity of parasitoids (Hymenoptera: Braconidae) of cereal aphids (Hemiptera: Aphididae) in winter and spring in southern England. *Bulletin of Entomological Research*, **76**, 491–504.

Vorley, W.T. & Wratten, S.D. (1985). A simulation model of the role of parasitoids in the population development of *Sitobion avenae* (Hemiptera: Aphididae) on cereals. *Journal of Applied Ecology*, **22**, 813–823.

Vorley, W.T. & Wratten, S.D. (1987). Migration of parasitoids (Hymenoptera: Braconidae) of cereal aphids (Hemiptera: Aphididae) between grassland, early-sown cereals and late-sown cereals in southern England. *Bulletin of Entomological Research*, **77**, 555–568.

Winder, L. (1990a). *Modelling the effects of polyphagous predators on the population dynamics of the grain aphid* Sitobion avenae *(F)*. PhD thesis, University of Southampton.

Winder, L. (1990b). Predation of the cereal aphid *Sitobion avenae* by polyphagous predators on the ground. *Ecological Entomology*, **15**, 105–110.

Winder, L., Carter, N. & Wratten, S.D. (1988). Assessing the cereal aphid control potential of ground beetles with a simulation model. *Proceedings Brighton Crop Protection Conference — Pests and Diseases*, **3**, 1155–1160.

Wratten, S.D. (1987). The effectiveness of native natural enemies. *Integrated Pest Management* (Ed. by A.J. Burn, T.H. Coaker & P.C. Jepson), pp. 89–112. Academic Press, London.

Wratten, S.D. & Thomas, M.B. (1990). Environmental manipulation for the encouragement of natural enemies of pests. *Crop Protection in Organic and Low Input Agriculture* (Ed. by D.M. Glen), pp. 87–92. British Crop Protection Council Monograph, Farnham.

Wratten, S.D., Bryan, K.M. Coombes, D.S. & Sopp, P.I. (1984). Evaluation of polyphagous predators of aphids in arable crops. *Proceedings of the British Crop Protection Conference — Pests and Diseases*, **1**, 271–276.

13. THE AGRO-ECOLOGY OF TEMPERATE CEREAL FIELDS: ENTOMOLOGICAL IMPLICATIONS

M.A. ALTIERI

Division of Biological Control, University of California, Berkeley, CA 94706, USA

INTRODUCTION

Given the serious environmental problems affecting modern agricultural systems today, there is a general consensus on the need for developing more sustainable agro-ecosystems based on low-input management (Altieri 1987). Vegetational diversification in time and space, recycling of biomass and nutrients, soil and water conservation features and built-in pest regulation mechanisms are some of the design characteristics that agro-ecologists emphasize to attain production stability (Gliessman 1989; Carroll, Vandermeer & Rossett 1990). Stability here is defined as the constancy of production of an agro-ecosystem subjected to a set of environmental stresses and/or disturbances (Conway 1985).

Within the range of world agricultural systems (Fig. 13.1) temperate cereal fields (herein considered as those composed of grain crops, excluding rice and maize) appear to require less inputs than modern orchards, field crops and vegetable cropping systems to achieve a certain level of desired stability. This greater stability apparently results from certain ecological and management attributes inherent to cereal cropping systems.

1   Cereals are usually grown as winter crops and therefore are rain-fed and depend on soil-stored moisture, not requiring supplemental irrigation water for optimal growth.

2   Cereals are usually grown or amenable to be grown in rotation with legumes or other crops, a practice with beneficial effects on pest control, soil fertility and conservation (Heichel 1978). In north-eastern USA corn/wheat, oats or barley/clover rotations are common, whereas in California wheat or barley/vetch or peas rotations are preferred.

3   In some Mediterranean areas, cereal fields are somewhat species and/or genetically diverse when composed of oat/barley mixtures or multilines, or when undersown with legumes or other grasses. Increased genetic diversity can significantly reduce the vulnerability of cereals to diseases (Browning & Frey 1969; Fischbeck, this volume, pp. 42–47).

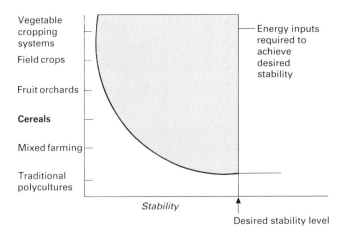

FIG. 13.1. Energy requirements to sustain a desired level of production stability in a range of farming systems.

4 In Europe, cereal fields are sometimes surrounded by hedgerows which increase plant diversity in field boundaries with important implications on the dynamics of insect pests and associated natural enemies (Potts & Vickerman 1974; Greig-Smith, this volume, pp. 360–362; Wratten & Powell, this volume, pp. 248–253).

5 Although cereals are colonized by a number of insect herbivores, these only erratically reach pest status (Andow 1983). Previously, in the US, total wheat losses due to pests were low, e.g. in 1976 losses were 1·6% of the total crop value. For this reason insecticide loads on cereals were low (about 0·03 kg/ha) when compared to cotton (0·50 kg/ha) or corn (0·40 kg/ha). Only 7·6% of the total US wheat acreage received insecticide treatments from 1967 to 1976 (Wilde 1981). Under such circumstances the prospects for the conservation and establishment of natural biological control agents were encouraging.

6 Cereal fields serve as habitat to a number of predators and parasitoids. More than 390 species of polyphagous predators have been recorded from cereal fields in England. Both naturally occurring polyphagous and aphid-specific predators have long been regarded as important control agents of cereal aphids and other insect pests (Edwards, Sunderland & George 1979).

ECOLOGICAL INTERACTIONS IN CEREAL FIELDS

The ecological dynamics of cereal fields in temperate regions vary widely depending on geographical location, crop species and genotypes, location and size of field, surrounding vegetation, spatial and temporal arrangement of crops, weediness and cultural management. Such variations determine the degree of diversity and management intensity of cereal fields in a region which may or may not benefit productivity of particular agro-ecosystems. Thus one of the main challenges facing agro-ecologists today is to identify the types of 'ecological designs' (either at the field or regional level) that will yield desirable agricultural results (i.e. pest regulation), given the unique environment and biota of each region.

This challenge can only be met by further analysing the multiple relationships between crops, weeds, arthropods, soils and cultural management. Our current understanding of key relationships affecting insect behaviour in cereals has been significantly advanced in the last two decades by numerous studies that have explored the ways in which crop and soil management influences the abundance and diversity of the herbivore and natural enemy community (Table 13.1).

From the available information it appears that several management techniques can minimize the incidence of cereal pests through direct effects of vegetation diversity and plant quality on colonization, establishment and reproduction of herbivores, or through increased pest mortality due to enhanced natural enemy abundance and effectiveness (Fig. 13.2).

INSECT PEST SITUATION IN NORTH AMERICAN CEREAL FIELDS

Historically, there were five major pest species of wheat in North America: the greenbug, *Schizaphis graminum* (Rondani) the grain aphid, *Sitobion avenae* F.; cereal leaf beetle *Oulema melanopa* L.; Hessian fly, *Mayetiola destructor* (Say) and wheat stem sawfly *Cephus cinctus* (Norton). Many of these pests constitute new introductions in certain cereal growing regions (i.e. the Hessian fly first recorded in the USA in 1778 did not reach Washington State until 1960 and Texas in 1978) and many are still spreading (Way 1988). Many changes in crop production technology (i.e. no tillage, increasing fertilizer use, new varieties, less rotations, etc.) undoubtedly affect the distribution and abundance of these species. For example, in Canada the wheat stem sawfly increased dramatically after 1910, favoured by changes in agricultural practices (Way 1988) such as:

1 summer fallowing which produced optimal wheat stems for oviposition and larval development;

2 introduction of early maturing cultivars which reduced parasitism;

3 stubble burning which killed parasites; and

4 change from ploughing to surface cultivation which killed fewer diapausing larvae.

TABLE 13.1. Relevant examples of crop and soil management techniques contributing to insect pest regulations in temperate cereal fields

Crop/soil management	Pest regulated	Effect or process	Reference
Crop temporal arrangement			
Crop rotation	Wheat stem sawfly (*Cephus pygmaeus (L.)*)	Breaks life cycle	Wilde (1981)
Planting date	Hessian fly (*Mayetiola destructor*)	Delay planting to avoid fly	Wilde (1981)
Late maturing cultivars	*C. pyramalus*	Favour sawfly parasites	Way (1988)
Crop spatial arrangement			
Crop density	Cereal aphids (*Metopolophium dirhodum*) *Rhopalosiphum maidis* (Fitch)	Population reduction in open plantings due to higher temperatures	Dixon (1987)
Undersowing with grass	Cereal aphids	Provides overwintering reservoir of aphid parasitoids	Powell *et al.* (1983), Dixon (1987)
Undersowing with clover	Cereal aphids	Enhanced abundance of carabid *Agonum dorsale* Pontoppidan	Luff (1987)
Trap cropping with brome grass	Wheat stem fly	Females preferentially lay eggs on brome grass where survivorship is low	Wilde (1981)
Weedy background	Cereal aphids and other pests	Enhanced abundance of carabids and other polyphagous predators	Speight & Lawton (1976)
Hedgerow management	Aphids and several insect pests	Refuge for polyphagous predators	Thiele (1977)
Sowing of host grasses by roads and ditches	Wheat stem sawfly	Favor sawfly parasites	Way (1988)
Control of volunteer wheat and other grass host plants	Russian wheat aphid	Elimination of RWA refuges	Baker & English (1988)
Soil management			
Tillage timing	Cereal leaf-beetle	By manipulating the soil temperature with different tillage techniques, emergence of adult parasites can be advanced or retarded	Wilde (1981)

| Low nitrogen fertilizer | Cereal aphids | Effects on plant quality which in turn decreases aphids' rate of increase | Dixon (1987) |
| Organic soil management | Cereal aphids | Lower aphid infestation due to increased resistance to aphid colonization attributed to changes in the proportions of non-protein to protein amino acids in the foliage | Kowalski & Visser (1983) |

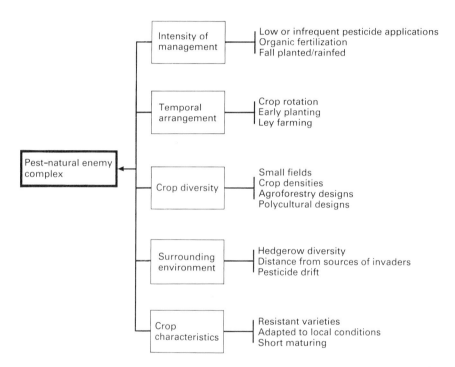

FIG. 13.2. Cultural management practices known to affect the dynamics of the pest–natural enemy complex in cereal fields.

The relatively low incidence of pests in North American cereals has changed recently. The Russian wheat aphid (*Diuraphis noxia* Mordvilko) was first recorded in Texas and Oklahoma in 1985. By 1986 *D. noxia* had spread into seven states and by the end of 1987 it was recorded in fourteen. In 1986 the Colorado Extension Service estimated a yield loss of $25 million in wheat. Around 0·5 million ha were treated with insecticides in 1987 and, despite these applications, yield losses amounted to 2500–3500 tonnes. In Texas, yield losses as high as 80% have been recorded with severe infestations. Already, millions of dollars have been spent in pesticide applications against the aphid, and it is likely that total expenditure in chemical control (currently about $12/ha) will continue to rise as the pest spreads throughout the 25 million ha of cereal potentially susceptible to infestation in the central United States (Anon. 1989).

Since *D. noxia* is an exotic pest in the USA it has been targeted as a primary candidate for classical biological control. This aphid originated in central Eurasia where a large number of predator species have been recorded attacking *D. noxia* or from plants infested with the aphid. These include aphid specific predators such as Coccinellidae, Syrphidae, Chrysopidae and Anthocoridae. Among insect parasitoids, *Aphidius colemani* Vierek and *A. uzbekistanicus* Luzhetskii are recorded from *D. noxia*, as is the cabbage aphid parasitoid, *Diaeretiella rapae* (McIntosh). In addition to these braconids, another, *Praon volucre* (Haliday), has been reared from *D. noxia* (Anon. 1989).

American scientists, in collaboration with the Commonwealth Institute of Biological Control (CIBC), initiated in 1988 an active foreign exploration programme in Turkey and Pakistan. Specimens collected in Turkey (i.e. *Aphidius matricariae* (Haliday)), *A. rhopalosiphi* (de Stephani), *Ephedrus plagiator* (Nees) and *P. volucre*) have been released in Idaho and other states, but so far none of these species could be catalogued as a successful biological control agent. The most aggressive *D. noxia* biological control programme in the USA has centred around large-scale releases of *Coccinella septempunctata* L. in various states. Evaluation of the impacts of such releases are still preliminary.

POSSIBILITIES OF IMPROVING BIOLOGICAL CONTROL IN CEREAL FIELDS

In large-scale cereal fields, native predators and parasitoids suffer from lack of alternate food sources and habitat (Speight & Lawton 1976). In many instances, monocultures offer enemy-free space to herbivorous pests, most of which exhibit rapid colonization and growth outstripping the ability of natural enemies to regulate them (Price 1981). A case in point is the outbreaks of cereal aphids attributed to the failure of predators, especially early in the season when aphids colonize fields (Chambers *et al.* 1987; Wratten & Powell, this volume, pp. 233–257).

A key to the enhancement of naturally occurring biological control is to improve the survival of predators in the off-season and also their colonization during early crop establishment (Powell *et al.* 1983). Despite the fact that the predator fauna of cereal fields has long been studied and known to be diverse, few manipulative methods to enhance their effectiveness have been implemented in real farming situations. Management of the composition and size of hedgerows, undersowing, rotations, weed control, and even intercropping are all cultural management techniques capable of creating conditions that may encourage colonization of cereal fields by predators. Burn (1987) summarizes several farming practices affecting predator effectiveness in English cereal fields. Effects of such practices vary according to geographical area, size of field, type of crop and/or variety, insect species complex, presence of non-crop vegetation, etc.

In regions where cereal fields are attacked by exotic pests, classical biological control has proved highly effective. A landmark example is the biological control of cereal aphids (*S. avenae* and *Metopolophium dirhodum* Walker) in the early 1970s in Chile. Despite the presence of resident natural enemies these aphids reached damaging levels which led to aerial applications of insecticides over 120 000 ha of wheat. In 1975, aphids and the barley yellow dwarf virus (BYDV) they transmit caused a loss of about 20% of the wheat production. In 1976 five aphidophagous predator species from South Africa, Canada and Israel, and nine species of parasitoids of the families Aphidiidae and Aphelinidae from Europe, California, Israel and Iran were introduced (Zuniga 1986). In 1976, more than 300 000 Coccinellidae were mass reared and released, and from 1976 to 1981 more than 4 million parasitoids were distributed throughout the cereal areas of the country. Such actions resulted in the long-term regulation of aphid populations below the economic threshold level. Although systematic evaluations of this biological control programme have not been done in Chile, it is estimated that if natural enemies had not been introduced, about 463 500 ha of cereals would have had to be chemically treated (twice per season with Dimethoate at 1500 cm^3/ha) at an annual cost of US$11 650 000 (Zuniga 1985). Today insecticides are rarely applied against wheat aphids in Chile, although aphids and BYDV are a sporadic problem in certain areas. This apparent success in Chile prompted Brazilian and Argentinian researchers to initiate similar programmes in the wheat growing regions of their countries (Crouzel 1983; Gassen 1983).

DESIGNING SUSTAINABLE
CEREAL CROPPING SYSTEMS

Among several options there are two main strategies on how to derive ecological information to design sustainable cereal cropping systems.

Comparative agro-ecological studies

It is obvious that much could be learned from a comparative study of cereal growing in different countries with temperate and Mediterranean types of climate, about the influence of biological, physical and management factors on the stability of farming systems. For example the introduction of the Mediterranean ley-farming system based on the use of annual, self-regenerating pasture legumes in rotation with cereals, may offer much to increase and stabilize crop production in other temperate areas with marginal soils, unpredictable water availability and lack of capital for fertilizers and pesticides. In selecting the legumes which are the basis of the ley-system, in addition to being adapted and productive, it may be important to encourage species that provide refuge and/or alternative food to natural enemies. A carefully designed rotation can enhance the survival of polyphagous predators by providing optimum cover and humidity as well as abundant prey availability (Burn 1987).

Clearly such rotations must be carefully designed based on the nature of the pest complexes and of the habitat. This is particularly relevant for wheat in England where out of thirty-nine insects reported as pests, thirty-three, including frit fly *Oscinella frit* L., are considered pests from the ley crop grown the year before (Gair 1975).

Using available information in cultural and biological control

There is sufficient information on certain forms of cultural and biological control probably applicable to virtually all pests of cereals of known biology. Based on such information, Perrin (1980) advanced a series of environmental management proposals to improve the control of insect pests affecting the cereal/rape system in southern England (Table 13.2). Although Perrin suggests some important changes in the design of cereal/rape systems, the protocol does not address some important dilemmas, e.g. whether it is desirable to have hedgerows removed or aerial spraying of insecticides. Nevertheless, Perrin's proposal is a step in the right direction in that it advances a regional approach where landscape diversity is manipulated in a co-ordinated manner by all agricultural sectors involved. The possibilities of such co-operation are not encouraging when antagonistic production and conservation views prevail. Such conflicts are well illustrated by the debate on hedges in England, where on the one hand their removal increases the efficiency with which the land can be farmed with modern machinery, but on the other may decrease the local diversity and abundance of birds, insects and plants. In these cases,

TABLE 13.2. Some proposals for environmental management in the cereal/rape system. After Perrin (1980)

Proposed practice	Potential effects on pests
Mixing cultivars within fields	Reduced spread of polycyclic diseases and aphids
Rotating cultivars between regions and years	Minimized build-up of resistance-breaking pathotypes/biotypes
Blocking of cereal types locally, e.g. winter wheat, spring barley in separate areas	Maximized 'edge-effects' diluted numbers of invading pests, easier aerial application of pesticide
Growing of insecticide-treated or resistant trap crops	Invading pests trapped before reaching main crop
Regional 'break' in cereals (i.e. synchronization of cropping sequence)	More effective starving-out of pests with poor powers of off-season survival
Removing all internal hedgerows and woodland	Minimized overwintering success of rape pests, fewer weed problems and less pigeon damage
Retaining and managing a few wild areas	Conservation of natural enemies for biological control
Geographically separating grassland and cereal areas as much as possible or changing grassland species composition	Fewer sources of cereal pests
Reducing sowing rates	Crops less attractive to some pests, but less competitive against weeds
Early autumn sowing	More vigorous crop to withstand weeds and wheat bulb fly, but more prone to disease

an agro-ecological approach must be developed so that economic, social and environmental goals are defined by the local rural community, and low-input technologies implemented to harmonize economic growth and environmental preservation (Fig. 13.3).

Within a sustainable development perspective, the argument for or against hedgerows must go beyond consideration of loss of yield and cost of conservation measures, to include aspects of sustainability and equity. This can be achieved by designing multiple-use hedgerows that provide numerous benefits and effects, in addition to those commonly perceived by farmers and environmentalists. Hedgerows that aid in conserving and regenerating the soil, modify the microclimate, provide useful products to the household and for the market, encourage wildlife, etc., can therefore provide economic benefits to farmers as well as help preserve the natural

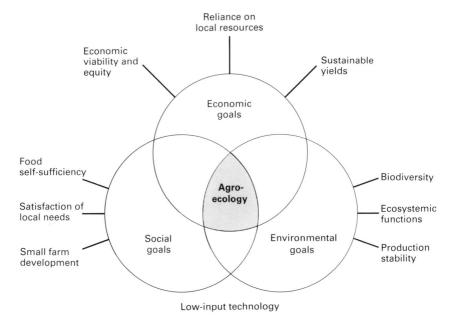

FIG. 13.3. The role of agro-ecology in satisfying social, environmental and economic goals in rural areas.

resource base that endows the local agricultural systems (Fig. 13.4). Advances along these lines have already been made in tropical countries where alley-cropping systems composed of hedges of fast-growing legumes (i.e. *Leucaena leucocephala*) have been developed to stabilize yields in marginal conditions (Nair 1984). Also in certain regions of China cereal fields are diversified with *Paulownia* spp. trees in various agroforestry designs. Such trees ameliorate the microclimate, and provide fuelwood, construction material and especially wood for small-scale furniture industries (Anon 1986).

CONCLUSIONS

The analysis of the ecology of insect populations in cereal fields indicates that insect abundance is influenced by:

1 the arrangement of crops in time and space;
2 the composition and abundance of non-crop vegetation within and around fields;
3 soil management (i.e. type of tillage, organic v. chemical fertilization);
4 intensity of management (i.e. pesticide applications, rotations, undersowing); and

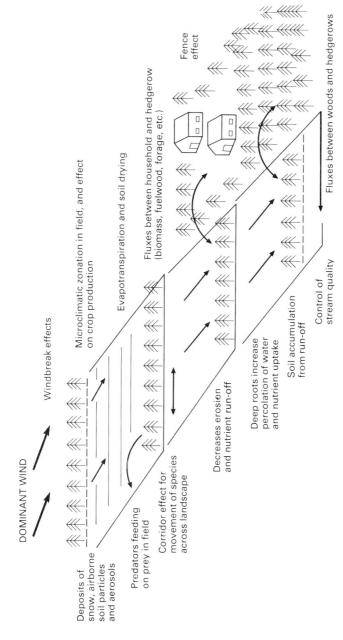

FIG. 13.4. Summary of major hedgerow functions within agro-ecosystems. After Forman & Baudry (1984).

5 location and size of field.

Insect populations in cereal fields may respond to environmental manipulations depending on their biology, phenological stage and degree of association with one or more of the vegetational components of the system. So far most of the research in cereal fields has concentrated on determining the predator and parasitic fauna of cereal fields, their feeding habits and their interactions with biotic and abiotic components of the agro-ecosystem. To advance beyond merely cataloguing, the mechanisms underlying pest/natural enemy responses to management and diversity must be further identified and elucidated.

What is difficult is that each agricultural situation must be assessed separately, since herbivore–enemy interactions vary significantly across regions, species-involved and management systems. One can only hope to elucidate the ecological principles governing insect dynamics in cereal fields, but the agro-ecosystem designs necessary to achieve pest regulation will depend on the agro-ecological conditions and socio-economic restrictions of each area. In this regard farmers' needs and preferences as well as the public's concerns must be fully considered if adoption of new sustainable designs is to occur. New designs will be most attractive if, in addition to pest regulation, agro-ecologically managed cereal fields offer benefits in terms of stable yields, increased soil fertility, provision of wildlife habitat, decreased weed competition and diseases as well as less dependence on chemical inputs.

In countries where high yield technology is emphasized, management strategies such as breeding for yield at the expense of various forms of pest resistance and the large increase in application of nitrogen fertilizers, can seriously jeopardize the implementation of alternative pest-control methods (Way 1988). In such situations, policies biased against low-input agriculture must be removed so that farmers are given incentives to adopt alternatives that sustain productivity while conserving the natural resource base.

REFERENCES

Altieri, M.A. (1987). *Agroecology: the Scientific Basis of Alternative Agriculture*. Westview Press, Boulder, Colorado.

Andow, D. (1983). The extent of monoculture and its effects on insect pest populations with particular reference to wheat and cotton. *Agriculture Ecosystems and Environment*, **9**, 25–36.

Anon. (1986). *Paulownia in China*. International Development Research Centre, Ottawa.

Anon. (1989). *Russian Wheat Aphid News*, **3**, (1–6). Department of Entomology, Colorado State University.

Baker, R.D. & English, L.M. (1988). *The Russian Wheat Aphid.* New Mexico State University Co-operative Extension Service Circular 528. Las Cruces.

Browning, J.A. & Frey, K.J. (1969). Multiline cultivars as a means of disease control. *Annual Review of Phytopathology*, **7**, 355–382.

Burn, A.J. (1987). Cereal crops. *Integrated Pest Management* (Ed. by A.J. Burn, T.H. Coaker & P.C. Jepson), pp. 209–256. Academic Press, London.

Carroll, R., Vandermeer, J.H. & Rossett, P.M. (1990). *The Ecology of Agricultural Systems.* McGraw Hill, New York.

Chambers, R.J., Sunderland, K.D., Stacy, D.L. & Wyatt, I.J. (1987). Control of cereal aphids in winter wheat by natural enemies: aphid-specific predators, parasitoids and pathogenic fungi. *Annals of Applied Biology*, **108**, 214–231.

Conway, G.R. (1985). Agroecosystem analysis. *Agricultural Adminstration*, **20**, 1–30.

Crouzel, I.S. (1983). El control biologico en la Argentina. *Informe Final IX CLAZ*, 169–174. Arequipa, Peru.

Dixon, A.F.G. (1987). Cereal aphids as an applied problem. *Agricultural Zoology Reviews*, **2**, 1–57.

Edwards, C.A., Sunderland, K.D. & George, K.S. (1979). Studies on the polyphagous predators of cereal aphids. *Journal of Applied Ecology*, **16**, 811–823.

Forman, R.T.T. & Baudry, J. (1984). Hedgerows and hedgerow networks in landscape ecology. *Environmental Management*, **8**, 495–510.

Gair, R. (1975). Cereal pests. *Proceedings of the 8th British Insecticides and Fungicides Conference*, **3**, 871–874. British Crop Protection Council, London.

Gassen, D.N. (1983). Controle biologico des pulgoes do trigo no Brasil. *Informe Agroprecuario*, **104**, 44–51.

Gliessman, S.R. (1989). *Agroecology: Researching the Ecological Basis for Sustainable Agriculture.* Springer Verlag, Berlin.

Heichel, G.H. (1978). Stabilizing agricultural energy needs: role of forager, rotations and nitrogen fixation. *Journal of Soil and Water Conservation*, **25**, 279–282.

Kowalski, R. & Visser, P.E. (1983). Nitrogen in a crop–pest interaction; cereal aphids 1974. *Nitrogen as an Ecological Factor* (Ed. by J.A. Lee, S. McNeill & I.H. Rorison), pp. 283–300. Blackwell Scientific Publications, Oxford.

Luff, M.L. (1987). Biology of polyphagous ground beetles in agriculture. *Agricultural Zoology Reviews*, **2**, 237–278.

Nair, P.K.R. (1984). *Soil Productivity Aspects of Agroforestry.* International Council on Research on Agroforestry, Nairobi.

Perrin, R.M. (1980). The role of environmental diversity in crop protection. *Protection Ecology*, **2**, 77–114.

Potts, G.R. & Vickerman, G.P. (1974). Studies on the cereal ecosystem. *Advances in Ecological Research*, **8**, 108–198.

Powell, W., Dewar, A.M., Wilding, N. & Dean, G. (1983). Manipulation of cereal aphid natural enemies. *Proceedings of the 10th International Congress of Plant Protection*, **2**, 780. British Crop Protection Council, Croydon.

Price, P.W. (1981). Relevance of ecological concepts to practical biological control. *Biological Control in Crop Production* (Ed. by G.C. Papavizas), pp. 3–19, Allarheld.

Speight, M.R. & Lawton, J.H. (1976). The influence of weed-cover on the mortality imposed on artificial prey by predatory ground beetles in cereal fields. *Oecologia (Berlin)*, **23**, 211–223.

Thiele, H.U. (1977). *Carabid Beetles in their Environments.* Springer Verlag, Berlin.

Way, M.J. (1988). Entomology of wheat. *The Entomology of Indigenous and Naturalized Systems in Agriculture* (Ed. by M.K. Harris & C.E. Rogers), pp. 183–206. Westview Press, Boulder, Colorado.

Wilde, G. (1981). Wheat arthropod pest management. *Handbook of Pest Management in Agriculture* (Ed. by D. Pimentel), pp. 317–328. CRC Press, Boca Raton, Florida.

Zuniga, E. (1985). Ochenta anos de control biologico en Chile. Revision historica y evaluacion de los proyectos desarrollados (1903–1983). *Agricultura Tenica*, **45**, 175–183.

Zuniga, E. (1986). Control biologico de los afidos de los cereales en Chile. I. Revision historica y lineas de trabajo. *Agricultura Tenica*, **46**, 475–477.

PART 4

CEREAL ECOLOGY — INTEGRATED PERSPECTIVES

Session Chairman: D.W. MACDONALD
Session Organizer: G.R. POTTS

Everything that we do is set against a background of increasing human pressure on the environment. The combination of population increase, technological innovation and increasing demand has progressively changed most of the natural world around us.

The consequence for ecologists is that they have, at least for the past 60 years, been running a race. They have, however, run it at two speeds. In the slow lane were those driven by a natural inquisitiveness about ecosystems. The methods were heuristic, the aims were to provide greater understanding and the final goal was an awareness of nature and of man's place in nature. As almost everyone has now realized, however, man was quickly changing the studied ecosytems in many ways, from obvious and deliberate to subtle and unwitting. Damage occurred and many ecologists were forced into a faster lane where the methods were pragmatic, sometimes even expedient; the aims were the solution of problems and the timetable set by events, usually unwelcome incidents of one sort or another. The inevitable result has been too much haste and too little science.

As we all know rather too well, ecosystems are not stable. In the long-term the norm is for equilibrium densities to drift, which is why monitoring is bound to give results which are increasingly 'noisy' with time. Here we have the core of the problem, noise, some of it natural, some of it anthropogenic. How do we know which is which? How do we know what is normal? Surely only by monitoring for long enough periods to embrace the full range of natural variation — yet how long is that? Some recent thinking, for example 'more time means more variation', has not been helpful because of the inbuilt anthropogenic trends. For example with the Common Bird Census of the British Trust for Ornithology much of the increased variation with time has been due to the preponderance of declines amongst the species monitored.

Against this background of uncertainty all sorts of measures need to be taken to manage the situation. For example, the European Commission in Brussels has passed over 200 laws on the environment (i.e. Regulations

& Directives excluding those for DGVi) and over fifty are pending. Decisions, sometimes hundreds of them per law, are being made but on data which are usually arguable in one way or another. Erring on the side of caution — the precautionary principle — is therefore a reasonable approach. Measures have thus been taken to limit damage to ozone by the CFCs long before science justified that course of action. Usually better data are needed on which to base decisions than are available, and that is what this session is all about.

As befits someone from Rothamsted, with its uniquely sustained record of work, Ian Woiwod eloquently argues that the historical component revealed by monitoring is an essential part of the answer to most current ecological questions. His data include almost the only quantatitive measures of the insect communities in the wider countryside from the pre-herbicide era. The aphid/aerial-plankton monitoring scheme he describes covers a huge geographical area and is thus capable of picking up general messages from the countryside. More intensive, though covering 62 km^2, is the cereal ecosystem monitoring scheme described by Nicholas Aebischer. Here the message is simple, most arthropod taxa are declining; the only question is why? Explanations or hypotheses have emerged but many of these need testing by experiment and that was the main purpose of the ADAS Boxworth experiment described by Peter Grieg-Smith. Effects were clearly detected on the 65 ha area subject to a high-pesticide regime though they were felt to be temporary in nature. The size of that area is, however, small — some individual fields in Aebischer's study area are over 65 ha and with larger areas the effects may not be so ephemeral. Meanwhile, Nick Sotherton and later Adel El Titi describe techniques, conservation headlands and an integrated farming system, respectively, that overcome many of the adverse effects of pesticides.

Here then are the three basic elements, monitoring, experimentation and application which ultimately will be linked with unifying theories. To complete the picture I outlined at the start, we can ask how the theories of the heuristic endeavours help the urgent necessarily pragmatic, reactions of the applied ecologists? David Macdonald and Helen Smith review this with a heroic effort to round off the session and with it, the whole volume.

14. THE ECOLOGICAL IMPORTANCE OF LONG-TERM SYNOPTIC MONITORING

I.P. WOIWOD

*AFRC Institute of Arable Crops Research, Rothamsted Experimental
Station, Harpenden, Hertfordshire AL5 2JQ, UK*

INTRODUCTION

Every year hundreds of thousands of agricultural experiments are carried out not by agricultural scientists but by farmers themselves, for this is in effect what they are doing every time they grow their crops. Procedures are based on previous research results, local conditions and experience, combined with economic considerations. At the end of each harvest the results of the experiments are soon apparent in terms of yield and profit. The natural vegetation succession is then arrested by cultivation or the application of herbicides ready for next year's experiment. It is perhaps because of this constant annual field testing that agriculture has been so fast to adopt new techniques, grow new cultivars or crops, and respond rapidly to changing financial situations. Agriculture, at least in temperate regions, is firmly tied to this yearly seasonal cycle and much of our improvement of agricultural practice also comes from annual field experimentation by scientists.

Traditionally ecology, when concerned with natural ecosystems, has had a different time perspective and the interest is often in studying populations or communities through time, usually over several generations of the dominant organisms. In these studies longer experiments and observations would be expected to be typical. Despite this a recent analysis of a random selection of 749 papers published in a prestigious American ecological journal from 1977 to 1987 showed that the most common length of an ecological research study was 1 year with the vast majority not lasting more than 3 years, and that many of the studies using more than 10-year runs of data were based on palaeo-ecological reconstruction not direct observation (Tilman 1987). It is perhaps surprising that by far the longest running experiments are still the agricultural Rothamsted 'classical' ones (Anon 1984), and that long-term studies of natural ecosystems or even single species populations are still few and far between.

For the cereal ecosystem case studies presented here (Aebischer, pp. 305–331; Greig-Smith, pp. 333–371; Sotherton, pp. 373–397; El Titi, pp. 399–411, this volume) there is a presumption that the normal 1–3 year experiment is inadequate as a basis for such work as the shortest study is 6 years (Sotherton, pp. 373–397, this volume) and the longest 20 (Aebischer, pp. 305–331, this volume). The importance of longer term studies is also implicit in several of the other contributions, notably for weeds where the pest species are known to interact with the crop over space and time (Firbank, pp. 209–231, this volume).

So why are there so few long-term sets of data available, why are they so important when considering ecological questions and what other scales of space and time need to be considered? These are all questions that have received increasing attention in ecology (Strayer *et al.* 1986; Likens 1987), notably in the USA where long-term ecological research programmes have recently been initiated because essential long-term data are lacking (Franklin 1987).

It is perhaps particularly appropriate that these questions should be reviewed here as one of the first and by far the longest running field experiment in the world, the Rothamsted Broadbalk experiment, is based on a temperate cereal ecosystem and I believe that this experiment still has useful lessons for those considering starting long-term studies.

How long is long term?

This is clearly a question for which it is impossible to provide a definitive answer. In ecology there are two equally valid perspectives (Strayer *et al.* 1986). One is that the study should last at least as long as necessary to include the generation time of the dominant organism or to include examples of all the important processes of the ecosystem. Thus with short-lived organisms in simple experimental ecosystems long-term studies can be carried out relatively quickly.

However, perhaps the more usual perspective is from the point of view of the human observer. In which case those lasting 1–3 years can definitely be classified as short term (Tilman 1987; Table 14.1) whereas anything of 5 years or longer starts to become long term, if only because of the relative rarity of such studies. However, as will be seen later, much longer runs of data are required to answer many ecological questions without ambiguity.

It is worth examining the numbers of different types of field experiments carried out at Rothamsted over the last 10 years in this respect (Table 14.1). Rothamsted is a large agricultural research station with a long history of field-based experiments, all of which go through a careful vetting process before experimental sites are assigned. At the same time it

TABLE 14.1. Field experiments carried out or started on Rothamsted Farm, 1980–1989

Type of experiment	Duration of experiment (years)	No. of experiments
Annual	1	746
Serial	2	5
	3	6
	4	4
	5	5
	6	1
	8	1
	20	1
		23
Rotational	>20	5
Classical	>130	8

is perhaps somewhat atypical in its well-known background of long-term studies.

As expected the vast majority (95%) of experiments are annual, although following good experimental practice usually replicated over 3 years on different sites, so the total given in Table 14.1 should be divided by 3 to give a more realistic idea of the number of different investigations. Serial experiments are those which require the same site, usually with the same crop, over several seasons. The number of years given are from the original proposals as many of these experiments are still in progress and some may be extended. These serial experiments are often ecological in outlook, indeed one is entitled 'farmland ecology' (5 years duration). Most of these studies are for a maximum of 5 years but it is interesting to see one experimental proposal concerning a 20-year study, a long-term investigation by any standard. This is a straw incorporation experiment where pest and disease build-up and straw breakdown are likely to be slow processes which need a longer-term perspective than available from annual field experiments. It is also a study which has become increasingly important in the last decade as straw has become surplus to requirements on arable farms and the practice of straw burning has become unacceptable.

Rotational experiments tend to be long-term in nature and most have been in existence for at least 30 years. The classical experiments are of course the really long-term studies instigated by Gilbert and Lawes in the last century. Two of these will be described in more detail later but they clearly represent the tail of the distribution when it comes to long-term experimentation and observation.

Why long-term studies are important in ecology

There are a number of areas of ecology where long-term observation or experiment are already accepted as essential and others where perhaps they should be (Likens 1983; Strayer *et al.* 1986).

Slow processes

The first and perhaps the most obvious reasons for long-term observation is to study slow processes. The usual example here is that of ecological succession, details of which are provided in every ecological textbook. In arable agriculture succession may not seem so important or relevant as the process is arrested at every harvest but the classical Rothamsted experiments of Broadbalk and Geescroft Wildernesses clearly demonstrate the inexorable change from arable land to woodland if cultivation is ever allowed to stop (Brenchley & Adam 1915) and there has been recent interest in the relevance of these observations to the problem of surplus agricultural land (Burnham 1989).

Another particularly relevant example of a slow process is that of soil formation which may occur at a rate of only 1 cm every 20–100 years (Buol, Hole & McCracken 1973).

Rare events

There are many events which are likely to effect ecological processes but which only occur sporadically and unpredictably. Obvious examples are climatic extremes of temperature, wind or rain but unusual migrations or outbreaks of pests and diseases are biological examples, although these in turn may be caused by chance weather effects. One important question is how often such events occur on average and hence how important they are ecologically, and this can only be ascertained by careful observation over long time periods.

Occasionally it is possible to study the effects of such rare events as part of experiments designed for other purposes. Thus the cereal aphid, *Metopolophium dirhodum* (Walker) had an exceptional outbreak in 1979 in southern Britain and also in continental Europe (Dewar, Woiwod & Choppin de Janvry 1980) but it was only by chance that the effect on cereal yields was able to be measured as a result of a multidisciplinary field experiment which happened to be in progress at the time (Prew *et al.* 1983). However, study of rare events is generally difficult within such closely controlled experimental procedures and structured long-term observations are often the only source of useful information.

Cyclical phenomena

Re-occurring patterns have often been looked for in population changes over time. There are several purposes in looking for such patterns. Practically they may be of use in prediction, and theoretically there are simple models which produce such cycles. Although powerful statistical techniques now exist for extracting regularly occurring patterns there have been several instances of cycles being reported which disappeared when longer data series became available. As new statistical technique becomes available the few good examples of long-term cyclical data that exist tend to be analysed repeatedly, just because of their rarity (e.g. Schaffer 1984).

Processes with high variability

This is a common situation with real field data where large annual fluctuations often mask more subtle trends. Again, long data runs are required to extract the 'signal' from the 'noise'. The analysis of the 'noise' itself is often of interest in its own right and also requires long-term observations for analysis.

To illustrate this point an example will be given later of trends in cereal aphids taken from the records of the Rothamsted Insect Survey.

Complex phenomena

Many ecological phenomena are the result of interacting factors. A useful way to assess the relative importance of such factors is by careful multivariate analysis of long-term quantitative observations.

Transient dynamics

A serious problem inherent in the interpretation of short-term experiments, particularly manipulative field experiments, has been highlighted by Tilman (1987). This is the question of transient dynamics where an ecosystem under study is changing as a result of the experimental procedure or other perturbation, and observations over short periods of time can lead to conclusions which are totally different from those which would be arrived at by long-term observations. Several examples of this phenomenon are given by Tilman (1987) and one of them from the Park Grass experiment at Rothamsted will be outlined later.

Testing and developing ecological theory

Many ecological phenomena are of types already mentioned, in particular slow processes, processes with high variability or complex phenomena. It follows from this that long sets of data are required to test the theories concerning them. Perhaps this is why there is a continual call in ecology for data to test theory, although the data are very rarely produced. Often formulation of theory in the form of mathematical models is relatively fast in comparison to data collection which is why much ecological theory is poorly founded in the real world. A good example of this is the theory of island biogeography originally put forward by MacArthur & Wilson (1967). This theory was very largely a theoretical construction but induced an enormous amount of interest and subsequent research, particularly in terms of the species–area relationship. However, the theory has not really stood the test of time well and has recently been assessed as trivial (Williamson 1989).

The process of testing ecological theory with data can be turned around so that theory naturally develops from an analysis of a large body of empirical and experimental data. Whether it is better to work from the theory to the data or vice versa is largely a matter of personal preference and both approaches have their adherents but either way long-term rather than short-term sets of data are likely to provide a firmer basis for ecological ideas.

Providing experimental arenas

Any area whose ecology is studied in detail for long periods is likely to become useful just because the body of knowledge about the site enables new investigations to be carried out more efficiently and many studies are only possible in such arenas. The classical experiments at Rothamsted have really survived for this reason and are still actively used as a result.

The temporal dimension

Ecology as a science has a very definite temporal component. In this respect it is similar to sciences such as geology, palaeontology and astronomy but different from some of the physical sciences (Taylor 1987).

Genetic selection can take place during the course of a series of observations so that it is impossible to repeat an experiment with the same organisms. It is also becoming increasingly apparent that communities are only loose associations of species which over historical time move in response to changes in physical environment at different rates and come back together in different combinations (e.g. Coope 1977). Soil is another important part of the ecosystem which has a definite historical component based on the underlying geology and past climatic and vegetational history of a site.

Therefore, in many respects it is not possible to repeat experiments in different places or at different times as many aspects of the environment or organisms will always be different. This historical component requires continuous or regular and prolonged observation to elucidate.

Problems in establishing long-term experiments and observations

As can be appreciated the nine previous reasons for doing long-term research cover nearly all aspects of ecological work at some stage. So why aren't all ecologists involved in long-term work? The answers are sometimes obvious and concern funding but other reasons are more complex and involve attitudes concerned with the conduct of science itself. The problems can be considered under three headings, scientific, political and personal.

Scientific

Many of our views on how science should be carried out are derived from the physical sciences and are based on the idea that experiments should be designed to answer specific theoretical predictions. It has recently been suggested that such a simplistic approach may not always be valid when applied to the more complex and historical situations presented to ecologists (Taylor 1987).

All long-term studies whether strictly experimental or more open-ended and observational are likely to require repeated quantitative observation over time. It is this repeated observation which is the life blood of such studies but which has in the past been far too often denigrated as 'mere' monitoring and assessed as uninteresting, low-grade or unimaginative by scientists unused to the difficulties of such work and the great creativity required in the analysis and interpretation of this type of data. This narrow view of what is good science has recently been comprehensively repudiated by Likens (1983), Strayer *et al.* (1986) and papers in

Likens (1987). There are indications in the last few years that attitudes are changing, partly driven by the need for long-term data sets to test some more interesting theoretical questions in ecology but also to answer many questions related to current concerns about global climatic change and pollution.

Political

Political decisions permeate science both locally and more widely in the way that the role of the scientist is perceived within society. Of crucial importance is the fact that the funding of much science has, at least historically, been provided through government sources. Other funding, for example through industrial sponsorship, tends naturally to be to provide answers to immediate questions and hence is nearly always short term in outlook, although funding through charitable trusts has often provided funds on a more stable basis.

Politicians themselves often lack scientific training and understandably have a rather short-term perspective (Cranbrook, this volume, pp. 26–28). They therefore require of scientists quick straightforward answers to simple questions. They are not impressed by the sort of answer from scientists that goes 'That is an interesting question and if you give us adequate funding for the next 20 years we should be able to provide a reasonable answer to it.' A commitment to long-term stable funding is becoming a very rare and valuable commodity in science but particularly important in ecological studies, and if politicians require quick yet reliable responses then the scientist must have already carried out much of the long-term background research before the question gets asked.

The short-term perspective of funding cycles is another well-known problem, whether it is the annual accountancy cycle or the 3-year postgraduate PhD grant. In British science there has been a dramatic increase in short-term, usually 3-year, contract work in recent years. This is supposed to improve flexibility and accountability in science. The effects on the possibility of carrying out serious long-term studies are dire.

Personal

The personal dimension should not be omitted from consideration. Recent analyses of long-term studies in ecology concluded that the majority of such studies are carried out as a result of the personal dedication and interest of individual scientists (Strayer *et al.* 1986; Taylor 1987). Not all scientists are equally interested in such studies, nor should they be expected to be. However, there are often direct disadvantages to individuals carrying out such work in terms of career prospects and personal advancement, curricula vitae often look better if they contain a wide range of experience which may not be possible to acquire when involved in detailed long-term studies.

Synoptic versus intensive

Apart from the important perspective of time there is another one that is often missing from ecological studies. This is the synoptic approach which implies a more general but less intensive study, really a case of not missing the wood by studying the trees too closely. Such studies tend also to imply a spatial perspective. The synoptic scale used in meteorological studies is familiar to everyone but there are very few ecological studies on similar scales. Current interest in the effects of climatic change make such studies vital, although there are also good ecological reasons for such work (e.g. Firbank, this volume, pp. 209–231).

A current interest in ecology is the question of scale and hierarchies for it is being realized increasingly that it is very difficult to predict the behaviour of ecological systems from detailed knowledge of small parts and that the scale of observation has to be right for the question being asked (Wiens 1989). It is only by studying systems at the correct scale that robust conclusions can be made.

EXAMPLES OF LONG-TERM AND SYNOPTIC STUDIES — THE ROTHAMSTED EXPERIENCE

Perhaps the best way to enlarge on some of these general points is by specific examples. Fortunately, working at Rothamsted one does not have to go far before finding good examples of most of the points mentioned above, and it would be almost impossible to discuss long-term experimentation in respect of the temperate cereal ecosystem without starting at the beginning with the well-known Broadbalk experiment and the related Park Grass experiment.

The Broadbalk continuous wheat experiment

I do not intend to cover in detail all aspects of this, the most famous of all the classical long-term experiments at Rothamsted, much of which is already well known and available elsewhere (Lawes & Gilbert 1895;

Bawden 1969) but there are several ways in which it demonstrates the difficulties and yet continuing value of long-term measurement and observation even after almost 150 years of continuous experimentation.

The experiment on Broadbalk was instigated by Sir John Bennet Lawes when, in 1843, the first experimental wheat crop was sown. This was just after he appointed the young chemist, J.H. Gilbert to help with this and other experiments on the Rothamsted estate. The general purpose of the experiment was to investigate what substances (inorganic chemicals) were necessary to supply to the soil to maintain its fertility (Lawes 1847).

So one of the first questions that needs to be asked is whether the experiment was envisaged as a long-term study by Lawes. Although it is not possible to say exactly what was in Lawes' mind when he set up the experiment it was certainly in response to immediate practical farming questions at that time. The first paper on the work was published after only 3 years of results (Lawes 1847) also suggesting a strong interest in the short-term applicability of the experiment.

A good case has recently been put forward to suggest that specific experimental objectives are inimicable to long-term research because as soon as these objectives are achieved the experiment is likely to be terminated (Taylor 1987). So how does Broadbalk fit into this idea? Did it have a theoretical background with a specific question to be answered? The answer rather surprisingly was yes, for one of the original purposes was to test Leibig's theory that plants obtain the nitrogen they required from the air (Leibig 1840). Lawes and Gilbert disagreed with this and Broadbalk, along with many of the other classical Rothamsted experiments, was initiated to settle this specific question (Lawes & Gilbert 1851), although more general questions about the other chemicals important for plant nutrition became increasingly important as the experiment proceeded.

The main point is that the theoretical question posed by Leibig's work was settled to everyone's satisfaction, except Leibig's, within 10 years (Lawes & Gilbert 1851) and many of the other main objectives of the experiment achieved by the 1860's (Taylor 1987). So why has the experiment survived for so long? The answer is not simple. First of all there was a deep personal antagonism between Leibig on the one hand and Lawes and Gilbert on the other. This undoubtedly kept Gilbert's interest alive long after the original point had been proved. Another factor was the stable working environment for Gilbert, a meticulous scientist who was able to continue his work at Rothamsted for over 50 years (Gilbert 1895), an impossible feat for any individual under current employment conditions. A further important feature of Broadbalk was its ideal layout as a demonstration plot, with the simple treatments running the length of the field which has a slope such that the results of the experiment are immediately visually apparent. This was particularly important in the early days of the experiment when farmers needed to be convinced of the importance of the use of artificial fertilizers to increase yields.

Despite these factors the continuation of the experiment after Lawes' death was only assured because the experimental fields had been rented to the Lawes Agricultural Trust securing their continued existence to the present day. Over the years the original experiment has been modified to retain its relevance to current agricultural practice and changing research requirements. Such changes have often added valuable new information but at the expense of ecological continuity.

One of the main justifications for the continuation of the experiment, apart from its historical importance, is that it has now become an increasingly valuable experimental arena suitable for many studies which could not have been envisaged or carried out at the time the original experiment was started. One example of such a use has been provided by Brookes, Wu and Ocio (this volume, pp. 95–111) where work on the turnover of microbial biomass requires soils where the history of chemical inputs is known and the soil organic matter is at equilibrium. A related study concerns the uptake of nitrogen into the wheat crop. This is a question which has important environmental as well as agricultural implications as there is currently a growing concern about the nitrate level in drinking water and how it can be reduced. To answer such questions we need to understand the turnover of nitrogen in the soil microbial biomass. Again Broadbalk has proved invaluable for these studies using ^{15}N-labelled fertilizer to see what proportion of nitrogen is taken up by the crop and how much is leached into the water table (Powlson *et al.* 1986; Jenkinson & Parry 1989). Although Lawes and Gilbert were very interested in the uptake of nitrogen fertilizer by crops the emphasis was very much on efficient application to improve crop yield. This emphasis has been maintained until very recently when the question has been turned on its head with the interest now being focused on the nitrogen being lost from the crop and soil into the ground water, and the latest analytical techniques have been brought to bear on this question. This is a theme that recurs again and again in long-term studies where the questions asked originally are replaced by further questions as theory, technology and current interests change in a way that is impossible to predict at the start of experiment.

A further example of this from the Broadbalk experiment concerns heavy metal pollution by cadmium. This work relied on the availability of stored samples of plant material and soil from the long-term experiments at

Rothamsted and the development of the analytical techniques of neutron activation analysis and graphite furnace atomic absorption spectrometry for detection of cadmium residues. It was shown that cadmium residues had increased in the Rothamsted soils since the 1850s (Jones, Symon & Johnston 1987) but that there was little evidence of a corresponding increase in cadmium concentrations in the crop (Jones & Johnston 1989). The availability of stored soil and crop samples were crucial to this particular study but maintaining these samples has always been a source of friction when storage space is at a premium (Taylor 1987). Yet they have proved an invaluable resource on several occasions since the 1950s as new analytical techniques have become available.

An example of this which also typifies the use of long-term data to study rare or unique events is the so-called thermonuclear 'bomb effect'. Thermonuclear explosions release neutrons which react in the air to produce $^{14}CO_2$ which mixes with naturally occurring $^{14}CO_2$ and then enters plants and subsequently soil organic matter. Thus a correction is required when radio-carbon dating is used to calculate the age of soil. Providing the pre- and post-bomb age of the soil is known and the organic matter is in a steady state throughout then the amount of organic matter entering the soil each year can also be calculated. The stored Broadbalk soil is uniquely suitable for this purpose and has enabled unrepeatable studies to be made (Jenkinson & Rayner 1977). There is no possible way that this use of the Broadbalk experiment could have been foreseen by the experiment's originators.

There are two other studies which are worth mentioning in relation to the Broadbalk experiment. An example of the use of long data runs to study complex phenomena is the analysis started by Fisher (1924) and continued by other workers into the effects of weather on yield on the Broadbalk experiment. Possibly because of secondary factors such as disease, the results from an immense amount of analysis were fairly meagre (Yates 1969).

The final example is an entomological one and concerns a remarkable run of observations carried out by H.F. Barnes on two species of wheat blossom midges (Cecidomyidae) on the Broadbalk experiment. The larvae of these species, *Contarinia tritici* (Kirkby) and *Sitodiplosis mosellana* (Géhin), were sampled from 500 ears of wheat on Broadbalk each year for 34 years from 1927 by Barnes and for a further 4 years after his death (Johnston, Lofty & Cross 1969). This record is exceptional in the entomological literature for its length and consistency of sampling methodology. These records were obtained initially to test the theory that major outbreaks of these species occurred every 5 years and that this cycle was driven by climate (Barnes 1958). The long run of data showed that local weather conditions sometimes disrupted this suggested cyclical pattern (Johnston, Lofty & Cross 1969).

A related study on one of the species *C. tritici* is also of note. This midge overwinters as a pre-pupal larva and it had been observed that the species could be dormant at this stage for several years. Larvae were taken from Broadbalk and kept outside in an insectary where some were observed to be emerging up to 18 years later when Barnes' sudden death brought the observations to an end, although a further emergence was observed at 19 years. The relevance of this observation to the population dynamics of the species and recent questions of metapopulation dynamics and the persistence of insect populations have never been investigated but may be very important. One could imagine the horror that would be created in funding agencies if a project was proposed to repeat and extend these observations under the present financial climate for long-term research.

The Park Grass experiment

This well-known Rothamsted classical experiment is not directly concerned with the cereal ecosystem but has several features which make it relevant to the question of long-term ecological observation and experimentation. The Park Grass experiment was laid down in 1856 on a field that had been under continuous grassland for at least 100 years previously. Its original purpose was similar to that of Broadbalk in that it was to determine the optimum combinations of fertilizers to obtain maximum yields of a crop, in this case hay. As this crop is now of less relevance in British agriculture this experiment has now become much more of an ecological rather than an agricultural one and is generally regarded as the longest-running ecological field experiment anywhere in the world (Tilman 1987). The ecological interest arises mainly from the large number of plant species present and the continuity of treatments, unlike many of the other Rothamsted classical experiments which are concerned with crop monocultures and have been modified over the years to maintain their agricultural relevance.

As with Broadbalk the Park Grass experiment soon showed how yield could be manipulated by fertilizers, however, after only 3 years Lawes & Gilbert (1859) noted that the plots looked more like a seed trial rather than a fertilizer experiment because the botanical composition of the plots had changed so much. As a result the botanical composition of the plots has been measured at intervals ever since, although gaps have arisen during periods where the continuing importance of the site was not appreciated

(Williams 1978). Here is perhaps a warning about the difficulty of keeping continuity of records in long-term studies as short-term scientific and political fashions change.

Park Grass has certainly demonstrated the slow process criterion for carrying out long-term work, for it took more than 100 years for intensive chemical treatment to reduce a mixed flora to a monoculture (Taylor 1987). It also provides good examples for at least three of the previously outlined criteria for long-term observation, namely transient dynamics, introducing the temporal dimension and testing ecological theory. It is of note that all these examples have been published in the last 20 years, over 100 years since the experiment started. This is not due to idleness on the scientists' behalf, for it should be remembered that evolutionary and ecological science did not exist at the time the experiment commenced and only relatively recently has the theoretical framework of these subjects reached a stage where the results from the Park Grass experiment have become relevant and often crucial.

An analysis of the dominance of plant species on the various plots has led to the discovery that nutrient addition can lead to long periods of transient dominance by species that are later displaced. Thus on a plot receiving full mineral fertilizers with nitrogen applied as ammonia, *Dactylis* dominated for 15 years before being briefly replaced by *Agrostis* until *Holcus lanatus* displaced all other species (Tilman 1982, 1987). This is a perfect example of transient dynamics where short-term studies would have led to erroneous conclusions.

The importance of the temporal dimension is well illustrated by the work of Davies and Snaydon who in a series of publications showed that selection had taken place on the experimental plots resulting in genetically distinct populations between plots with very sharp boundaries (e.g. Davies & Snaydon 1973; Snaydon & Davies 1976). In some respects then the Park Grass experiment has actually modified the organisms that determine its ecology, a process that would be impossible to reverse and arrive at the original experimental starting point in 1856.

Finally, Park Grass has provided tests for several important aspects of ecological theory, including resource competition and succession (Tilman 1982), and more recently it has been able to provide one of the few examples of ecological stability (Silvertown 1987) after the existence of communities exhibiting such stability had been questioned (Connell & Sousa 1983). As Silvertown points out, the theory is exceedingly difficult to test and is only possible with this data because of the detailed studies continued over long periods of time. Ecological theory is often difficult to test for just these reasons.

The Rothamsted Insect Survey — a synoptic study

As mentioned previously, synoptic studies in ecology are rare. One of the few which has a direct relevance to the cereal ecosystem is the Rothamsted Insect Survey. This was instigated in the early 1960s by Roy Taylor when he set up a network of standard Rothamsted light traps run by volunteers throughout Great Britain to monitor moth populations (Taylor 1986). In 1963, following the publication of Rachel Carson's book *Silent Spring*, government funds became available for environmental projects to reduce insecticide usage and this enabled the work to be extended to include a nationwide network of 12·2 m high suction traps for monitoring aerial aphid populations (Taylor 1987). These two trapping networks became known as the Rothamsted Insect Survey.

The original idea behind the work was the fundamental one of introducing the spatial element, and particularly migration, into insect population dynamics. However, as the project developed a multitude of other studies have been made covering the full range from fundamental to applied (Taylor 1986). In particular the migration of pest aphids has been studied and forecasting systems developed to give advance warning of crop infestation and spread of virus diseases (e.g. Woiwod, Tatchell & Barrett 1984; Tatchell 1985; Tatchell, Plumb & Carter 1988; Camell, Tatchell & Woiwod 1989; Harrington 1989). Moths and aphids have been used in fundamental studies on the spatial and temporal population dynamics of species (e.g. Taylor & Woiwod 1980; Taylor, Woiwod & Perry 1980). Moths have been studied to measure changes in insect fauna in relation to land use, particularly with respect to agriculture and urbanization (Taylor, French & Woiwod 1978). Basic data and annual summaries have also been published at regular intervals (e.g. Taylor *et al.* 1981; Taylor *et al.* 1982; Woiwod & Dancy 1987; Woiwod *et al.* 1989).

As with many long-term studies the Insect Survey was not designed originally to be long-term as the spatial aspects of the work were considered to be more important. However, it was always realized that the data would become more valuable as time went on and this has proved to be the case. Requests for access to the data are increasing and with current interest in faunal changes as the results of possible global climatic warming, the Survey's database has become valuable in a way that certainly was not envisaged at its inception.

Practical forecasts that were originally developed to provide information for pest control are often based on weather parameters and trap samples collected over many years. These are already producing some of the first quantitative predictions of the long-term consequences of possible climatic change (Harrington 1989). Examples of two such forecasts are shown in Figs 14.1a and b for the cereal aphid *Sitobion avenae* (Fabricius), one forecasting the time of first flight in the year from preceding winter temperatures (Fig. 14.1a) and the other the relation between the size of the aerial population and the corresponding mean temperature (Fig. 14.1b). The important feature here is that each point on the graph represents a single year's data and long runs of such data are required to provide adequate temperature ranges to make such forecasts reliable. Once the relationship is established it can be verified and updated each year but only if monitoring continues.

Although the Survey covers many species and the data have been used in diverse studies there are two aspects where it can be used to answer synoptic questions relevant to the cereal ecosystem. These are in investigations concerning cereal aphids which are amongst the most important group of pest insects on cereals (Carter & Plumb, pp. 193–208, Wratten & Powell, pp. 233–257, this volume) and more generally to provide some information on the effect of agricultural intensification on non-pest species.

The data have already been used extensively to study the migrations of cereal aphids and related questions in cereal virus epidemiology (e.g. Dewar, Tatchell & Turl 1984; Tatchell, Plumb & Carter 1988) but a question that is synoptic and yet slightly removed from the normal agricultural perspective is to consider the total annual aerial populations and ask if numbers of cereal aphids have increased or decreased since adequate monitoring began. Here is a good example of an analysis of a process with high variability as aphids have very large potential rates of increase and year-to-year variation can be considerable. Twenty years is about the minimum run required to extract trends although longer sets may be necessary for certain types of time-series analysis. Figures 14.2a – g show logarithmic plots of annual totals for the seven cereal aphid species for the eight sites throughout Britain which have run consistently for the 20 years from 1969 to 1988.

A visual inspection of these plots reveals some immediate points of interest. Thus *M. dirhodum* (Fig. 14.2a) is considerably more variable in numbers than the other species, with populations fluctuating over at least four orders of magnitude. This ties in with the known erratic nature of the pest status of this species and the outbreak years of 1979 in East Anglia and 1982 in Scotland can be seen clearly. These outbreaks are thus examples of rare events and the two occurrences since monitoring began have enabled tentative explanations to be provided, although further observation will be required for confirmation and as a basis for forecasting (Dewar, Tatchell & Turl 1984).

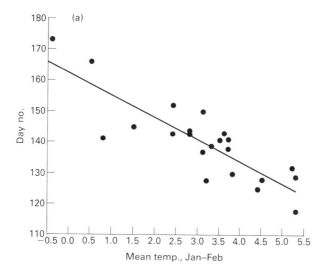

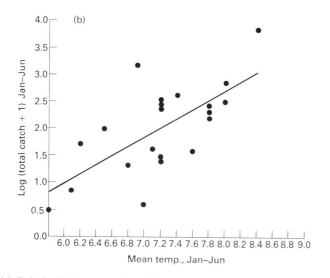

FIG. 14.1(a) Relationship between time of first flight of *Sitobion avenae* and the preceding January and February mean temperature at Rothamsted 1968–89. 70% of the variation is accounted for by the regression. (b) Relationship between total catch of *Sitobion avenae* in the Rothamsted suction trap between January and June and the corresponding mean temperature 1968–88. 45% of the variation is accounted for by the regression.

Another interesting feature is revealed in the graph for *Rhopalosiphum maidis* (Fitch) (Fig. 14.2d). This is the rarest of the British cereal aphids and is a characteristic species of southern Europe. It shows a curious alternating annual cycle with a much higher degree of synchronization between sites than is apparent in most of the other species. Very little is known about the ecology of this species in Britain because it is not considered of economic importance and its low population levels make it difficult to study, however, its status could change if climatic warming occurred. Similar patterns have been observed in the migrant moth species *Autographa gamma* (Taylor 1986) and it is tempting to suggest that it may be a feature of species unable to overwinter in Britain, which immediately poses a question about the overwintering status of *R. maidis* for which we presently have no answer. Perhaps this is a good example of a long-running synoptic dataset producing interesting questions to be answered by more detailed studies.

Analysis of variance and analysis of parallelism show that all species have significant site and year effects but that only three species have significant trends across the years. In one species *Rhopalosiphum insertum* (Walker) (Fig. 14.2c), the trend is curvilinear but this appears to be caused by a linear downward trend between 1969 and 1976 followed by a period with a higher mean but no trend after that date.

The other two species with linear trends are *Metopolophium festucae* (Theobald), Fig. 14.2b, where the upward trend is slight and only just significant ($P < 0.05$). However, *Sitobion avenae* (Fabricius), Fig. 14.2f, exhibits a highly significant linear upward trend ($P < 0.001$) and analysis of parallelism fails to detect any significant difference between sites in this trend. This result is of particular interest because *S. avenae* is the main pest of cereal crops in summer and control measures are regularly applied against it. The regression equation for the observed trend reveals that on average *S. avenae* was approximately twice as numerous in 1988 as it was 20 years earlier corresponding to an annual average rate of increase of 3·5%.

Two further points are worth noting about this result. Firstly 20 years is probably about the minimum viable length of time required to start asking questions about such long-term trends, the *R. insertum* example providing a warning about the danger of studying trends on short data runs. The second point is the one of the scale of investigation and comes out of the fact that trends in numbers of aphids over 20 years from the North Farm study in Sussex (Aebischer, this volume, pp. 314–315) are in the opposite direction to the ones described here from the Insect Survey suction trap network. This difference is very intriguing and requires further analysis. A

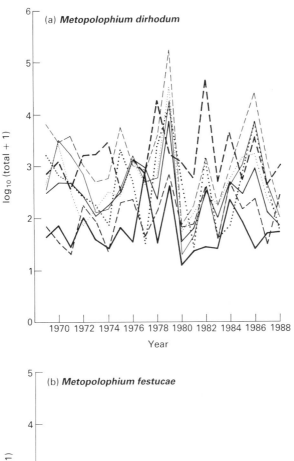

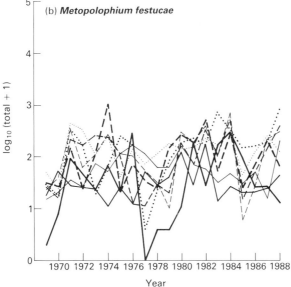

FIG. 14.2. Annual totals of seven species of cereal aphids from eight Survey suction traps (represented by the different lines) from 1969 to 1988. (a) *M. dirhodum;* (b) *M. festucae;* (c) *R. insertum;* (d) *R. maidis;* (e) *R. padi;* (f) *S. avenae;* (g) *S. fragariae.*

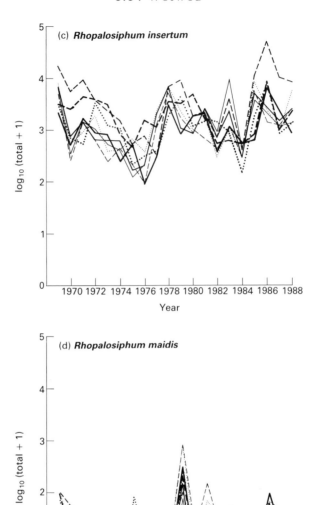

FIG. 14.2. (*Cont.*)

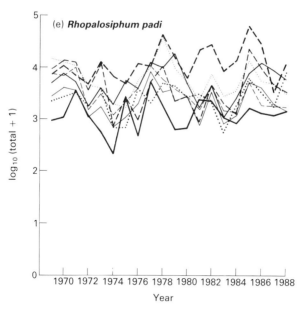

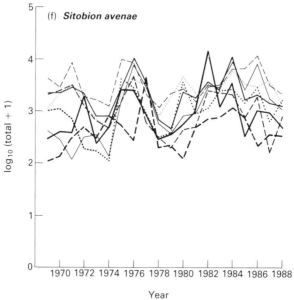

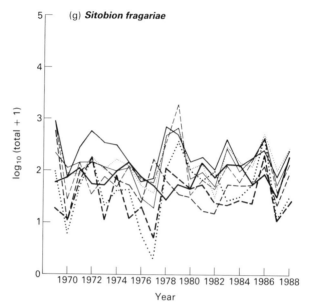

FIG. 14.2. (*Cont.*)

possible explanation is provided by the changes in agricultural practice which are outlined by Aebischer, for example, the big change from spring-sown to winter-sown cereals. This may provide larger areas of crops suitable for aphid multiplication, but is also likely to have altered the phenological relationship between the crop and the pest. The samples in suction traps integrate aerial aphid populations over a large area and are thus representing a different scale of sampling. As mentioned previously it is very difficult to predict from one scale to another but studies at all scales are required to gain a full understanding of how ecosystems function.

Synoptic studies often feature samples taken spatially on an irregular grid. The distance and direction of samples from each other is often relevant but very difficult to include in formal analyses. An example of this problem is apparent in Figs 14.2a – f, where the spatial element is inherent in the data but impossible to disentangle by inspection of the graphs. One approach that has proved useful for Survey data is the use of computer-produced density distribution maps (Woiwod & Tatchell 1984; Cammell, Tatchell & Woiwod 1989). An example of such maps is given in Fig. 14.3 which show *M. dirhodum* for 10 of the 20 years used in Fig. 14.2a. These maps cover both of the outbreak years 1979 and 1982 and different spatial aspects of these outbreaks are clearly visible from these maps.

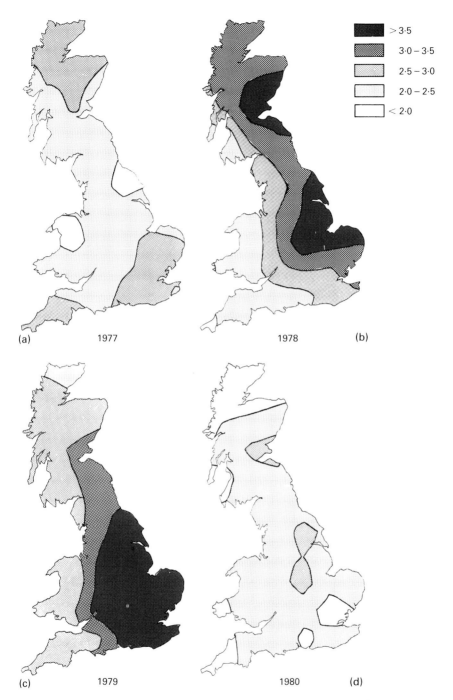

FIG. 14.3. Annual density distribution maps ($\log_{10}$ numbers trapped per year) of *Metopolophium dirhodum* for the 10 years 1977–86.

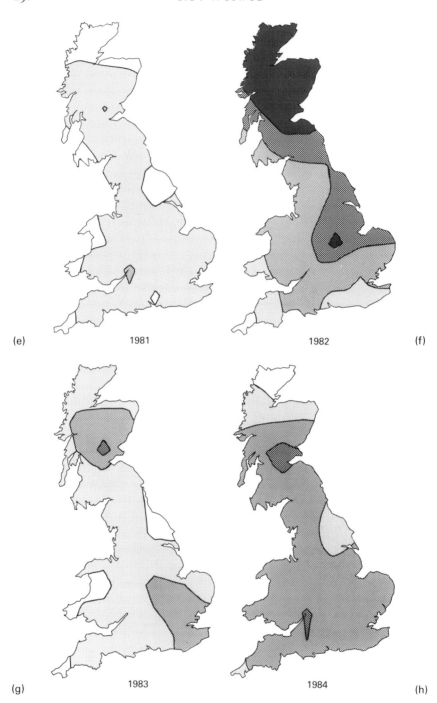

(e)　　　　　　　　1981　　　　　　　　1982　　　　　　　　(f)

(g)　　　　　　　　1983　　　　　　　　1984　　　　　　　　(h)

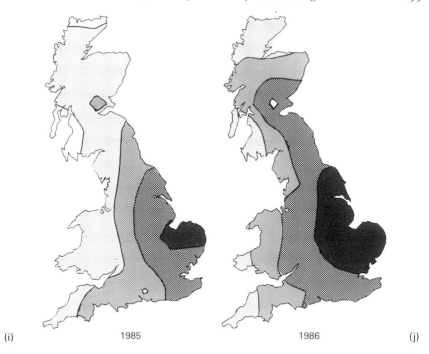

(i)   1985   1986   (j)

A final question that concerns the effect of agricultural intensification on non-pest insect species can be addressed by data from the Insect Survey. For this we need to examine some of the moth data gathered by the Survey's light-trap network. Generally moths are only sporadic pests of arable crops so only a small proportion of the Survey's light-trap samples have been in arable situations. However, one of these was the first trap of the network and was placed by the side of one of Rothamsted's classical experiments, Barnfield, in 1960 where a similar trap had been operated by C.B. Williams from 1933 for 4 years and again for a 4-year period in the 1940s. This set of data is a unique sample covering as it does a large group of insects over a period of 57 years.

The changes in moth diversity at this site are shown in Fig. 14.4 together with two other long-running sites on the Rothamsted estate, Geescroft Wilderness, a stable woodland site, and Allotments, a site subjected to large changes in local environment (Taylor 1986). The most interesting feature of this set of data is the rather surprising observation that the main changes in the Barnfield catches took place in the 1950s. Unfortunately no sampling was carried out in this period enabling us to follow changes in detail and again because of the historical perspective it is

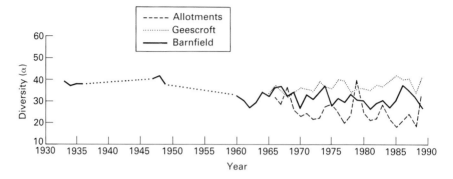

FIG. 14.4 Annual diversity (α from log-series distribution) of the macrolepidoptera from three long-running light traps at Rothamsted.

impossible to go back and repeat the study. All we can hope for is better vigilance in respect of studying future changes.

The changes in the 1950s are reflected in species richness and total moth populations; pre-1950 levels only appear to be found now in the farm woodland site. It might be argued that the Rothamsted result is atypical of the agricultural situation more generally and it is true that there is no spatial replication for this particular result. However, Table 14.2 gives geometric mean annual numbers of moths caught for all the Survey data from arable areas. For this purpose a sample is only considered to be arable if the trap is either on a field margin or within a few metres of an arable field. As can be seen, arable field samples throughout Britain and post-1960 Barnfield samples are not significantly different ($P > 0.10$) in their annual productivity suggesting that the Barnfield results may not be too atypical. Grassland sites tend to be significantly more productive

TABLE 14.2. Average annual catches of macrolepidoptera from the Insect Survey light trap network

Type of site	No. of sites	Geometric mean catch
Arable sites	20	1176
Grassland sites	13	2105
All sites	314	2638
Barnfield 1933–1950	1 (7 years)	3910
Barnfield 1960–1989	1 (30 years)	1280

All differences are significant ($P < 0.05$) except arable and post-1960 Barnfield, and grassland v. all sites.

but not significantly below the average for all sites. However, pre-1950 Barnfield was consistently more productive than the average ($P < 0.05$). It will be very interesting to see what changes will occur if agriculture becomes less intensive in the future.

CONCLUSIONS

I hope from what has been said and the examples given that the vital importance of long-term and spatially replicated studies is apparent. I would also like to think that the idea that long-term studies can be replaced by short-term experimental work will soon be a thing of the past. The two approaches are complementary rather than mutually exclusive and specific experimental studies centred around long-term studies seem to be particularly productive in ecology (Taylor 1987).

Long-term studies will never be easy to start, fund, or manage but without them whole areas of ecological ignorance will remain. It is perhaps an encouraging sign that in Britain, at least, ecological monitoring has jumped from low to high priority in just a few years and that there are proposals to set up long-term ecological monitoring sites in Britain with Rothamsted's classical experiments as part of this network. It is also appropriate to remind readers again that such studies started almost 150 years ago, and that studies on the cereal ecosystem played an important part in their inception.

REFERENCES

Anon (1984). *Guide to the Classical Field Experiments.* Lawes Agricultural Trust, Harpenden.

Barnes, H.F. (1958). Wheat blossom midges on Broadbalk, Rothamsted Experimental Station, 1927–1956. *Proceedings X International Congress of Entomology,* **3**, 367–374.

Bawden, F.C. (1969). *The Broadbalk Wheat Experiment.* Rothamsted Experimental Station Report for 1968, Part 2. Lawes Agricultural Trust, Harpenden.

Brenchley, W.E. & Adam, E. (1915). Recolonisation of cultivated land allowed to revert to natural conditions. *Journal of Ecology,* **3**, 193–210.

Buol, S.W., Hole, F.D. & McCracken (1973). *Soil Genesis and Classification.* The Iowa State University Press, Ames.

Burnham, P. (1989). Returning set-aside land to nature. *Ecos,* **10**, 13–17.

Cammell, M.E., Tatchell, G.M. & Woiwod, I.P. (1989). Spatial pattern of abundance of the black bean aphid, *Aphis fabae*, in Britain. *Journal of Applied Ecology,* **26**, 463–472.

Connell, J.H. & Sousa, W.P. (1983). On the evidence needed to judge ecological stability or persistence. *The American Naturalist,* **121**, 789–824.

Coope, G.R. (1977). Fossil coleopteran assemblages as sensitive indicators of climatic changes during the Devensian (last) cold stage. *Philosophical Transactions of the Royal Society of London,* B, **280**, 313–340.

Davies, M.S. & Snaydon, R.W. (1973). Physiological differences among populations of

Anthoxanthum adoratum L. collected from the Park Grass Experiment, Rothamsted. I. Response to calcium. *Journal of Applied Ecology*, **10**, 33–45.

Dewar, A.M., Tatchell, G.M. & Turl, L.A.D. (1984). A comparison of cereal-aphid migrations over Britain in the summers of 1979 and 1982. *Crop Protection*, **3**, 379–389.

Dewar, A.M., Woiwod, I.P. & Choppin de Janvry, E. (1980). Aerial migrations of the rosegrain aphid, *Metopolophium dirhodum* (Wlk.), over Europe in 1979. *Plant Pathology*, **29**, 101–109.

Fisher, R.A. (1924). The influence of rainfall on the yield of wheat at Rothamsted. *Philosophical Transactions of the Royal Society*, B, **213**, 89–142.

Franklin, J.F. (1987). Importance and justification of long-term studies in ecology. *Long-term Studies in Ecology* (Ed. by G.E. Likens,), pp. 3–19. Springer Verlag, New York.

Gilbert, J.H. (1895). Agricultural investigations at Rothamsted, England, during a period of fifty years. *USAD Bulletin No. 22*, 1–316.

Harrington, R. (1989). Greenhouse greenfly? *Antenna*, **13**, 169–172.

Jenkinson, D.S. & Parry, L.C. (1989). The nitrogen cycle in the Broadbalk wheat experiment: a model for the turnover of nitrogen through the soil microbial biomass. *Soil Biology and Biochemistry*, **21**, 535–541.

Jenkinson, D.S. & Rayner, J.H. (1977). The turnover of soil organic matter in some of the Rothamsted classical experiments. *Soil Science*, **123**, 298–305.

Johnson, C.G., Lofty, J.R. & Cross, D.J. (1969). Insect pests on Broadbalk. *The Broadbalk Wheat Experiment* (Ed. by F.C. Bawden), pp. 141–156. Lawes Agricultural Trust, Harpenden.

Jones, K.C. & Johnston, A.E. (1989). Cadmium in cereal grain and herbage from long-term experimental plots at Rothamsted. *Environmental Pollution*, **57**, 199–216.

Jones, K.C., Symon, C.J. & Johnston, A.E. (1987). Retrospective analysis of an archived soil collection. II. Cadmium. *Science of the Total Environment*, **67**, 75–89.

Lawes, J.B. (1847). On agricultural chemistry. *Journal of the Royal Agricultural Society of England*, **8**, 226–260.

Lawes, J.B. & Gilbert, J.H. (1851). Agricultural chemistry, especially in relation to the mineral theory of Baron Leibig. *Journal of the Royal Agricultural Society of England*, **12**, 1–40.

Lawes, J.B. & Gilbert, J.H. (1859). Report of experiments with different manures on permanent meadow land. *Journal of the Royal Agricultural Society of England*, 1st Series, **19**, 552–573.

Lawes, J.B. & Gilbert, J.H. (1895). The Rothamsted experiments over fifty years. *Transactions of the Highlands and Agricultural Society of Scotland Series 5*, 11–354.

Leibig, J. (1840). *Organic Chemistry and Physiology*. Taylor & Walton, London.

Likens, G.E. (1983). A priority for ecological research. *Bulletin of the Ecological Society of America*, **64**, 234–243.

Likens, G.E. (Ed.) (1987). *Long-term Studies in Ecology, Approaches and Alternatives*. Springer Verlag, New York.

MacArthur, R.H. & Wilson, E.O. (1967). *The Theory of Island Biogeography*. Princeton University Press, Princeton.

Powlson, D.S., Pruden, G., Johnston, A.E. & Jenkinson, D.S. (1986). The nitrogen cycle in the Broadbalk Wheat Experiment: recovery and losses of ^{15}N-labelled fertilizer applied in spring and inputs of nitrogen from the atmosphere. *Journal of Agricultural Science, Cambridge*, **107**, 591–609.

Prew, R.D., Church, B.M., Dewar, A.M., Lacey, J., Penny, A., Plumb, R.T., Thorne, G.N., Todd, A.D. & Williams, T.D. (1983). Effects of eight factors on the growth and nutrient uptake of winter wheat and on the incidence of pests and diseases. *Journal of Agricultural Science, Cambridge*, **100**, 362–382.

Schaffer, W.M. (1984). Stretching and folding in lynx fur returns: evidence for a strange attractor in nature? *American Naturalist*, 124, 798–820.

Silvertown, J. (1987). Ecological stability: a test case. *The American Naturalist*, 130, 807–810.

Snaydon, R.W. & Davies, M.S. (1976). Rapid population differentiation in a mosaic environment. IV. *Anthoxanthum odoratum* L. at sharp boundaries. *Heredity*, 36, 9–25.

Strayer, D., Glitzenstein, J.S., Jones, C.G., Kolasa, J., Likens, G.E., McDonnell, M.J., Parker, G.S. & Pickett, S.T.A. (1986). *Long-term Ecological Studies*. Institute of Ecosystem Studies, New York.

Tatchell, G.M. (1985). Aphid control advice to farmers and the use of aphid monitoring data. *Crop Protection*, 4, 35–50.

Tatchell, G.M., Plumb, R.T. & Carter, N. (1988). Migration of alate morphs of the bird cherry aphid (*Rhopalosiphum padi*) and implications for the epidemiology of barley yellow dwarf virus. *Annals of Applied Biology*, 112, 1–11.

Taylor, L.R. (1986). Synoptic dynamics, migration and the Rothamsted Insect Survey. *Journal of Animal Ecology*, 55, 1–38.

Taylor, L.R. (1987). Objective and experiment in long-term research. *Long-term Studies in Ecology* (Ed. by G.E. Likens), pp. 20–70. Springer Verlag, New York.

Taylor, L.R. & Woiwod, I.P. (1980). Temporal stability as a density-dependent species characteristic. *Journal of Animal Ecology*, 49, 209–224.

Taylor, L.R., French, R.A. & Woiwod, I.P. (1978). The Rothamsted Insect Survey and the urbanization of land in Great Britain. *Perspectives in Urban Entomology* (Ed. by G.W. Frankie & C.S. Koehler), pp. 31–65. Academic Press, New York.

Taylor, L.R., French, R.A., Woiwod, I.P., Dupuch, M.J. & Nicklen, J. (1981). Synoptic monitoring for migrant insect pests in Great Britain and Western Europe. I. Establishing expected values for species content, population stability and phenology of aphids and moths. *Rothamsted Experimental Station. Report for 1980*, Part 2, 41–104.

Taylor, L.R., Woiwod, I.P. & Perry, J.N. (1980). Variance and the large scale spatial stability of aphids, moths and birds. *Journal of Animal Ecology*, 49, 831–854.

Taylor, L.R., Woiwod, I.P., Tatchell, G.M., Dupuch, M.J. & Nicklen, J. (1982). Synoptic monitoring for migrant insect pests in Great Britain and Western Europe. III. The seasonal distribution of pest aphids and the annual aphid aerofauna over Great Britain 1975–80. *Rothamsted Experimental Station. Report for 1981*, Part 2, 23–121.

Tilman, D. (1982). *Resource Competition and Community Structure*. Princeton University Press, Princeton.

Tilman, D. (1987). Ecological experimentation: strengths and conceptual problems. *Long-term Studies in Ecology* (Ed. by G.E. Likens), pp. 136–157. Springer Verlag, New York.

Wiens, J.A. (1989). Spatial scaling in ecology. *Functional Ecology*, 3, 385–397.

Williams, E.D. (1978). Botanical composition of the park grass plots. *Rothamsted Experimental Station, Report for 1977*, Part 2, 31–36.

Williamson, M. (1989). The MacArthur and Wilson theory today: true but trivial. *Journal of Biogeography*, 16, 3–4.

Woiwod, I.P. & Dancy, K.J. (1987). Synoptic monitoring for migrant insect pests in Great Britain and Western Europe. VII. Annual population fluctuations of macrolepidoptera over Great Britain for 17 years. *Rothamsted Experimental Station, Report for 1986*, Part 2, 235–262.

Woiwod, I.P. & Tatchell, G.M. (1984). Computer mapping of aphid abundance. *1984 British Crop Protection Conference — Pest and Diseases*, 675–683. British Crop Protection Council, Farnham.

Woiwod, I.P., Tatchell, G.M. & Barrett, A.M. (1984). A system for the rapid collection, analysis and dissemination of aphid-monitoring data from suction traps. *Crop Protection*, 3, 273–288.

Woiwod, I.P., Tatchell, G.M., Dupuch, M.J., Macaulay, E.D.M., Parker, S.J., Riley, A.M. & Taylor, M.S. (1989). *Rothamsted Insect Survey Twentieth Annual Summary*. Lawes Agricultural Trust, Harpenden.

Yates, F. (1969). Investigations into the effect of weather on yield. *The Broadbalk Wheat Experiment* (Ed. by F.C. Bawden), pp. 46–49. Lawes Agricultural Trust, Harpenden.

15. TWENTY YEARS OF MONITORING INVERTEBRATES AND WEEDS IN CEREAL FIELDS IN SUSSEX

N.J. AEBISCHER

The Game Conservancy Trust, Fordingbridge, Hampshire SP6 1EF, UK

INTRODUCTION

Farming methods are in a state of rapid change. Over the past 20 years, the use of fertilizers and pesticides has increased, traditional ley farming and crop rotations have been largely abandoned, minimum cultivation or direct drilling have come and mostly gone again, hedgerows have been removed and fields made larger in response to greater mechanization (Sturrock & Cathie 1980; Davies 1984; Barr *et al.* 1986; Rands, Hudson & Sotherton 1988). A number of arable farms have changed from a patchwork of small fields in diverse crops to an intensive monoculture of winter wheat spanning hundreds of hectares (Pollard, Hooper & Moore 1974). Such dramatic changes must have had many effects upon the distribution and abundance of wild plants and animals — either vertebrate or invertebrate — which coexist with agriculture and, in many ways, have become dependent upon it (Bunyan & Stanley 1983; Chancellor, Fryer & Cussans 1984; Edwards 1984; O'Connor & Shrubb 1986).

In order to detect and quantify such changes, it is necessary to have data collected in the same way over a sufficient number of years to encompass the major phases of agricultural change. The Game Conservancy has monitored levels of invertebrate abundance and various other aspects of the cereal ecosystem, as well as crop management, every year from 1970 onwards in West Sussex (e.g. Potts & Vickerman 1974; Potts 1984). This ongoing study thus now spans exactly 20 years. Based on the monitoring, this paper highlights changes that have taken place in the cereal ecosystem from 1970 to 1989 inclusive; it investigates relationships within the system and the effects of changes in farming techniques.

METHODS

Study area and data collection

The Sussex study area lies across 62 km^2 of the South Downs between the river Arun in the west and the river Adur in the east (Potts & Vickerman

1974). Every year, The Game Conservancy has monitored invertebrates and weeds in the second half of June from approximately 100 cereal fields spread across fourteen farms. To ensure comparability between years, analyses in this paper were restricted to a core area of five farms (28 km²) which were consistently and comprehensively sampled over all years (Table 15.1).

The invertebrate samples were obtained using a 'D-vac' vacuum insect sampler (Dietrick 1961); they consisted of five subsamples, each of 0·092 m², taken at 5 m intervals along a diagonal transect into the field. The invertebrates in the samples were subsequently sorted into taxonomic groups in the laboratory. At the same time as the invertebrates were sampled, records were taken of crop type, amount of grass and of broad-leaved weeds present in the crop (on an ordinal scale of 0–5), species of grass and of broad-leaved weeds in the crop, and any other factors (such as undersowing) which were relevant.

TABLE 15.1. Median sampling date, total number of fields sampled for invertebrates, and numbers of fields sampled for invertebrates on the five farms in the core Sussex study area from 1970 to 1989

Year	Sampling date	Total fields	Core fields	Fields sampled per farm				
				F1	F2	F3	F4	F5
1970	17/6	143	123	28	9	55	23	8
1971	22/6	107	86	11	9	41	19	6
1972	28/6	99	78	19	10	31	13	5
1973	28/6	155	107	34	11	38	14	10
1974	22/6	154	100	30	13	33	13	11
1975	23/6	139	94	29	11	28	14	12
1976	21/6	121	67	27	6	16	8	10
1977	20/6	82	69	16	7	31	6	9
1978	19/6	94	78	27	6	28	9	8
1979	18/6	96	82	22	9	33	13	5
1980	18/6	100	83	28	7	27	13	8
1981	17/6	106	89	27	6	38	10	8
1982	18/6	96	83	23	6	29	17	8
1983	20/6	104	84	23	11	30	13	7
1984	19/6	99	80	22	14	25	12	7
1985	17/6	100	82	22	14	24	12	10
1986	18/6	99	86	26	17	20	16	7
1987	18/6	100	83	26	14	17	16	10
1988	16/6	100	82	23	13	20	17	9
1989	16/6	85	79	23	13	16	19	8

Data analysis

In 1970 and 1971, the invertebrates were sorted from the D-vac samples by eye for identification under a binocular microscope; in subsequent years, a microscope was used for sorting as well as for identification. To compensate for the less efficient sorting technique, mean invertebrate densities for the two early years were amplified by a correction factor (varying between taxonomic groups) calculated from nine randomly selected samples sorted using both methods. Correction factors and standard errors were derived from these samples for twenty invertebrate groups covering all taxa sampled; the correction factors varied from 1·05 for ground beetles, which were large and obvious (95% sorting efficiency), to 2·29 for apterous aphids (44% efficiency). All samples were preserved in alcohol for future reference. In particular, this enabled the annual numbers of springtails, not counted originally, to be assessed retrospectively as far back as 1972 by inspection of fifty randomly selected samples per year.

Because of considerable individual variation between samples, analyses were based on arithmetic averages of the samples by farm and by year. The use of the arithmetic mean maintained the additivity of the data (even if a correction factor had been used), thereby facilitating comparisons between adults and larvae or between species, and enabling them to be grouped if necessary. For the sake of clarity in the figures, we have plotted means without confidence limits. An idea of their accuracy may be obtained from Table 15.2.

Thus the accuracy of means calculated for the more abundant species was good; that for less common ones was acceptable. Arithmetic means are

TABLE 15.2. The average ratio of standard error over mean for aphids, one of the most abundant groups discussed in this paper, and for *Demetrias atricapillus*, one of the least numerous species mentioned

	Average density	Average SE	Ratio SE/mean
Based on all farms pooled			
Aphids	137.3	16.2	0.118
Demetrias atricapillus	0.253	0.057	0.225
Based on individual farms			
Aphids	148.3	33.2	0.224
Demetrias atricapillus	0.239	0.112	0.469

more likely to be affected by outlying values than geometric ones, and the accuracy of any mean depends upon the number of samples from which it was calculated. For the purposes of regression analyses and analyses of variance, the means calculated here were therefore transformed to logarithms (base 10) and weighted by sample size. Consistent upward or downward trends in the means through time were detected using linear regression (Kendall 1976). In the case of log-transformed means, the slope of the regression line gave the instantaneous rate of change (r); the corresponding annual rate of change was calculated as $10^r - 1$.

RESULTS

Crops and farming methods

Crops

Since 1970, seven different types of cereal crop were sampled for invertebrates in the Sussex study area: winter wheat, spring wheat, winter barley, spring barley, winter oats, spring oats and triticale. As cereals were sampled according to their availability (with the exception of maize, present in some years in one or two fields but not suitable for D-vac sampling), the changes in the proportions of each cereal in the annual samples reflected the changes in the proportions of each cereal grown in the study area (Fig. 15.1). Spring oats and triticale have been omitted from Fig. 15.1 because they constituted less than 1% of cereals sampled. The changes in the proportions of the other cereals since 1970 have, for some, been dramatic. Spring barley was the dominant cereal grown in the early 1970s (c. 60% of cereal fields), but declined to under 25% of cereal fields by the late 1980s. The situation with winter wheat was the opposite: since 1986, over 60% of cereal crops were winter wheat, compared to around 25% at the start of the study. Winter barley constituted less than 10% of cereal fields for the first 5 years of the study, then rose to around 20% from 1979 to 1985 before declining again. Winter oats averaged around 5% of cereals grown until 1983, but has not been grown on the study area since. Spring wheat has always been an uncommon crop, sown on under 4% of cereal fields.

These changes are similar to the national picture: winter wheat was grown on 35% of arable land in the United Kingdom in 1974, and the proportion increased to 50% in 1987; the area of winter barley likewise increased from 7 to 25%, whereas during the same period spring barley declined nationally from 49% of arable land to 22% (Anon 1986, 1989).

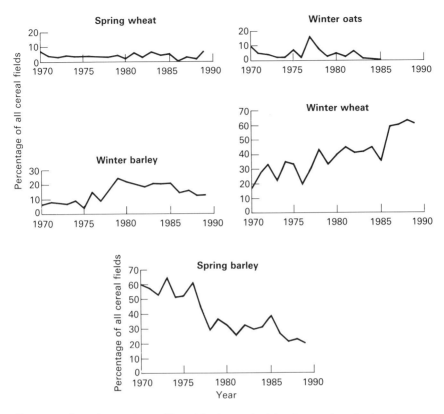

FIG. 15.1. Annual percentages of invertebrate samples taken from spring wheat, spring barley, winter wheat, winter barley and winter oats in the core Sussex study area from 1970 to 1989.

Crop management

The farming practices adopted by the five farms on the core study area have also changed since 1970 (Table 15.3). Farm 1 (856 ha) has sustained a modern system of non-rotating arable and grass, with large-scale hedge removal and field enlargement taking place at three distinct periods. Farm 2 (420 ha), originally managed traditionally, switched to arable and grass from 1973 onwards, with a period of minimum cultivation in the early 1980s. Farm 3 (856 ha) reverted in the mid 1970s from its original all-arable system of farming to one of non-rotating arable and grass. Farm 4 (393 ha) kept up a traditional ley rotation until the end of the 1980s. It is now given over entirely to winter wheat and has the highest pesticide input on the

TABLE 15.3. Major changes in farming practices on the five farms in the core Sussex study area over the period 1970–89. The negative percentages given correspond to hedgerow removal

	Farm 1	Farm 2	Farm 3	Farm 4	Farm 5
Start of study					
Kilometres of hedgerow per km^2	6·0	5·8	6·2	6·9	4·3
Average field size (ha)	11·0	17·5	10·9	12·5	12·1
1970	AG	TL	AA	TL	TL
1971	AG −7%	TL	AA	TL	TL
1972	AG	TL	AA	TL	TL
1973	AG	TL	AA	TL	TL
1974	AG	AG	AA	TL	TL
1975	AG	AG	AA	TL	TL
1976	AG	AG	AG	TL	TL
1977	AG −18%	AG	AG	TL	TL
1978	AG	AG	AG	TL	TL
1979	AG	AG*	AG	TL	TL
1980	AG	AG*	AG	TL	TL
1981	AG	AG*	AG	TL	TL
1982	AG	AG*	AG	TL	TL
1983	AG	AG*	AG	AG −7%	TL
1984	AG	AG	AG	AG	TL
1985	AG −5%	AG	AG	AG −3%	TL
1986	AG	AG	AG	MW	TL
1987	AG	AG	AG	MW −8%	TL
1988	AG	AG	AG	MW −9%	TL
1989	AG	AG	AG	MW	TL
End of study					
Kilometres of hedgerow per km^2	4·3	5·8	6·2	5·2	4·3
Average field size (ha)	14·3	17·5	12·4	15·3	13·5

* Minimum cultivation or zero tillage.
AA, all arable crops with no grass, stubbles burnt.
AG, arable and grass not necessarily rotating, no undersowing (or not undersown traditionally), winter-wheat stubbles burnt, others baled.
MW, monoculture winter wheat, stubbles burnt.
TL, traditional ley rotation with undersowing and bastard fallow, no stubble burning.

study area; removal of hedgerows and grass banks has been considerable in recent years. This is the one farm which, since 1984, has used the sterile strip system, i.e. maintained a strip of bare ground round each field by, in this instance, herbicide use. Farm 5 (257 ha) has maintained a traditional outlook throughout the period of the study, with mixed farming and crop rotations; however, it was the farm with the least hedgerows. The five farms have therefore differed widely in their manner of farming, both within and between farms, taking in most of the important changes in arable farming over the past 3 decades (Raymond 1984).

The use of cereals as a nurse crop for grass, known as undersowing, typifies the agricultural changes that have taken place on the study area since 1970 (Fig. 15.2). Although practised originally on four out of the five farms, it is now carried out only on Farm 5, where it has been consistent throughout the study. On the other farms, the decline of undersowing has been complete: none has taken place since 1985. The overall result is that only 2% of cereal fields on the study area are now undersown, compared to about 10% at the start of the study.

Arable weeds

The overall abundances of grass weeds and of broad-leaved weeds in the crop were scored on a scale of 0 (no weeds) to 5 (completely dominated by weeds). The average annual abundances of these two major types of arable weed are given in Fig. 15.3. Despite a coincidence in peaks in 1971 and in 1985, and in troughs in 1970 and in 1976, there was no significant correla-

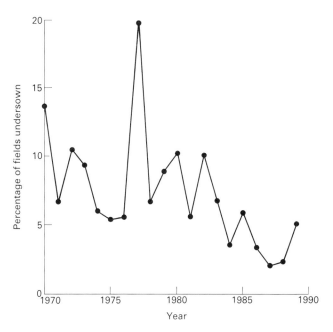

FIG. 15.2. Annual percentage of sampled cereal fields in the Sussex study area which were undersown, from 1970 to 1989.

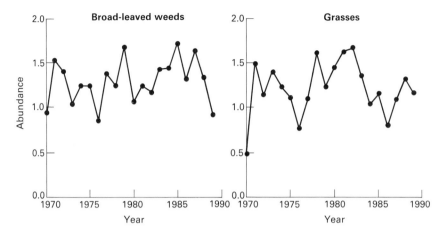

FIG. 15.3. Annual mean index of abundance for broad-leaved weeds (left) and for grass weeds (right) in cereal crops in the core Sussex study area from 1970 to 1989.

tion between the year-to-year abundances of the two weed types. There was no evidence that the abundance of grass weeds had increased or decreased over the 20 years of the study. Likewise, the overall abundance of broad-leaved weeds did not show any particular trend during the study. Bearing in mind that the Sussex study began after the introduction and widespread use of herbicides, this is broadly in line with the results of Whitehead & Wright (1989) who noted that the most abundant weeds have remained unchanged for over 20 years.

The measures of abundance used above covered a wide range of weed species, and a lack of overall pattern did not imply that the abundance of individual species had not varied in the course of the study. Although no abundance index was available for individual grass or broad-leaved weed species, a list of species present at each sampling site was recorded. This showed that, for example, the percentage occurrence of barren brome *Bromus sterilis* in the sampled fields had increased dramatically (Fig. 15.4). Barren brome had been extremely rare in cereal fields — though common in hedgerows — until 1979, but it is now present in over 30% of cereal fields in the study area. None of the infested fields belonged to the traditionally managed Farm 5. A broad-leaved weed which has increased markedly, though less so than barren brome, was cleavers *Galium aparine* (Fig. 15.4). During the first years of the study, it was present on an average of 3% of cereal fields; it now occurs in over 14% of them. The infestation was half as severe on the traditionally managed farm as on the rest of the

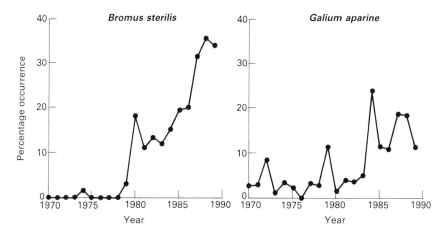

FIG. 15.4. Annual percentages of cereal fields where barren brome *Bromus sterilis* (left) and cleavers *Galium aparine* (right) were recorded during invertebrate sampling in the core Sussex study area from 1970 to 1989.

area. Again, these were changes which have been observed on a national scale, and related to changes in tillage, straw-burning, and a trend towards early-sown autumn cereals (Froud-Williams & Chancellor 1982; Froud-Williams 1983; Wilson & Froud-Williams 1988).

Total invertebrates

Overall figures for invertebrates sampled in cereals in Sussex were available from 1972 onwards (no measure of springtail — Collembola — abundance in 1970/71), and excluded mites (Acari) whose abundance had not been assessed. The overall abundance of invertebrates (excluding mites) showed little change between 1972 and 1989 (Fig. 15.5). However, almost a third (32%) of this total was made up of springtails, whose annual density increased significantly ($F_{1,16} = 5.19$, $P < 0.05$) in the course of the study (Fig. 15.5). The numbers of springtails recorded per sample collapsed during the dry summers of 1975 and 1976; the pattern of their abundance in subsequent years suggests that this group of invertebrates took 3–5 years to recover. After deducting springtails from the overall invertebrate total, the latter showed a significant decline of 4·2% per year ($F_{1,18} = 6.61$, $P < 0.05$), equivalent approximately to a halving of abundance over 20 years (Fig. 15.5).

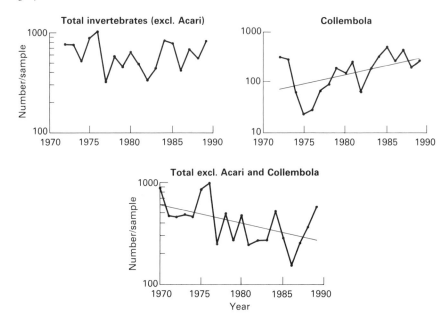

FIG. 15.5. Annual mean totals per sample (logarithmic scale) for all invertebrates excluding Acari (top left), for Collembola (top right) and for invertebrates excluding Acari and Collembola (bottom) from 1970 to 1989 in the Sussex study area. Collembola density increased significantly in the course of the study, with a slope of 0·036 ± 0·016 (r_{16} = 0·495, $P < 0·05$); after deducting Collembola, the density of remaining invertebrates (excluding Acari) declined significantly with a slope of −0·019 ± 0·007 (r_{18} = −0·518, $P < 0·05$).

Aphids

Annual changes in abundance

The mean density of aphids sampled in the third week of June declined significantly between 1970 and 1989 ($F_{1,98}$ = 25·0, $P < 0·001$), with no detectable between-farm difference in the annual rates of decline which averaged 8·4% (Fig. 15.6). Moreover, the same general pattern of year-to-year variation could be observed across the five farms: high numbers were sampled throughout the core study area in 1970, 1975, 1976, 1980, 1984 and 1989, and low numbers in 1979 and 1986. Exactly the same pattern of highs and lows was observed when the data were split according to crop type (Fig. 15.7); again, the rates of decline were similar for winter wheat, winter barley and spring barley, averaging 7·9% per year.

The decline rates calculated above imply that, on average, the present-day abundance of aphids is about a fifth of that recorded at the start of the

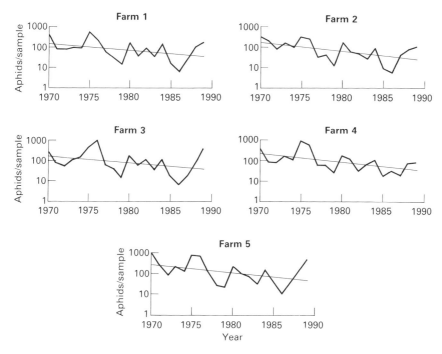

Fɪɢ. 15.6. Annual mean numbers (logarithmic scale) of aphids per sample from 1970 to 1989 on each of the five farms in the core Sussex study area. Aphid density decreased significantly in the course of the study, with no between-farm differences in the regression lines; the pooled slope was -0.038 ± 0.008 ($r_{98} = -0.450$, $P < 0.001$).

study. The decline was not caused by an increase in the use of aphicidal sprays shortly before sampling: the use of summer aphicides in the study area has been relatively uncommon, and far from consistent (Table 15.4). At most, 16% of cereal fields in the study area were treated with aphicide, in 1984, and 24% in 1989 (the latter concentrated on Farm 4). None was used in over half of the years of the study, including the 4 recent years 1985–88.

The next step is to examine the two major groups of invertebrate predators known to feed on aphids, as well as the group of wasps which parasitize aphids.

Aphid predators

The first group of predators are aphid-specific species, taken here to include the adults and larvae of ladybirds (Coleoptera: Coccinellidae),

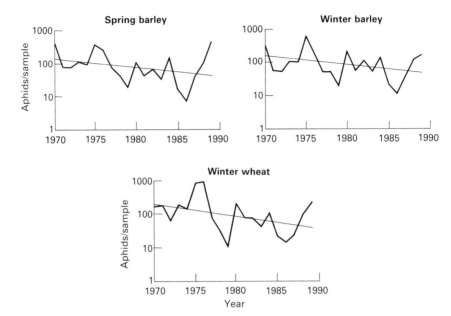

FIG. 15.7. Annual mean numbers (logarithmic scale) of aphids per sample from 1970 to 1989 in each of the three main cereal crops grown in the core Sussex study area (spring barley, winter barley, winter wheat). Aphid density decreased significantly in the course of the study, with no differences in the regression lines between crops; the pooled slope was -0.035 ± 0.010 ($r_{58} = -0.430$, $P < 0.001$).

soldier beetles (Coleoptera: Cantharidae), lacewings (Neuroptera) and hoverflies (Diptera: Syrphidae). On all farms, the numbers of aphid-specific predators caught per sample showed no significant trend (Fig. 15.8).

The second group is that of the polyphagous predators, comprising spiders (Araneae), harvestmen (Opiliones), ground beetles (Coleoptera: Carabidae), rove beetles (Coleoptera: Staphylinidae), earwigs (Dermaptera) and predatory flies (Diptera: Dolichopodidae, Empididae, Scathophagidae). Here the picture is quite different (Fig. 15.9): polyphagous predators have declined significantly on all farms ($F_{1,98} = 31.7$, $P < 0.001$). There was no difference in the rate of decline between farms, which averaged 3.9% per year.

The rate of change in aphid abundance was more pronounced than that of either group of aphid predators. As a result, it would be expected that the average aphid : predator ratio in the samples would decrease during the course of the study. Figure 15.10 presents the results for both groups of predators, combining data from the five farms for conciseness. For each predator type, the aphid : predator ratio declined significantly and at a

TABLE 15.4. Summer aphicides: total percentage of cereal fields treated on the Sussex study area from 1970 to 1989, and percentage treated before the fields were sampled for invertebrates

Year	Fields treated (%)	
	Throughout summer	Before sampling
1970	0	0
1971	<1	0
1972	0	0
1973	1	0
1974	0	0
1975	12	1
1976	7	4
1977	0	0
1978	0	0
1979	0	0
1980	1	1
1981	0	0
1982	1	1
1983	1	1
1984	16	16
1985	0	0
1986	0	0
1987	0	0
1988	0	0
1989	24	24

similar rate on each farm: on average by 6·9% per year for aphid-specific predators ($F_{1,98} = 20·6$, $P < 0·001$), and by 4·8% per year for polyphagous predators ($F_{1,98} = 7·13$, $P < 0·01$).

A decline in aphid : predator ratio is equivalent to an increase in the numbers of predators per aphid, i.e. in the predation pressure exerted upon the aphid population. Is this increase in predation pressure the cause of the aphid decline? If such were the case, aphid numbers would be negatively related to predator numbers. In actual fact, the relationship between aphids and aphid-specific predators was a significant positive one (Fig. 15.11), implying that aphid-specific predators were responding numerically to aphid abundance by the third week of June. The relationship between annual aphid density and polyphagous predators was also a positive one (Fig. 15.11), but not as clear-cut as for aphid-specific predators: it was significant only when calculated across the five farms instead of the pooled data ($F_{1,98} = 10·4$, $P < 0·01$). A forward stepwise multiple regression of aphid density against the densities of the six major taxa of polyphagous predators picked out only one as being significantly linearly

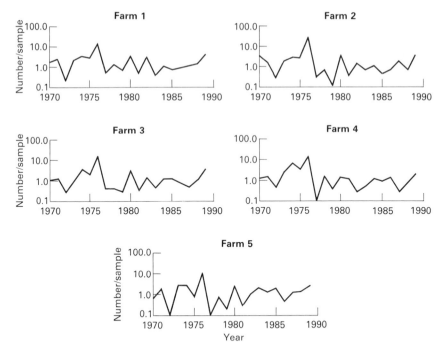

FIG. 15.8. Annual mean numbers (logarithmic scale) of aphid-specific predators per sample from 1970 to 1989 on each of the five farms in the core Sussex study area.

related to aphid density, namely spiders ($F_{1,98} = 21\cdot4$, $P < 0\cdot001$); again, the slope of the relationship was positive. There was therefore no evidence that any of the taxa of polyphagous predators were depressing aphid numbers by the second half of June.

Parasitoid wasps

The changes in the annual abundance of parasitoid wasps (Parasitica: Hymenoptera) are summarized in Fig. 15.12. Numbers decreased significantly across all farms, at the same average rate of $5\cdot9\%$ ($F_{1,98} = 11\cdot9$, $P < 0\cdot001$). Aphids which had been mummified were identified and counted separately from healthy aphids, and the annual percentages of mummified aphids varied from 0% in 1977 to 20·5% in 1974. The numbers of parasitoid wasps recorded in the samples each year were closely and positively correlated with the numbers of mummified aphids (Fig. 15.12). As mummified aphids were themselves correlated with total numbers of aphids ($r_{98} = 0\cdot715$, $P < 0\cdot001$), it was possible that the previous relationship was merely a consequence of both insect groups responding in the

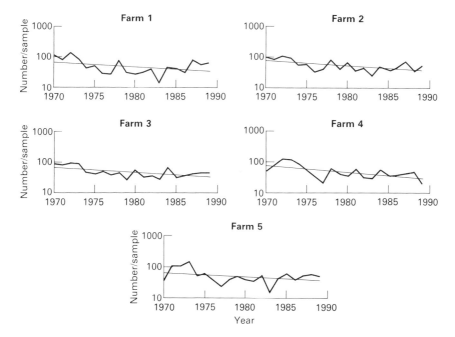

FIG. 15.9. Annual mean numbers (logarithmic scale) of polyphagous predators per sample from 1970 to 1989 on each of the five farms in the core Sussex study area. Polyphagous-predator density decreased significantly in the course of the study, with no between-farm differences in the regression lines; the pooled slope was -0.017 ± 0.003 ($r_{98} = -0.494$, $P < 0.001$).

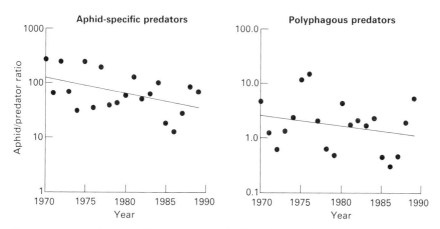

FIG. 15.10. Annual mean aphid-to-predator ratios (logarithmic scale) from 1970 to 1989 for aphid-specific predators (left) and polyphagous predators (right) based on samples from the core Sussex study area. The data for the five farms were combined for clarity; analysis of the full dataset showed that for both groups of predators the ratios declined in the course of the study, with no between-farm differences in the regression lines; the pooled slopes were -0.031 ± 0.007 ($r_{98} = -0.417$, $P < 0.001$) for aphid-specific predators, and -0.021 ± 0.008 ($r_{98} = -0.261$, $P < 0.01$) for polyphagous predators.

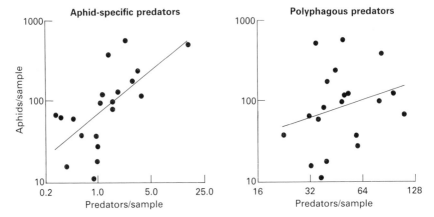

FIG. 15.11. Annual mean aphid densities (logarithmic scale) in relation to the annual mean predator densities (logarithmic scale), for aphid-specific predators (left) and polyphagous predators (right), based on samples from the core Sussex study area from 1970 to 1989. The data for the five farms were combined for clarity; analysis of the full dataset revealed no between-farm differences in the regression lines; the pooled equations were $\log (y) = 0.654 \log(x) + 1.86$ ($r_{98} = 0.572$, $P < 0.001$) for aphid-specific predators, and $\log (y) = 0.761 \log (x) + 0.62$ ($r_{98} = 0.310$, $P < 0.01$) for polyphagous predators.

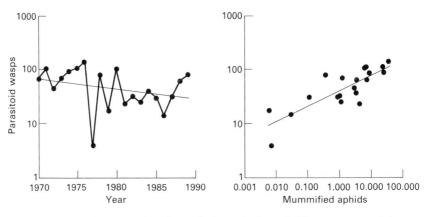

FIG. 15.12. Annual mean numbers (logarithmic scale) of parasitoid wasps per sample from 1970 to 1989 in the core Sussex study area (left), and the relationship between the annual mean densities of mummified aphids (logarithmic scale) and annual mean parasitoid densities (logarithmic scale) for the same years (right). The data for the five farms were combined for clarity; analysis of the full dataset revealed no between-farm differences in the regression lines. The pooled slope (left) was -0.021 ± 0.006 ($r_{98} = -0.330$, $P < 0.001$); the pooled regression equation (right) was $\log (y) = 1.90 \log (x) - 2.94$ ($r_{98} = 0.751$, $P < 0.001$).

same way to an extrinsic annual factor such as weather. However, after removal of the effect of year, there remained a significant positive within-year relationship between the numbers of mummified aphids and the numbers of parasitoids ($F_{1,79} = 4.78$, $P < 0.05$). It seems likely that the relationship between mummified aphids and parasitoids was causal, for instance the parasitoids caught during sampling may have emerged from the mummified aphids.

Aphid-specific predators

General changes in the abundance of aphid-specific predators have already been presented in the context of their relationship with aphids. In particular, it was observed (Fig. 15.11) that instead of the negative correlation which would have been expected if the predators had been reducing the numbers of their prey, the actual correlation was significantly positive. Further analysis showed that after removing the year effect, there remained a significant within-year relationship between aphids and predators across the different farms ($F_{1,79} = 5.49$, $P < 0.05$). This ruled out the possibility that the correlation was caused by year-to-year climatic effects. It seems therefore that the numbers of aphid-specific predators were a direct response to the abundance of aphids.

The numbers of aphid-specific predators recorded in the samples were made up of ladybirds (56%), soldier beetles (7%), lacewings (17%) and hoverflies (20%). Ladybirds were thus numerically the most important aphid-specific predator to be picked up in the sampling; the vast majority of ladybirds sampled were larvae (87%).

Considering ladybirds alone, the species most often recorded in the samples was *Propylea 14-punctata* (40% of adults). The response of ladybirds to large numbers of aphids lay not so much in an increase in numbers of adult beetles feeding on the aphids, but in increased breeding in the cereals (Fig. 15.13). The peaks in ladybird larvae were much more marked in years of high aphid abundance, especially 1975, 1976, 1980 and 1989, than were the peaks in adult ladybirds. Moreover, there was a significant positive relationship between the ratio of larvae to adult ladybirds in each year and aphid density ($F_{1,18} = 8.82$, $P < 0.01$).

Polyphagous predators

Polyphagous predators were shown earlier to have declined over the whole of the core study area. They are a very heterogeneous group, comprising

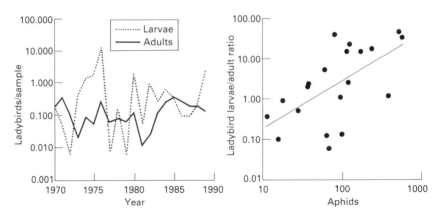

FIG. 15.13. Annual mean densities (logarithmic scale) of ladybird adults and larvae from 1970 to 1989 (left), and the relationship between annual mean ladybird larvae : adult ratios (logarithmic scale) and aphid densities (logarithmic scale) for those years (right). Densities of adults were low, so all samples from the core Sussex study area were pooled. The regression equation (right) was $\log(y) = 1\cdot09 \log(x) - 1\cdot74$ ($r_{18} = 0\cdot593$, $P < 0\cdot01$).

ground beetles (3% of polyphagous predators sampled), rove beetles (50%), spiders (21%), harvestmen (<1%), predatory flies (25%) and earwigs (<1%). It is examined below how numbers within each constituent group have varied in the course of the study, and whether they have all varied in the same way. Earwigs were not considered because of their scarcity in the samples; for the same reason, harvestmen were included with spiders.

Ground beetles

The mean numbers of ground beetles (adults and larvae) sampled each year showed no consistent trend, although the average number of beetles per sample appears to be slightly higher now than at the start of the study (Fig. 15.14). Certain species, for instance *Agonum dorsale*, *Bembidion lampros* and *Demetrias atricapillus*, are known to disperse from their over-wintering habitats into adjacent cereal crops by crawling; *D. atricapillus* in particular has been found to be more abundant in fields surrounded by hedgerows than in fields surrounded by fence-post and wire boundaries (Coombes & Sotherton 1986). In view of the different types of field-boundary management on the five farms on the core study area, it would be expected that the abundance of these species in the samples would vary accordingly. However, such did not seem to be the case: a comparison of mean numbers of beetles of each species on the five farms during the first

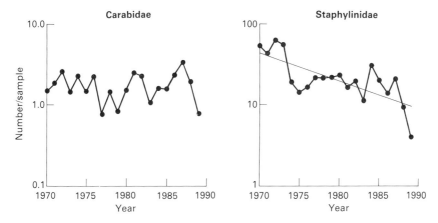

FIG. 15.14. Annual mean numbers (logarithmic scale) of Carabidae (left) and of Staphylinidae (right) per sample from 1970 to 1989 in the core Sussex study area. The data for the five farms were combined for clarity; analysis of the full dataset showed that staphylinid density declined in the course of the study, with no between-farm differences in the regression lines; the pooled slope was -0.035 ± 0.004 ($r_{98} = -0.658$, $P < 0.001$).

and the last five years of the study showed a statistical between-farm difference for *A. dorsale* only (Table 15.5), which was probably an artefact of very low abundance. Despite the shortage of hedges on Farm 5 in both time periods, and despite the systematic removal of hedgerows on Farm 4 in recent years, there was no evidence of a corresponding reduction in numbers of any of the three species on Farm 5 in the two time periods, or on Farm 4 in the most recent period (Table 15.4). The clearest effect in Table 15.4 was the three-fold increase in overall mean numbers of *B. lampros* between 1970–74 and 1985–89, an increase which took place to varying degrees on all farms. It should be noted, however, that these species are not well sampled by D-vac, thereby giving rise to low apparent densities and large sampling errors (Potts & Vickerman 1974).

Rove beetles

The mean numbers of rove beetles (adults and larvae) sampled each year have decreased by over two-thirds since the start of the study (Fig. 15.14). The decline rate was the same on all five farms, and averaged 7·8% per annum ($F_{1,98} = 74·8$, $P < 0.001$). The genus *Tachyporus* constituted 65% of all rove beetles sampled, and is examined in greater detail below.

The annual mean number of *Tachyporus* adults and larvae in the samples have both gone down significantly in the course of the study (Fig.

TABLE 15.5. Mean numbers of *Agonum dorsale*, *Bembidion lampros* and *Demetrias atricapillus* on each of the five main farms in the Sussex study area, during 1970 to 1974 and during 1985 to 1989 (first and last 5 years of the study, respectively). As the analysis was performed on logarithmic data, the values presented here are back-transformations of the logarithmic means, with approximate standard errors

Farm	A. dorsale		B. lampros		D. atricapillus	
	1970–74	1985–89	1970–74	1985–89	1970–74	1985–89
F1	0·03 ± 0·02	0·03 ± 0·01	0·07 ± 0·06	0·26 ± 0·20	0·32 ± 0·08	0·28 ± 0·16
F2	0·06 ± 0·02	0·06 ± 0·03	0·14 ± 0·10	0·17 ± 0·11	0·15 ± 0·09	0·34 ± 0·25
F3	0·07 ± 0·08	0·03 ± 0·01	0·08 ± 0·14	0·28 ± 0·15	0·32 ± 0·20	0·27 ± 0·11
F4	0·06 ± 0·03	0·06 ± 0·04	0·07 ± 0·07	0·29 ± 0·25	0·39 ± 0·21	0·22 ± 0·24
F5	0·13 ± 0·06	0·06 ± 0·01	0·11 ± 0·05	0·21 ± 0·10	0·23 ± 0·13	0·24 ± 0·13
Overall	0·05 ± 0·02	0·04 ± 0·01	0·08 ± 0·04	0·25 ± 0·07	0·30 ± 0·07	0·27 ± 0·08
Statistical tests						
Time period, $F_{1,40}$		2·88 NS	12·6 $P < 0·001$			0·13 NS
Farm, $F_{4,40}$	3·39 $P < 0·05$		0·02 NS			0·18 NS
F4 v. others, $F_{1,40}$	0·95 NS		0·00 NS			0·02 NS
F5 v. others, $F_{1,40}$	2·99 NS		0·02 NS			0·32 NS

None of the interactions between time period and any of the farm factors was statistically significant. NS, not statistically significant ($P > 0·05$).

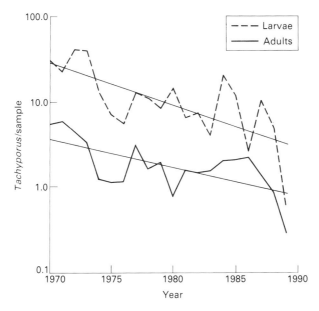

FIG. 15.15. Annual mean numbers (logarithmic scale) of *Tachyporus* adults and larvae per sample from 1970 to 1989 in the core Sussex study area. The densities of both adults and larvae declined in the course of the study, with no differences in the slopes of the regression lines; the pooled slope was -0.042 ± 0.007 ($r_{37} = -0.684$, $P < 0.001$).

15.15). There was no difference in the rates of decline of the two age classes (average 9.3% per year), so that the ratio of adults to larvae in the samples remained approximately constant at 1 adult to 6 larvae. Although the larvae had not been identified to species, the adults belonged to the following four species: *T. chrysomelinus*, *T. hypnorum*, *T. nitidulus* and *T. obtusus*. The mean numbers of adults of all four species declined significantly during the study ($F_{1,75} = 26.3$, $P < 0.001$), with again no difference in their rates of decline despite varying overall levels of abundance of the different species (Fig. 15.16).

Spiders and harvestmen

Spiders and harvestmen together accounted for 21.3% of all polyphagous predators sampled in the core Sussex area. The mean numbers of spiders and harvestmen caught per sample per year have also declined in abundance since the study began (Fig. 15.17). Again, there was no difference between farms, and the overall annual decline rate was 4.1% ($F_{1,98} = 25.0$, $P < 0.001$). On average, therefore, the abundance of these two groups in

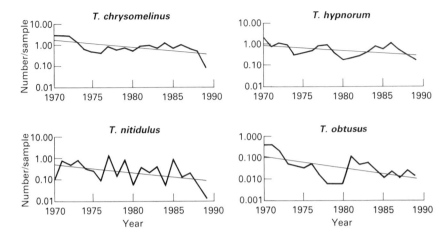

FIG. 15.16. Annual mean numbers (logarithmic scale) of adults of four species of *Tachyporus* (*T. chrysomelinus, T. hypnorum, T. nitidulus, T. obtusus*) per sample from 1970 to 1989 in the core Sussex study area. The densities of all four species declined in the course of the study, with no differences in the slopes of the regression lines; the pooled slope was -0.038 ± 0.007 ($r_{75} = -0.509, P < 0.001$).

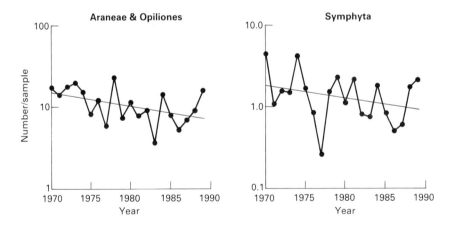

FIG. 15.17. Annual mean numbers (logarithmic scale) of Araneae and Opiliones (left) and of Symphyta (right) per sample from 1970 to 1989 in the core Sussex study area. The data for the five farms were combined for clarity; analysis of the full dataset showed that the densities of both groups declined in the course of the study, with, in either case, no between-farm differences in the regression lines. The pooled slopes were -0.018 ± 0.004 ($r_{98} = -0.451$, $P < 0.001$) for Araneae & Opiliones, and -0.020 ± 0.007 ($r_{98} = -0.270, P < 0.01$) for Symphyta.

cereals in the core study area is now less than half of that at the start of the study. It was found earlier, in a stepwise multiple regression analysis of aphid density against the different groups of polyphagous predators, that spiders were the only group to be selected. The above suggested that the reason for the selection of spiders was simply that aphids and spiders were both groups which had gradually declined in abundance over the past 19 years. Rove beetles, whose abundance had also gone down, were not selected because of dual positive correlations with aphids and spiders.

Predatory flies

Predatory flies, comprising the Dolichopodidae, Empididae and Scathophagidae, made up about a quarter of the polyphagous predators to have been sampled. Figure 15.18 shows that the densities of Dolichopodidae and Empididae, which together account for 89% of the predatory flies in the samples, varied roughly in parallel through time. After a stable period lasting 6 years, their abundance dropped by nine-tenths between 1975 and 1979, then gradually recovered to their former levels; the pattern was

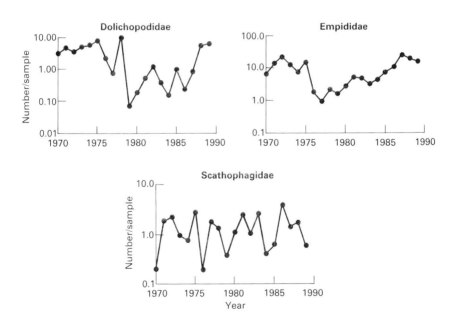

FIG. 15.18. Annual mean numbers (logarithmic scale) of Dolichopodidae, Empididae, and Scathophagidae, the three families of predatory Diptera sampled from 1970 to 1989 in the core Sussex study area.

similar, but not as clear, for the Dolichopodidae, numerically less common than the Empididae. It is possible that both groups were affected by the hot dry summers of 1975 and 1976. Numbers of Scathophagidae, on the other hand, fluctuated apparently at random, with no clear pattern or trend.

Sawflies

Sawflies (Symphyta: Hymenoptera) as a group are an important source of insect protein for gamebird chicks foraging in cereal crops (Potts 1986). The mean numbers of sawflies recorded per sample annually are shown in Fig. 15.17. Numbers declined significantly at an average annual rate of 4·4% ($F_{1,98} = 7·73$, $P < 0·01$).

Time-series analysis using auto-regression techniques showed that the annual mean numbers of sawflies on each farm were significantly related to amount of undersowing, summer temperature and summer rainfall, all in the previous year; there was also a strong carry-over effect, i.e. density dependence, from one year to the next (Aebischer 1990). These variables, undersowing in particular, completely explained the decline in annual mean sawfly abundance.

DISCUSSION

The most unexpected result to have come out of the computerization of the Sussex study is that for each of a wide variety of different invertebrate taxa, the pattern of change in abundance from one year to the next was remarkably similar on each of the five farms in the core study area. Such a result was obtained despite different attitudes towards farming and conservation on each farm, and different changes in crop management technique on each farm in the course of the study. In particular, the degree of changes in field boundaries, which differed markedly on the five farms, seem to have had little or no effect on the densities of a number of major groups of invertebrates sampled in the cereals in the second half of June. Thus the ground beetle *Demetrias atricapillus*, one of the most likely species to have responded to such changes in field boundaries (Coombes & Sotherton 1986), showed no between-farm or between-year variation which might have been attributed to boundary differences. It is possible that the sampling methodology was inappropriate to detect a response in this case, but the observed declines in the abundance of aphids, polyphagous predators and sawflies all occurred synchronously on the five farms. The implication is that the scale of spatial change was so large that even a study area measuring 16 km across could not detect it according to accepted statistical criteria. That leaves two main options to explain the scale of the observed changes: (a) invertebrate densities are determined to such a large extent by climate that they are relatively unaffected by differences in crop management, or (b) the impact of modern farming upon the abundance of invertebrates living in cereals has been so extensive that their density has declined everywhere, including on 'islands' where traditional farming is still practised. The climatic option is the easiest one to consider, and the next step in the analysis of the Sussex data is to examine in detail how climate in the first half of the year may relate to the abundance of the various invertebrate taxa sampled in the second half of June. As regards a possible overwhelming impact of modern farming, extensive experimentation would be the most conclusive, if not the most feasible, way of testing the hypothesis.

The declines in abundances of aphids, polyphagous predators and sawflies are now considered more specifically. In the case of aphids, there was no evidence that the reduction in density in the yearly samples was related to an increase in numbers of predators. However, the aphid : predator ratio increased during the period of the study, so although the density of aphids in the second half of June may be independent of the density of predators at that time, it is possible that increasing predation pressure may be having an effect on the subsequent height or timing of the annual peak in aphid density, or the probability of an outbreak at all. Indeed, Dixon (1987) reports a tendency for peaks in aphid abundance in East Anglia to occur later each year, consistent with this hypothesis.

The overall abundance of polyphagous predators declined over the past 20 years, owing mainly to the declines in abundance of rove beetles and spiders, which together constituted 71% of polyphagous predators in the samples. In turn, much of the decline in densities of rove beetles was attributable to that of *Tachyporus* adults and larvae. As well as being predatory, *Tachyporus* feed on a variety of fungi, mainly mildews and rusts (Sunderland 1975; Dennis, Wratten & Sotherton 1991). It is possible that the reduction in *Tachyporus* density is a consequence of a reduction in the availability of fungal foods through an increase in the use of foliar fungicides (Potts 1986). Indeed, the level of mildew and rust-infecting cereals in Sussex have dropped considerably, in line with the increasing use of foliar fungicides on all farms since the start of the study (Aebischer & Potts 1990). A shortage of pathogenic fungus could not have affected spiders directly, although their prey numbers might have been reduced. Otherwise, we cannot isolate any particular type of farming practice which might be at the root of the decrease in spider density, other than the general intensification of agriculture.

In the case of sawflies, the reduction in mean numbers sampled each year was explained by linking changes in sawfly density to the amount of undersowing and the mid-summer climate in the previous year (Aebischer 1990).

This study has quantified a number of hypotheses concerning the relationship between density of invertebrates in cereals and the management of the crops in which they live. Because of the complexity of the database, so far answers have only been sought to some very basic questions concerning trends and patterns in abundance. Areas which require further investigation include the relevance of climate before sampling, carrry-over effects from one year to the next, community analysis through time, and spatial distribution through time. Meanwhile, the long-term downward trends in the densities of a wide range of cereal invertebrates are worrying, particularly since the causes are not yet established.

ACKNOWLEDGMENTS

We are particularly indebted to the Sussex farmers who have allowed us to collect samples from their cereal fields throughout the course of the study. We thank the many people who have helped with the sampling, sorting and identification of the invertebrates, especially G.P. Vickerman, K.D. Sunderland, S.J. Moreby and S.J. Duffield.

REFERENCES

Aebischer, N.J. (1990). Assessing pesticide effects on non-target invertebrates using long-term monitoring and time-series modelling. *Functional Ecology*, **4**, 369–373.

Aebischer, N.J. & Potts, G.R. (1990). Long-term changes in numbers of cereal invertebrates assessed by monitoring. *Proceedings of the Brighton Crop Protection Conference — Pests and Diseases 1990*, pp. 163–172. British Crop Protection Council, Farnham.

Anon. (1986). *Agricultural Statistics United Kingdom 1984*. Ministry of Agriculture, Fisheries and Food. HMSO, London.

Anon. (1989). *Agricultural Statistics United Kingdom 1987*. Ministry of Agriculture, Fisheries and Food. HMSO, London.

Barr, C.J., Benefield, C., Bunce, R.G.H., Riddesdale, H. & Whittaker, M.A. (1986). *Landscape Changes in Britain*. Institute of Terrestrial Ecology, Monks Wood, Huntingdon.

Bunyan, P.J. & Stanley, P.I. (1983). The environmental cost of pesticide usage in the United Kingdom. *Agricultural Ecosystems and Environment*, **9**, 187–209.

Chancellor, R.J., Fryer, J.D. & Cussans, G.W. (1984). The effects of agricultural practices on weeds in arable land. *Agriculture and the Environment* (Ed. by D. Jenkins), pp. 89–94. Institute of Terrestrial Ecology Symposium No. 13, Natural Environment Research Council, Cambridge.

Coombes, D.S. & Sotherton, N.W. (1986). The dispersal and distribution of polyphagous predatory Coleoptera in cereals. *Annals of Applied Biology*, **108**, 461–474.

Davies, D.B. (1984). Trends in mechanization in the lowlands. *Agriculture and the Environment* (Ed. by D. Jenkins), pp. 44–48. Institute of Terrestrial Ecology Symposium No. 13, Natural Environment Research Council, Cambridge.

Dennis, P., Wratten, S.D. & Sotherton, N.W. (1991). Mycophagy as a factor limiting aphid predation by staphylinid beetles in cereals. *Bulletin of Entomological Research*, **81**, 25–31.

Dietrick, E.J. (1961). An improved backpack motorised fan for suction sampling of insects. *Journal of Economic Entomology*, **54**, 394–395.

Dixon, A.F.G. (1987). Cereal aphids as an applied problem. *Agricultural Zoology Reviews*, **2**, 1–57.

Edwards, C.A. (1984). Changes in agricultural practice and their impact on soil organisms. *Agriculture and the Environment* (Ed. by D. Jenkins), pp. 56–65. Institute of Terrestrial Ecology Symposium No. 13, Natural Environment Research Council, Cambridge.

Froud-Williams, R.J. (1983). The influence of straw disposal and cultivation regime on the population dynamics of *Bromus sterilis*. *Annals of Applied Biology*, **103**, 139–148.

Froud-Williams, R.J. & Chancellor, R.J. (1982). A survey of grass weeds in cereals in central southern England. *Weed Research*, **22**, 163–171.

Kendall, M.G. (1976). *Time-Series*, 2nd edn. Griffin, London.

O'Connor, R.J. & Shrubb, M. (1986). *Farming and Birds*. Cambridge University Press, Cambridge.

Pollard, E., Hooper, M.D. & Moore, N.W. (1974). *Hedges*. Collins, London.

Potts, G.R. (1984). Monitoring changes in the cereal ecosystem. *Agriculture and the Environment* (Ed. by D. Jenkins), pp. 128–134. Institute of Terrestrial Ecology Symposium No. 13, Natural Environment Research Council, Cambridge.

Potts, G.R. (1986). *The Partridge: Pesticides, Predation and Conservation*. Collins, London.

Potts, G.R. & Vickerman, G.P. (1974). Studies on the cereal ecosystem. *Advances in Ecological Research*, **8**, 107–197.

Rands, M.R.W., Hudson, P.J. & Sotherton, N.W. (1988). Gamebirds, ecology, conservation and agriculture. *Ecology and Management of Gamebirds* (Ed. by P.J. Hudson & M.R.W. Rands), pp. 1–17. Blackwell Scientific Publications, Oxford.

Raymond, W.F. (1984). Trends in agricultural land use: the lowlands. *Agriculture and the Environment* (Ed. by D. Jenkins), pp. 7–13. Institute of Terrestrial Ecology Symposium No. 13. Natural Environment Research Council, Cambridge.

Sturrock, F. & Cathie, J. (1980). *Farm Modernisation and the Countryside*. Occasional Paper No. 12. University of Cambridge Department of Land Economy, Cambridge.

Sunderland, K.D. (1975). The diet of some predatory arthropods in cereal crops. *Journal of Applied Ecology*, **12**, 507–515.

Whitehead, R. & Wright, H.C. (1989). The incidence of weeds in winter cereals in Great Britain. *Proceedings of the Brighton Crop Protection Conference — Weeds 1989*, pp. 107–112. British Crop Protection Council, Farnham.

Wilson, B.J. & Froud-Williams, R.J. (1988). The effect of tillage on the population dynamics of *Galium aparine* (L.) (cleavers). *VIIIème Colloque International sur la Biologie, l'Ecologie et la Systématique des Mauvaises Herbes*, 81–90, Columa, Paris.

16. THE BOXWORTH EXPERIENCE: EFFECTS OF PESTICIDES ON THE FAUNA AND FLORA OF CEREAL FIELDS

P.W. GREIG-SMITH

Central Science Laboratory, Ministry of Agriculture, Fisheries & Food, Tangley Place, Worplesdon, Surrey GU3 3LQ, UK

INTRODUCTION

Modern cereal growing creates an environment which differs in three major respects from the majority of 'natural' terrestrial environments. First, cereal fields have sharply-defined boundaries, at which there are abrupt changes in the occurrence of flora and fauna, in microclimate, soil structure and chemistry, and most other aspects of the environment. This feature applies to many man-made habitats, but it is particularly marked in the case of arable crops. Second, the habitat undergoes major changes in structure from year to year, as the pattern of cropping follows a rotation from cereals to break crops such as beans or rape. Third, the environment is subject to sudden, often severe perturbations during each year, caused by cultivations, pesticide applications, and other husbandry operations.

One consequence of this peculiar environment is that the resident fauna and flora are likely to be limited to species equipped to survive under these constraints. The attributes which might allow animals and plants to persist in these conditions include high mobility and life cycles which offer an ability to 'escape' the effects of sudden changes, or to recover afterwards. It seems probable that many of the species likely to be successful in these areas will be generalists, but it is also possible that the cereal field may offer particular opportunities to certain specialist species.

The 'Boxworth Project' (Hardy 1986; Greig-Smith 1989a) provides an unusually good opportunity to examine in detail the effects of one type of perturbation, the application of pesticides to crops in cereal fields. The project is one of very few large-scale experimental studies that have examined pesticide effects over several years with the aim of identifying the ecological implications of intensive chemical inputs (see also El Titi, this volume, pp. 399–411).

The range of possible side-effects of pesticide applications includes some that are *short term* (generally direct toxic effects of exposure to the active ingredient at lethal or sublethal levels), and some *long term*, including the continuing effects of a single exposure, the cumulative effects of repeated exposures to one or more chemicals, and indirect changes such as depletion of prey or removal of vegetation cover. Figure 16.1 provides a general framework to identify the possible consequences of pesticide use for wild animals and plants.

The severity of short-term effects depends partly on physiological susceptibility to the toxic effects of the pesticide, but much more importantly, from an ecological point of view, on *exposure* to the chemical. Exposure itself is determined jointly by the pattern of use of the product and its subsequent distribution and persistence in the environment, and by the ecology and behaviour of the species, affecting contact with it. We may expect the exposure of non-target organisms to vary in space and time. Thus, species or individuals whose habits allow 'escape' from the site of application, by movement to and from other areas, or by occupying protected parts of the field habitat, should experience less-damaging exposures than others. Similarly, the ability to recover from population reductions, by immigration from adjacent habitats, recruitment from seed-banks or reproduction, will help to overcome short-term effects. Therefore, mobility and breeding seasonality should be important determinants of the severity of short-term adverse effects, in conjunction with the timing and scale of local use of the pesticide.

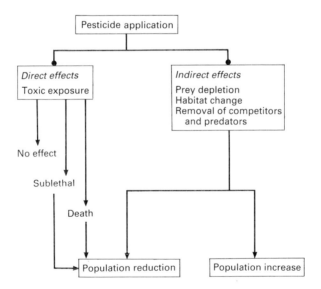

FIG. 16.1. Outline of the main routes by which pesticide applications can cause changes in wild populations.

The same properties are also relevant to long-term toxic effects of pesticides. However, possible indirect effects, in the form of prey depletion and habitat change, are likely to be influenced more strongly by other aspects. In particular, dietary specialists might be at risk from reductions in the density of their prey, whereas generalist feeders are likely to be less vulnerable if they can switch to alternative prey.

It is generally assumed that the side-effects of pesticide use are detrimental to the environment. However, if the community of animals or plants is altered by direct or indirect effects on vulnerable species, there may be new opportunities for certain other species. These are likely to be poor competitors, normally held at low density by predators or superior competitors, but able to increase in numbers when these pressures are relaxed.

Evidently, there is a broad range of possible consequences of pesticide use on the ecology of cereal fields. This paper uses information obtained in the course of the Boxworth Project to examine: (i) how the animal and plant communities differ between areas of high pesticide use and lower-input areas; (ii) the effects on individual species of the full annual pesticide 'package'; and (iii) the mechanism of impact of particular pesticide applications on selected species. These aspects are considered for each major group studied; birds, mammals, soil invertebrates, crop invertebrates and plants.

THE BOXWORTH PROJECT

The project was initiated in the late 1970s, to investigate the possibility that the sustained use of a wide range of pesticides for prophylactic crop protection was causing harmful environmental effects. Although individual products were evaluated to assess their environmental safety before approval for commercial use was granted, there had then been no study of the total combined consequences of the entire pesticide 'package'. A particular aim of the Boxworth Project was therefore to examine such overall effects, and to do so over a period of several years, so as to reveal any changes that might develop gradually after the imposition of a high-input regime.

The aims, design and timetable of the Project have been outlined elsewhere (Hardy 1986; Jarvis 1988; Greig-Smith 1989a), and will be fully described in the Final Report of the work (Greig-Smith, Frampton & Hardy 1991). The following paragraphs briefly explain the main features relevant to this paper.

The basis of the project was an experimental comparison of matched blocks of fields, subject to high or reduced inputs of pesticides, but otherwise similar in crops, husbandry, fertilizers and other farming operations. The study was carried out on 130 ha at MAFF's Boxworth Experimental Husbandry Farm, a few miles north of Cambridge. This area was divided into three parts (Fig. 16.2), representing contrasting approaches to pesticide use:

1 A *Full Insurance* programme (on 67 ha) comprised prophylactic applications of herbicides, fungicides and insecticides, designed to eliminate any possibility of pest, weed or disease problems.

2 In a *Supervised* treatment (46 ha), levels of pests, weeds and diseases were monitored regularly, and pesticides applied only if thresholds known to correspond to economically serious damage were exceeded.

3 An *Integrated* regime (22 ha) employed similar monitoring to determine when pesticides were needed, but also involved less damaging practices where possible, such as the use of more specific pesticides, and disease-resistant varieties to reduce the need for fungicides. In many respects, the Supervised and Integrated treatments were similar, and can be compared with the Full Insurance area as a combined 'reduced-input' area.

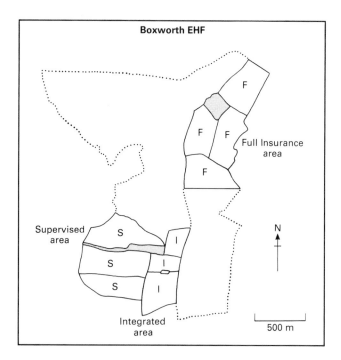

FIG. 16.2. Map of Boxworth Experimental Husbandry Farm, showing the location of fields in the three experimental treatment areas.

The fields were sown with winter wheat, except for periodic break crops of oil-seed rape. So far as possible, everything apart from pesticide use was kept similar in all fields. Nitrogen and phosphate fertilizers were applied at rates appropriate to the crop, regardless of pesticide treatments. Also, husbandry operations were matched so that each of the Full Insurance fields had 'partners' in the Supervised and Integrated areas, which were sown, cultivated, harvested, etc. at the same time. In this way, it was possible to eliminate many of the variations in sowing conditions, for example, that might otherwise have confounded differences due to the contrasts in pesticide use.

The investigation began with 2 years of 'baseline' data collection (Oct 1981–Sept 1983), to provide information on the plant and animal populations present on the project fields at the outset, while farmed under previous, moderate pesticide inputs. Thereafter, the three contrasting pesticide regimes were applied for 5 years (1984–88 harvests), during which changes in the density of animals and plants in the fields and their margins were monitored.

Because all the fields within each of the three treatment areas were next to one another in a block, in order to maximize the chances of identifying large-scale effects, the treatments were not replicated in the formal sense that would allow conventional statistical testing of differences between treatments. A replicated trial of small plots was incorporated into the project to provide support for the interpretation of some differences observed at a field scale (see Greig-Smith 1989a). However, the principal value of monitoring populations is to trace changes from densities in the baseline years in the Full Insurance area, relative to changes in the Supervised and Integrated areas.

Research studies in the project were carried out by scientists from the MAFF Central Science Laboratory (birds, soil invertebrates), the AFRC Long Ashton Research Station (plants), Cambridge University (small mammals, invertebrate predators of cereal pests), Southampton University (invertebrate populations in fields) and Reading University (biochemical effects on birds and mammals). A supplementary study of pesticide spray drift was carried out within the project fields by the NERC Institute of Terrestrial Ecology.

For the purposes of this paper, the information obtained by monitoring animal and plant populations was used in two ways. Differences between the Full Insurance area and the Supervised/Integrated areas at the end of the treatment phase provide an indication of the set of species able to persist under the high pesticide-input regime. The densities of the species then present, and the relative numbers of various ecological types (predators, parasites, herbivores, detritivores, etc.; annual and perennial plants) also give an indication of the functional attributes of the resulting communities. The data are also used to identify effects of the total pesticide programme on individual species of interest. Although the study was not designed to attribute significance to particular applications, the monitoring data suggest at least the time of year at which the most critical exposures occurred. Some of those cases were pursued in more detail by specific research directed at identifying the mechanisms of adverse effects on selected species of birds, mammals and invertebrates.

PESTICIDE USE AT BOXWORTH

The planned programme for crop protection in the Full Insurance area included a comprehensive annual input of seven insecticide and molluscicide applications, against slugs (Sept/Nov), autumn aphids carrying barley yellow dwarf virus (Nov), frit fly (Oct/Nov), yellow cereal fly (Feb/Mar), summer aphids (June/July) and, if considered necessary, against wheat bulb fly (Jan/Mar). Most of these treatments involved organophosphorus or carbamate pesticides, which are of short persistence but high toxicity to some non-target animals.

The prophylactic approach also included provision for seven herbicide applications, and five of fungicides. These were considered less likely than the insecticides to cause direct side-effects on animals, but herbicides affect non-target plant species, and can also cause indirect effects on invertebrates by modifying habitat structure.

In practice, fewer applications than those outlined above were made each year in the Full Insurance area (5·2 insecticides, 5·2 herbicides and 4·0 fungicides on average). The Supervised approach to crop protection problems on the other two areas reduced inputs even further, to averages of 0·8 insecticides, 3·4 herbicides and 2·6 fungicides in the Supervised area, and 0·9, 2·8 and 2·4 in the Integrated area. The similarity between these two areas justifies combining data from them in several of the later analyses.

Figure 16.3 shows the seasonal pattern of pesticide use, as well as the other principal husbandry operations. Clearly, there was a pattern of two major periods of pesticide use during the year, broadly identifiable as 'summer' and 'winter'. In the Full Insurance programme, each field received at least one application per month on average in October, November and December, and two–three applications per month in the period from April to June. However, there was substantial variability in timing, so that individual fields sometimes had up to six applications within a month. For the insecticide use that was expected to be most ecologically

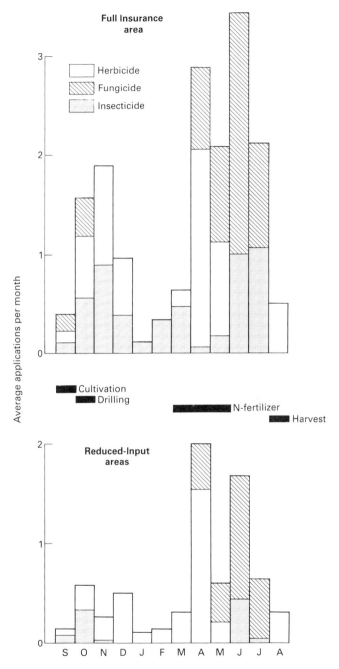

FIG. 16.3. Seasonal patterns of pesticide use in the Full Insurance area and the Supervised/Integrated areas. Columns show the average number of applications of herbicides, fungicides and insecticides per field per month, over the period 1984–88.

damaging, the principal periods of risk were November and June/July.

Although there were two periods of heavy pesticide use in the Supervised and Integrated areas, these were more restricted than in the Full Insurance area for insecticides and fungicides, but not for herbicides.

The precise timing of applications may be important for certain species, depending on whether treatments occurred before or after a critical point in the annual cycle.

Because the project was designed to minimize all differences other than pesticide use, the seasonal patterns of cultivation (September) drilling (October), nitrogen fertilization (March/May) and harvest (August/September) were similar in both parts of the farm in most years.

EFFECTS OF PESTICIDES ON BIRDS

The bird fauna at Boxworth is typical of farmland in eastern England. With the exception of one species, the skylark *Alauda arvensis*, which prefers to nest in fields (O'Connor & Shrubb 1986), all the common resident birds nest in hedges or woods. For the majority of birds, therefore, exposure to pesticides is likely to be through spray drift from the fields into adjacent habitats, or (for those that forage in fields) through contamination of vegetation or prey. The latter species might also suffer indirect effects of prey depletion, making the area less suitable for them.

The occurrence and density of eleven common breeding species was monitored by M.R. Fletcher, using a version of the 'Common Birds Census' (Marchant *et al.* 1990), which involves plotting the locations of territories during the spring. This method is likely to reveal indirect effects of pesticides affecting habitat quality, but would not identify short-term toxic effects occurring during the breeding season.

Table 16.1 compares the numbers of territories in the two areas in the baseline years, and in a 2-year period at the end of the project. There was an overall decline in numbers in both areas, but this was proportionately similar for the totals of all species combined. Only one species, the starling *Sturnus vulgaris*, showed a markedly greater decrease in the Full Insurance area than in the Supervised/Integrated area. This pattern (Fig. 16.4) appears to be consistent with an adverse effect of the high pesticide-input programme.

Starlings have a generalized invertebrate diet in the breeding season, in which leatherjackets, the larvae of crane-flies (Tipulidae), are particularly important (Tinbergen 1981). These and other prey declined in the Full Insurance area (see later) and it is possible that a change in prey availability might have caused starlings to breed elsewhere than in the woods of

TABLE 16.1. Average densities (territories per 50 ha) of eleven common breeding bird species in the Full Insurance and Supervised/Integrated areas at Boxworth EHF in the baseline years (1982–83) and at the end of the treatment period (1987–88)

	Full Insurance area		Supervised/Integrated area	
	1982–83	1987–88	1982–83	1987–88
Wren	8·6	7·5	8·8	7·7
Dunnock	7·5	5·2	11·0	5·5
Robin	3·7	4·1	7·0	5·1
Blackbird	6·0	7·1	9·2	6·2
Song thrush	4·8	4·1	6·6	3·3
Blue tit	7·5	9·0	9·2	8·1
Great tit	2·6	6·7	6·2	8·4
Starling	38·4	18·7	43·4	30·9
Tree sparrow	21·6	19·8	16·2	11·0
Chaffinch	4·5	6·3	9·9	9·2
Yellowhammer	9·3	5·2	8·4	6·6
	114·6	93·7	136·0	102·2

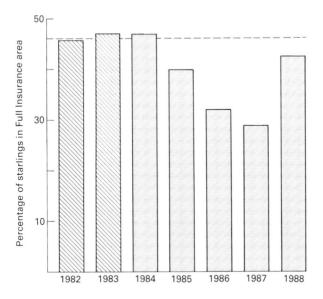

FIG. 16.4. Change in the distribution of breeding starlings in the Full Insurance and reduced-input areas at Boxworth. The dashed line indicates the average of the two baseline years (1982–83). The relative decline in the Full Insurance area after 1984 was greater than expected by chance.

that part of the farm. However, the picture is complicated by a number of other factors, including removal of trees and hedges, and changes in the presence of cattle in fields near to the project area. Starlings are well known to favour feeding around livestock (Feare 1984), and the removal of cattle from the area in 1987 may have affected their numbers. No single factor was closely associated with the change in starling density, however, and the reasons for it remain unclear.

Short-term pesticide effects in the breeding season were studied by examining nesting success. The species in Table 16.1 include hole-nesters (starling, tree sparrow *Passer montanus*, house sparrow *P. domesticus*, blue tit *Parus caeruleus* and great tit *P. major*) and open nesters. Because the four hole-nesting species readily used nest-boxes, their breeding activities could be monitored in detail. Success was very variable from year to year, but there were no major consistent differences between the Full Insurance and Supervised/Integrated areas that could be confidently ascribed to effects of pesticide use (Table 16.2).

This lack of evidence for any short-term effects, covering a time when summer insecticides were applied, may mean that breeding birds 'escaped' exposure by switching to feed elsewhere. Detailed studies were carried out on tree sparrows to assess this possibility by examining changes in behaviour and diet, and measuring exposure by biochemical monitoring (Greig-Smith 1989b; Hart *et al.* 1990).

Tree sparrows are particularly likely to be exposed to summer aphicides, because they feed their young chiefly on invertebrates gathered from cereal fields near to their nests in woods and hedges. Automatic recorders were fitted to many nests by A.D.M. Hart, allowing all visits to the nest to be registered for several days. This provided an ability to detect any changes in parental care coinciding with pesticide applications or other events (cf. Grue, Powell & McChesney 1982; Powell 1984). However, at those nests

TABLE 16.2. Production of young by birds breeding in the Full Insurance area at Boxworth in the baseline years (1982–83) and in the final 2 years of the treatment period (1987–88). Values are given as the number of young produced per pair, and (in brackets) as a percentage of production in the reduced-input areas. There were no consistent trends for lowered breeding success in the Full Insurance area in these species

	Starling	Tree sparrow	Great tit	Blue tit
1982	3·2 (126%)	5·2 (92%)	5·7 (75%)	5·6 (63%)
1983	3·5 (110%)	5·8 (148%)	3·8 (112%)	3·8 (55%)
1987	2·8 (64%)	3·6 (63%)	4·3 (86%)	6·0 (120%)
1988	2·2 (99%)	2·7 (153%)	2·7 (159%)	2·7 (41%)

for which data were obtained before and after aphicide applications, no consistent changes in behaviour were detected. Apart from initial disturbance during spraying, tree sparrows continued to provision their young at rates undiminished by the sprays. Direct observation also showed that they visited the fields to gather food to about the same extent after applications (about 50% of visits on average, although individual nests varied greatly). However, there was a change in diet. Examination of faecal samples produced by nestling tree sparrows revealed a diverse diet in which beetles were a major element in the period before spraying of aphicides. However, immediately after applications of demeton-S-methyl, the proportion of beetles in the diet fell, associated with a large increase in the numbers of cereal aphids eaten (Fig. 16.5), implying that the birds concentrated their foraging on the ready supplies of dead or moribund aphids.

This evidence indicates that for several days after application, nestling tree sparrows were fed on insects liable to carry residues of the pesticide. To determine what effects this form of exposure might have, samples of blood were taken from nestlings, and assayed by H.M. Thompson for the level of activity of serum cholinesterase. This enzyme is inhibited by organophosphorus pesticides such as demeton-S-methyl (Ludke, Hill & Dieter 1975), and the results revealed depressions of activity in all the nests studied at the time of the aphicide application (H.M. Thompson, unpublished data). However, neither the growth-rates nor the survival of nestlings was adversely affected (Hart *et al.* 1990). It was not feasible to catch adult tree sparrows at the time of application, but analyses of house

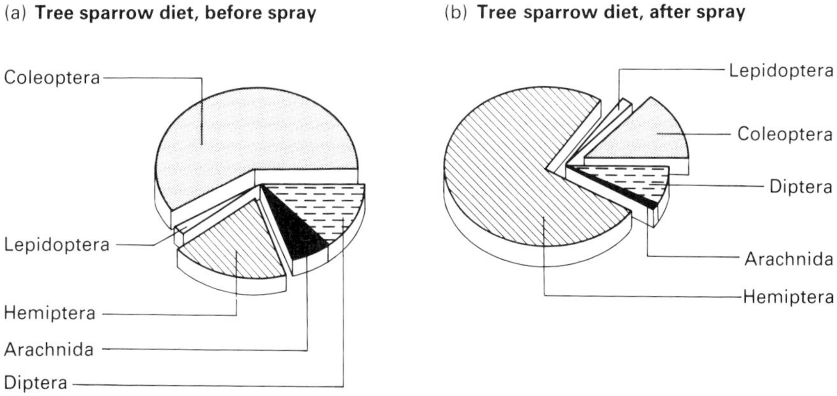

(a) **Tree sparrow diet, before spray**   (b) **Tree sparrow diet, after spray**

FIG. 16.5. Composition of the diet of nestling tree sparrows within 3 days before and 3 days after spraying of fields with an organophosphorus aphicide in 1988.

sparrow cholinesterase levels revealed evidence of inhibition in apparently healthy birds (H.M. Thompson & K.A. Tarrant, unpublished data).

The information gathered in these investigations of a species considered to be particularly liable to exposure to an insecticide suggests substantial resilience of the tree sparrow population. This can be attributed partly to behavioural avoidance of exposure by adult birds at the time of spraying, and partly to a physiological ability to withstand the levels of exposure experienced while foraging on contaminated prey during the next few days. Even the effects of a more toxic pesticide might be diluted by the variability of exposure among nests, where the parents' foraging activity ranged from 0 to 70% of visits into the crop. The distribution of breeding territories, and the availability of alternative habitats in which to forage, are likely to be the major determinants of this variability. In a diverse mosaic of fields and non-crop habitats, severe effects may be restricted to a small number of individual birds.

Effects on some other species that use fields more intensively, and do not have the opportunity to switch their foraging to other habitats (e.g. skylark and corn bunting *Miliaria calandra*), could be more serious. However, for the majority of species, which spend less time in the cereal fields or have mixed diets, effects are likely to be less extreme. Therefore it is not surprising to find that the whole bird community was resilient to the use of summer insecticides.

SMALL MAMMALS

Most of the mammals that occur in and around cereal fields are rodents and shrews. At Boxworth, regular trapping, carried out in the project fields by R. Hare, J.R. Flowerdew and I.P. Johnson, revealed the presence of field voles *Microtus agrestis*, bank voles *Clethrionomys glareolus*, wood mice *Apodemus sylvaticus*, common and pygmy shrews *Sorex araneus* and *S. minutus*, house mice *Mus musculus* and harvest mice *Micromys minutus*. However, only wood mice were abundant residents in cereal fields throughout the year, as they are elsewhere (Green 1979). Attention was therefore focused on the effects of pesticides on this species.

Mark–release–recapture studies revealed that numbers fluctuated strongly from year to year, and showed a marked annual cycle, with peak density in late summer (July/August) (Fig. 16.6). This seasonal pattern is rather different to that occurring in woodland habitats, where density continues to rise through the autumn (e.g. Crawley 1970). The reason is probably connected with the physical disturbance of harvest and cultivation, causing mice to move out of fields, rather than a true change in

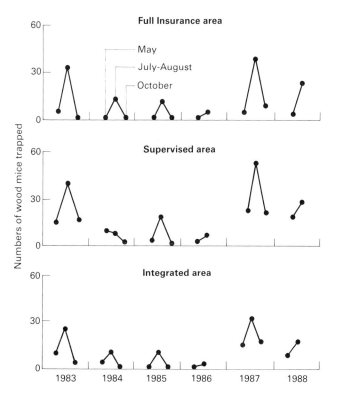

FIG. 16.6. Density of wood mice on wheat fields in the three treatment areas, determined by capturing mice in 2 ha grids of live-traps at three times of year.

numbers. The results revealed similar trends in all three treatment areas at Boxworth, with no evidence for any effect of pesticide use on density.

This overall pattern conceals one striking effect of a particular pesticide application. Molluscicide pellets were used in the Full Insurance area for autumn slug control, and were also applied in some years in the Supervised and Integrated fields. For most of the project, pellets containing the active ingredient methiocarb were broadcast on the surface of the fields shortly after drilling. On each occasion, there was a sudden drop in the number of wood mice captured immediately after application (Table 16.3). Numbers recovered quickly, so that the capture rate returned to its previous level after 7 days, but few of the animals caught and marked before treatment were recaptured afterwards, and a predominantly adult population was replaced largely by juveniles. This implies that adult wood mice living in or visiting fields just after pellets were spread were killed, and the area recolonized by juvenile immigrants from adjacent habitats. Wood mice

TABLE 16.3. Effects of the surface application of molluscicide pellets containing methiocarb on populations of wood mice in winter wheat fields at Boxworth

	Before application	After application	
		2–4 days	7–27 days
Number of wood mice trapped per 100 trap-nights	4·33	0	4·58
Ratio of adults : juveniles	5·8 : 1	—	0·6 : 1
Recapture of previously-marked mice:			
Treated field	—	1/38 (3%)	
Untreated field	—	8/45 (18%)	

feed on invertebrates as well as grain (Watts 1968), so the animals might have been exposed to the chemical either by ingestion of the pellets, which are formed with a nutritious bran base, or by eating poisoned slugs and earthworms. These possibilities cannot be firmly distinguished, for laboratory studies have shown that wood mice will ingest pellets (Tarrant & Westlake 1988) and will eat slugs killed by methiocarb (Johnson, Hare & Flowerdew 1990). However, the fact that the hazard to wood mice at Boxworth was removed by drilling pellets into the soil along with the seed (Johnson, Hare & Flowerdew 1991) suggests that the route of exposure may be chiefly through ingestion of pellets.

Studies elsewhere have failed to show a similar impact of broadcast molluscicides on wood mice (Greig-Smith & Westlake 1988; Tarrant *et al.* 1990). It is probable that their use of fields, and the willingness of rodents to eat pellets, depend on the availability of food in alternative habitats as well as on fields. Thus, after harvest of a cereal crop, the field is likely to offer supplies of spilt grain that may attract large numbers of wood mice to the field, but diminish their need to feed on pesticide pellets. Exposure to such pesticides is therefore a complex issue, in which the risk of adverse effects is related both to environmental conditions and to the characteristics of the product.

The ability of rodent populations to recover from pesticide-induced mortality also depends heavily on the habitats surrounding cereal fields. At Boxworth the hedges and woods provided a source for rapid recolonization which may not be present in many other arable areas.

The two species of shrews at Boxworth are entirely insectivorous, feeding particularly on some of the insects that were affected by the high pesticide-input programme. They might therefore be vulnerable to reduction in prey availability in the Full Insurance area. The diet of shrews was studied by I.P. Johnson, based on examination of prey remains in the guts of shrews found dead in pitfall traps in spring and summer. Results for common shrews showed that many of the major diet components were similar in all treatment areas, but leatherjackets were taken much less frequently in the Full Insurance area (Fig. 16.7). This matches the fact that the density of leatherjackets declined during the treatment period. The smaller number of pygmy shrew guts examined produced no clear differences between areas.

There was also some evidence that the pesticide regimes influenced the suitability of fields for shrews. As the project progressed, a greater proportion of the shrews captured in the Full Insurance area were found in hedgerows, rather than in the open fields. This accords with the decline in many invertebrate prey species under the high pesticide input (see later).

Rabbits *Oryctolagus cuniculus* are common in most types of farmland, including cereal fields. Their exposure to summer aphicides was investigated at Boxworth by trapping rabbits in fields just before and just after demeton-*S*-methyl applications in 1987 and 1988. Assays of blood and brain enzymes by H.M. Thompson, and histological studies by K.A. Tarrant, revealed evidence of sublethal exposure, perhaps by grooming, in

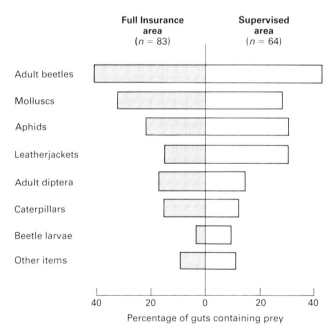

FIG. 16.7. Diet composition of common shrews trapped in the Full Insurance and Supervised areas.

most of the animals. Inhibition of enzyme activity was greater in those rabbits trapped after 2 or 3 days than in those trapped immediately after the application. This implies that exposure was accumulated over a period of several days, and raises the possibility of substantial variation between individuals, according to the time they spend in the crop.

SOIL FAUNA

Approximately 90% of soil biomass consists of microbial life. No attempts were made to investigate this component of the system during the treatment years of the Boxworth Project, because of the difficulties in interpretation of short-term changes in microflora and microfauna relative to the normal wide fluctuations due to 'natural' causes (Somerville & Greaves 1987). However, a programme of sampling was introduced to monitor populations of the soil microfauna in the three treatment areas. Soil cores were taken regularly and invertebrates were extracted using modified Tullgren funnels (MacFadyen 1961). This provided samples of springtails (Collembola) and mites (Acari) that were identified and counted by microscopical examination. The heavy clay soil at Boxworth harbours only very small numbers of earthworms, and no attempt was made to monitor them.

Figure 16.8 summarizes some of the major patterns found in the results. Overall, there were no significant differences between the Full Insurance and reduced-input areas in the density of total Collembola, nor of total mites. However, several individual species, and other taxonomic categories, did show indications of differences between areas, or of trends through the project. For example, *Folsomia quadrioculata* was present at high density in some Full Insurance fields during the baseline period, but not during the treatment years. In contrast, *Isotomiella minor* showed the opposite pattern, becoming more prevalent in the Full Insurance area relative to the reduced-input areas as the project progressed. Possibly this might be the result of a reduction in the numbers of predators or competitors of this species, but too little is known of its ecology to be certain.

One of the most striking features of Fig. 16.8 is the variability between fields, even though samples were collected at the same times. This was also evident within fields, where there was extreme patchiness of distributions. A further complication lies in the differences between treatment areas that were present during the baseline years. In order to help overcome these problems, soil samples were also taken from a set of small replicated plots sited in one of the Full Insurance fields. These plots (each 12 × 24 m)

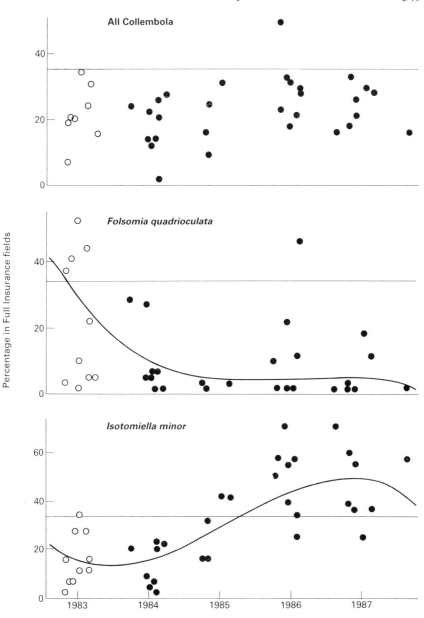

FIG. 16.8. Trends in the density of Collembola (springtails) in soil cores collected from the project fields. Each point represents the numbers recorded in samples from one Full Insurance field, compared to those in matched samples from one Supervised and one Integrated field. The lines are fitted polynomial regressions.

included replicates of the Full Insurance and Supervised programmes, and a 'Minimum use' treatment which differed in minor ways from the Integrated treatment.

Patterns observed in these samples largely supported what was seen on a field scale. There were significant differences between treatments in a few cases (Table 16.4), but many comparisons were inconsistent, or dominated by large numbers recorded on a few occasions. Differences between species are to be expected, because some are better protected from pesticides by subterranean habits than others.

Overall, the study of soil fauna suggests a number of adverse effects of the high pesticide-use programme, but no attempt was made to attribute the declines of some species to particular pesticides. Whatever their cause, the changes were highly variable in time and in space, perhaps partly because the scale of distribution of these poorly mobile Collembola was close to the scale on which there was major variation in pesticide application rates.

This part of the work also raises the possibility of interactions between species, causing pesticide-related changes to animal communities. The rise in density of *I. minor* in the Full Insurance area is consistent with a reduction in its competitors and/or predators. Evidence to support this possibility was not obtained for Collembola, but data on certain other groups of crop invertebrates also suggested an indirect effect through reduced competition or predation (see later).

TABLE 16.4. Results of paired comparisons between the density of soil invertebrates in Full Insurance plots and that in reduced-input plots in a replicated trial at Boxworth, 1984–88

	Number of comparisons	Full Insurance higher than reduced*	Full Insurance lower than reduced*
Total mites	13		1
Total Collembola	13		4
Onychiuridae	12		2
Sminthuridae	1		1
Neanuridae/Hypogastruridae	12		2
Neelidae	8	1	
Lepidocyrtus cyaneus	6		3
Folsomia quadrioculata	13		10
Isotoma spp.	13	1	
Isotomiella minor	11	1	

* Comparisons significant at $P < 0.05$.

CROP INVERTEBRATES

The range of invertebrate species inhabiting cereal crops is very broad, including pests and their natural enemies; herbivores, predators and detritivores; resident and mobile forms, and so on. Not surprisingly, the effects of pesticides on these groups at Boxworth were also highly varied.

By virtue of their intended purpose, most insecticides are liable to affect non-target arthropods as well as the pests themselves, particularly if the chemicals are formulated as broad-spectrum pesticides. The full range of pesticide effects on invertebrates is described elsewhere by the researchers who carried out studies at Boxworth (Burn 1988a,b, 1989, 1990; Vickerman 1988, 1991). The following sections provide a summary of the principal findings for each of the major trophic categories.

Populations were monitored principally by two methods; pitfall trapping (which provides an index reflecting both the numbers and activity of surface-dwelling invertebrates) and the use of a Dietrick vacuum insect net (which collects animals resting on vegetation or ground to which the suction nozzle is applied). In each year of the project, samples were taken from fields regularly throughout the spring and summer, and (less frequently) in the autumn and winter. A variety of additional techniques was employed to investigate predation (see later).

Herbivores

Virtually all the herbivorous insects in the samples were of species that feed on the wheat plants. They included thrips (Thysanoptera), flies (Diptera) and aphids and plant bugs (Hemiptera). Densities were similar in the three treatment areas in the baseline years. The general pattern thereafter was for a sudden decline in numbers in the Full Insurance area compared to the reduced-input areas, occurring in the first of the treatment years. However, the extent of the change varied greatly from species to species (Table 16.5), and there were also major differences from year to year (Vickerman 1991). For example, field populations of the grain aphid *Sitobion avenae*, one of the major target pests, were low at the end of spring, and increased during June at an exponential rate in some years, but more slowly in others. Applications of aphicides checked this growth only in some years, and density often continued to rise as a result of immigration by winged aphids. Generally, densities were lower in Full Insurance fields, but in 1986, peak numbers of this species, and of the rose-grain aphid *Metopolophium dirhodum*, were substantially *higher* in the Full

TABLE 16.5. Changes in the density and species-composition of the principal herbivorous invertebrates in the Full Insurance area at Boxworth, from the baseline years (1982–83) to the treatment period (1984–88)

	Density in the Full Insurance area as a percentage of reduced input area		Percentage of all herbivores in the D-vac samples	
	1982–83	1984–88	1982–83	1984–88
Thysanoptera (thrips)	90·0	24·9	24·5	21·8
Hemiptera (aphids & plant bugs)	80·6	76·3	35·8	54·3
Diptera (flies)	91·5	30·9	39·4	22·5

Insurance area than the Supervised/Integrated areas. This illustrates the unpredictability of chemical control of cereal aphids, and of the effects of insecticides on non-target species.

Overall, the total density of herbivores in the Full Insurance fields was reduced by 50%. This was accompanied by a change in species composition, so that by the end of the treatment period, Hemiptera (chiefly aphids) formed a larger proportion of the fauna in the Full Insurance area than they did in the other areas (Vickerman 1991).

Detritivores

Almost all of the detritivores recorded were either Coleoptera (mainly rove beetles and fungus-feeding species of several other families) or Diptera. Overall, the density of detritivores was hardly affected by the Full Insurance programme, remaining similar to the Supervised/Integrated areas throughout the project, although a few species showed evidence of modest reductions or increases in abundance (Vickerman 1991).

Predators

Among the insectivorous species were many predators and parasitoids that are 'natural enemies' of pests. The predominant predators were beetles (Coleoptera), flies (Diptera) and money-spiders (Arachnida; Linyphiidae). Although similar in the baseline years, densities in the Full Insurance area fell on average to about 47% of those in the reduced-input areas, though in some years the effect was much greater. Coleoptera appeared to be less affected than either of the other main groups, although certain species were adversely affected. As with the herbivores, the composition of the predatory fauna changed as a result of the Full Insurance regime; Coleoptera formed a progressively larger part than in the Supervised/Integrated areas, where Arachnids predominated (Vickerman 1991).

The predatory beetles provide an example of the reasons for differences between species in susceptibility to insecticides. It is to be expected that populations will be more resilient if individual insects can 'escape' the short-term toxic effects of pesticides by occupying protected parts of the field environment (e.g. below the soil surface, or under a plant canopy), or if their life cycles are such that adult emergence and reproduction occur after the principal insecticide applications. Similarly, population recovery is likely to be greater and more rapid in species that have several generations each year, or have good powers of dispersal. Burn (1988a) has classified predators into four groups on the basis of these characteristics: (i) poor dispersers of which the adults overwinter in fields; (ii) moderate dispersers that overwinter in field boundaries; (iii) good dispersers, colonizing fields from distant non-crop habitats; and (iv) species of moderate dispersive ability that overwinter on fields as larvae.

Table 16.6 lists the main groups of predatory beetles at Boxworth, showing that although rove beetles (Staphylinidae) and ladybirds (Coccinellidae) were reduced in numbers by the Full Insurance programme, numbers of ground beetles (Carabidae) and soldier beetles (Cantharidae) were relatively higher than in the reduced-input areas

TABLE 16.6. Changes in the density and species-composition of the principal four groups of predatory beetles in the Full Insurance area, from 1982–83 to 1984–88

	Density in the Full Insurance area as a percentage of reduced input area		Percentage of all Coleoptera in the D-vac samples	
	1982–83	1984–88	1982–83	1984–88
Staphylinidae (rove beetles)	95·8	74·5	74·1	57·4
Carabidae (ground beetles)	89·7	104·1	22·7	36·3
Cantharidae (soldier beetles)	63·0	123·3	2·8	6·0
Coccinellidae (ladybirds)	91·4	9·4	0·4	0·3

(Vickerman 1991). These contradictory trends are the result of differences in the effects of pesticides on individual species.

Figure 16.9 shows how the numbers of four species of ground beetles caught in pitfall traps varied through the project. *Bembidion obtusum* is an example of a species in which adults are present in the fields during winter, living on the surface. These insects are therefore very vulnerable to the effects of autumn and winter insecticides, and the species was virtually eliminated from the Full Insurance area early in the treatment years of the project. In contrast, *Agonum dorsale* overwinters in field boundaries, where the adults are not directly exposed to winter insecticides. The crop is then recolonized in spring. Accordingly, effects of the Full Insurance regime were much less on this and similar species, whose populations did not decline until several years after the start of the treatment phase of the project.

Another common species that overwinters in the fields, *Trechus quadristriatus*, was able to overcome the potential hazards of winter insecticide use because most individuals were present as larvae, beneath the soil surface. The population was not only protected from mortality, but actually increased to higher levels in the Full Insurance area in all years but

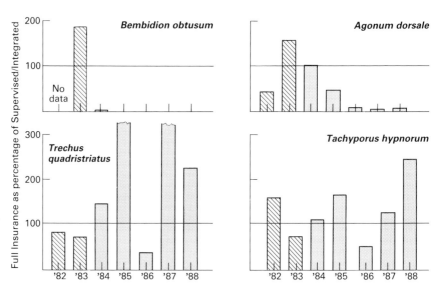

FIG. 16.9. Changes in the density of four selected species of ground beetles during the course of the project. Columns represent the numbers recorded in the Full Insurance fields compared to the reduced-input fields each year. Each species is an example of a different response to the high pesticide inputs. See text for details.

1986 (Fig. 16.9). In that year, poor development of the crop canopy may have increased the exposure to summer insecticides of this and other species.

Some species are highly dispersive, and colonize the crop each year from distant sources far removed from any effect of winter insecticides. Activity of these species, such as *Tachyporus hypnorum*, was generally unaffected by the Full Insurance regime, probably because the protective canopy of the crop plants normally sheltered them from direct exposure to summer insecticides. Even the large treatment areas at Boxworth are probably too small to reveal any long-term effects on such species from widespread insecticide use. In long-term monitoring of cereal invertebrates in Sussex, *T. hypnorum* populations declined substantially as pesticide use increased (Aebischer & Potts 1990; Aebischer, this volume, pp. 323–326).

Changes in density under a high pesticide regime may not be caused solely by these variations in exposure and recovery from short-term toxic effects. Some species were probably reduced in numbers because their prey was depleted. In particular, polyphagous predators that feed on aphids in summer rely on the availability of alternative prey such as Collembola earlier in the year. The importance of these effects to the beetles *A. dorsale* and *Loricera pilicornis* was studied by A.J. Burn.

In the baseline years, there were no differences between areas in the occurrence of Collembola in the guts of *A. dorsale* and *L. pilicornis*. However, from 1984 onwards, there was a significant reduction in Collembola in the guts of beetles in the Full Insurance area (Table 16.7), in line with the lower density of these insects on the ground surface. Also, this lack of availability of prey may have led to lowered fecundity, at least in *L. pilicornis* (Burn 1989, 1991), which would tend to add to the effects of direct pesticide-induced mortality on the population.

Parasitoids

Almost all of the insect parasitoids recorded in the Boxworth fields were Hymenoptera, particularly wasps in the family Braconidae and super-family Chalcidoidea. As with the herbivores and predators, total numbers in the Full Insurance area relative to the Supervised/Integrated areas were reduced to about half of their levels in the baseline years. For some groups the reduction was much more severe; for example, by the end of the project, the density of ichneumon wasps was only 8% of their abundance in the baseline years (Vickerman 1991). Only one group of parasitoids, the fairy flies (Mymaridae), was more numerous in the Full Insurance area.

TABLE 16.7. Indirect effects of the Full Insurance regime on the beetles *Agonum dorsale* and *Loricera pilicornis*, acting through reduction of prey density. For each species, 1 baseline year is compared to 1 year from the treatment period; within each year, asterisks in column 1 indicate a significant difference between Full Insurance and reduced-input data, and means sharing the same letter in column 2 are not significantly different

	Percentage of insects examined with Collembola in the gut	Average number of eggs per gravid female
Agonum dorsale		
Full Insurance 1983	12	$3 \cdot 1^a$
Supervised 1983	23	$4 \cdot 6^b$
Integrated 1983	19	$4 \cdot 9^b$
Full Insurance 1985	7^*	$3 \cdot 8^a$
Supervised 1985	48	$5 \cdot 2^b$
Integrated 1985	34	$4 \cdot 0^{ab}$
Loricera pilicornis		
Full Insurance 1983	78	$7 \cdot 1^{ab}$
Supervised 1983	81	$5 \cdot 2^a$
Integrated 1983	74	$7 \cdot 9^b$
Full Insurance 1988	35^*	$5 \cdot 2^a$
Supervised 1988	83	$6 \cdot 6^b$
Integrated 1988	84	$7 \cdot 4^b$

PREDATION AND PARASITISM OF CEREAL APHIDS

A majority of the polyphagous predators shown to have declined as a result of the Full Insurance regime are predators of the aphids *Sitobion avenae* and *Metopolophium dirhodum*. Studies were carried out by A.J. Burn to determine the extent of any consequent reduction in predation pressure on these pests in the Full Insurance area.

Small exclusion plots (less than $100 \, m^2$) were established in some fields, by sinking plastic barriers into the ground, and removing ground-dwelling predators from within them (Burn 1988a). The plots were protected from summer aphicide applications. Comparisons could then be made of the build-up of aphid populations inside plots with predators excluded and control plots with predators present. Results showed that in some years there was a difference between matched exclusion and control plots in the Supervised and Integrated fields, but not in the Full Insurance fields (Fig. 16.10). This indicates that only the community of predators in the reduced-

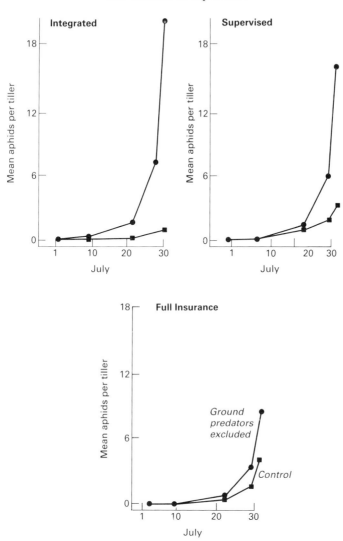

Fɪɢ. 16.10. Examples of the effects of predator exclusion from barriered plots on the development of aphid populations in the three treatment areas. Data are for 1986, a year in which there was a substantial predatory effect in the Supervised and Integrated fields.

input areas was still effective in reducing summer aphid populations. However, this pattern was seen only in years when aphid populations built up slowly; in outbreak years, predator exclusion had no effect on population growth, either in Full Insurance or in reduced-input areas, perhaps because the density of predators was too low to influence the large numbers of aphids.

Predation pressure was measured more directly by (i) placing cards bearing known numbers of *Drosophila* pupae in the fields, and examining them after 24 hours to evaluate the number attacked, and (ii) establishing small artificial 'colonies' of aphids on crop plants. The first of these methods revealed no consistent difference between the high-input and reduced-input areas, but predation on artificial colonies supported the conclusions from exclosure experiments, that significant predation occurred in some years where predator populations had not been affected by insurance pesticide treatments.

Effects of pesticides on predation may be complex due to the elimination of certain alternative prey species, and can thereby cause a *higher* level of predation on pests in an area of high pesticide use. Evidence for this was obtained at Boxworth by the use of an immunoassay to determine whether individual predators had been feeding on *S. avenae*. For example, beetles of the species *Trechus quadristriatus*, a generalist predator, showed a disproportionately high rate of feeding on aphids in the Full Insurance area (Burn 1989).

Parasitism of cereal aphids is difficult to assess, because of sampling biases caused by changes in the behaviour of parasitized aphids. However, approximate measures of parasitism were made by determining the proportions of mummified aphids during visual counts at times of highest aphid population density. Although some differences between areas were detected, there was substantial variation from year to year and from field to field, and no clear long-term effects attributable to pesticide use were apparent.

Overall, this part of the work shows that predation was reduced in some circumstances, but was very variable, depending on the species of predators present and the rate of population increase of aphids. Investigations of predation on slugs (Burn 1988b) also revealed evidence for complex combined effects of direct and indirect pesticide impacts.

RECOVERY OF INVERTEBRATE POPULATIONS

Following the 5-year treatment phase of the project, it was clear that many groups of invertebrates had been severely reduced in density by the Full Insurance regime. In order to assess whether, and how quickly, these species would recover when pesticide inputs were reduced, the project was continued for a further 3 years, in modified form.

After the 1988 harvest, five of the ten fields were retained under experimental control. One Full Insurance field was kept as a prophylactic, high-input standard, while two other Full Insurance fields were switched to a Supervised programme, and compared to two of the previous Supervised fields under similar management.

At the end of the first continuation year, sampling by G.P. Vickerman showed that previously affected ground beetles (*Agonum dorsale*, *Bembidion obtusum*, *Notiophilus biguttatus* and *Pterostichus* spp.) had not increased in the fields recovering from a high input. Similarly, numbers of *Trechus quadristriatus*, which was favoured by the high-input regime, showed no signs of returning to their previous level. However, among the spider fauna, the Tetragnathidae had recovered to near the level recorded in the Supervised fields, while Lycosidae and Linyphiidae both showed signs of increase. The Linyphiidae (money-spiders), in particular, are likely to be rapid colonists, by virtue of their habits of 'ballooning' through the air. Ground-walking colonists evidently take longer to repopulate an area, and it is to be expected that for many susceptible taxa, recovery from an adverse pesticide regime will take longer than the original reduction in numbers. This applies both to short-term recovery within a season (cf. Jepson & Thacker 1990) and to changes on a longer time scale.

Densities of several pest species (e.g. the grain aphid *Sitobion avenae* and the orange wheat blossom midge *Sitodiplosis mosellana*) were higher in the fields recovering from the Full Insurance programme. This is consistent with the fact that predator populations had not recovered, and with the use of a more selective aphicide (pirimicarb), in place of demeton-S-methyl, which provided greater control of pests under the previous Full Insurance regime.

PLANTS

The flora of cereal fields and their boundaries includes some species regarded as weeds, and some that are not damaging to the growth of the crop. Both categories were comprised of grasses and broad-leaved species. Much of the investigation of plants at Boxworth, carried out by E.J.P. Marshall, was directed at assessment of weed densities, in order to aid decision-making on the use of herbicides. This involved periodic measurements of the densities of several species of grasses (based on inflorescences per m^2) and of dicotyledonous species in the crop (using the density of plants, weighted by a correction factor related to the competitiveness of the species; see Wilson 1986). Weed populations during the treatment years were rather lower in the Full Insurance area than the Supervised/ Integrated area, corresponding to the greater use of herbicides (Marshall

1991). Nevertheless, certain species (particularly meadow brome *Bromus commutatus*) continued to be recorded at high density in Full Insurance fields.

To investigate the field weed flora further, soil cores were taken and the density of viable seed was assessed by allowing germination to occur over a 2-year period. This revealed a higher seed density, and a greater number of species, in the soil seed-banks of the Supervised and Integrated fields. However, there was no baseline information, and the lack of any trend over the course of the 5 treatment years indicates that the difference probably existed before 1984. There was a strong pattern, however, for a higher density of seed in fields following oil-seed rape, illustrating the importance of factors other than pesticide use.

Surveys of field margins were conducted to determine any effects of pesticide drift. Analysis of the species composition of the hedge-bottom flora identified a variety of different boundary types, but there were no consistent differences between the three treatment areas (Marshall 1991). The average numbers of species at each sample point were 13·9 (Integrated area), 14·7 (Supervised) and 13·6 (Full Insurance). However, there was substantial variation from field to field, reflecting aspects of hedge structure, particularly the width of the boundary and the closeness of cultivation. There was also a high density of seed in soil from the boundaries compared to open field samples (Fig. 16.11), probably as a result of cultivations and previous pesticide use on the fields. These results suggest that pesticide drift had no significant effect on plants in the field margins, although it is possible that the species there may be those that are tolerant of low concentrations of drift.

The relationship between field and boundary floras is important, because of the possibility that weed problems are associated with the spread of weeds from the field margin (Marshall & Smith 1987). The surveys conducted at Boxworth and elsewhere have shown that this is not the case; levels of particular weed species in fields were not correlated with their densities in the margins. Indeed, most species occur either in the hedge or the crop, but few are present in both habitats. Marshall (1989a) has defined four categories, covering species confined to the boundaries, species at highest density in the open fields, species that favour the transitional area of the headland between crop and boundary, and the few species, such as cleavers *Galium aparine*, that do spread into the crop as weeds. Examples of these patterns are shown in Fig. 16.12.

Overall, therefore, the studies of plants at Boxworth suggest that there were few consequences of the Full Insurance regime on species composition or density, at least over a 5-year period. It is likely that the existence of

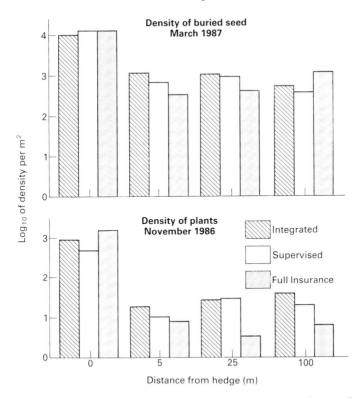

FIG. 16.11. Density of broad-leaved plants, and of buried seed, at various distances from the field boundary in the three treatment areas in 1986–87.

seed-banks protected from pesticide effects provides a stable reservoir of colonists that would only decline after a long period of sustained high inputs. Such effects would be partly checked by the opposite influence of break crops on plant density and seed production.

DISCUSSION

Research in the Boxworth Project has revealed a wide range of consequences arising from a particular high-input, prophylactic approach to chemical crop protection for cereals. The broad scope of the ecological investigations carried out demonstrated different combinations of short- and long-term toxic effects, and indirect changes, for birds, small mammals, invertebrates and plants. Table 16.8 summarizes some of the principal patterns.

Variability of pesticide-induced effects is attributable to three general

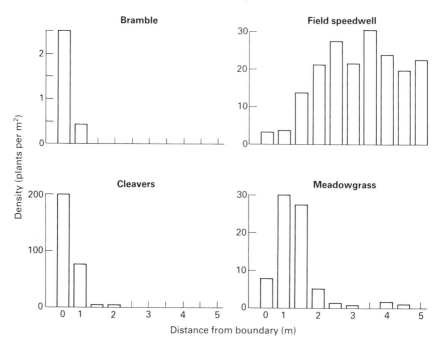

FIG. 16.12. Examples of plant distribution patterns close to the field boundary. The data are average densities along transects up to 5 m into the field, for bramble *Rubus fruticosus*, field speedwell *Veronica persica*, cleavers *Galium aparine*, and meadowgrass *Poa trivialis*.

properties of plants and animals, at both the individual and population levels: (i) the ability to '*escape*' exposure to the pesticide; (ii) inherent *susceptibility* to the doses experienced, changes in prey availability, or other effects; and (iii) the capacity to *recover* from adverse effects. It is the combination of these three aspects that determines which species are able to survive in an environment created by sustained high pesticide usage.

Escape from exposure

Resilience to pesticides can be achieved in a variety of ways. For example, 'escape' from exposure may be a result of behaviour allowing individual animals to move away from the place of application. This is probably the major route by which mobile vertebrates such as birds and rabbits avoid the most severe toxic effects of exposure, as they leave fields during the disturbance of agricultural operations. However, animals may then return, to experience exposure that depends largely on the persistence of chemical residues on ground, foliage or prey. Studies of rabbits suggested such a delay in exposure, and observations of tree sparrows also demonstrated that the birds failed to avoid areas recently treated with summer aphicides. Thus 'escape' by a behavioural response is often incomplete and short-lived.

In addition to short-term departure from the site of pesticide use, there are many animals whose migratory habits ensure that they are absent at the time of some potentially damaging applications. Examples are small mammals that retreat from fields on a daily basis, and some of the polyphagous predators that spend the winter in field margins, usually recolonizing fields only after winter insecticides have been applied. These habits are less certain to protect animals from exposure than is direct avoidance behaviour, particularly if applications are sometimes delayed until after the animals have recolonized.

For other species, it is their selection of microhabitats that determines the level of exposure. For example, invertebrate species that are wholly subterranean, such as many springtails, or have underground larval stages (e.g. *Trechus quadristriatus*), are able to avoid direct exposure to certain chemical applications. Others spend time on the ground, or low in vegetation, where spray applications are intercepted by the dense crop canopy, reducing exposure.

Species may also escape in time, if they have seasonal breeding cycles in which emergence from protected sites, or recruitment, occurs after potentially harmful applications have taken place. This was seen in a number of predatory beetles at Boxworth (Burn 1989), and in annual plant species which emerge from buried seed.

Susceptibility

Even among animals that receive a particular level of exposure, variation in effects is to be expected as a result of inherent differences between species in susceptibility. This issue was not addressed in the Boxworth Project, and few data are available on the comparative toxicity of agricultural chemicals to wild species (e.g. Smith 1987; Marshall 1989b). Insecticides are more likely than are herbicides or fungicides to affect non-target invertebrates directly, although the overall indirect effects of herbicides may be more serious. Also, the toxicity of insecticides to vertebrates is generally lower than to invertebrates, because of screening for selectivity against pests. Understanding of the reasons for species-differences in pesticide toxicity is poor (e.g. Ware 1980; Greig-Smith, Walker & Thompson, in press) although some differences can be related to taxonomic trends in enzyme systems (Walker 1983).

Table 16.8. Summary of some of the major effects of the Full Insurance pesticide regime at Boxworth on the fauna and flora of winter wheat fields

| | Changes in the density of species | | | |
	Direct	Indirect	Community effects	Critical pesticide applications
Birds	None	Reduction in starling density (cause unknown)	None	Summer aphicides affected diet and esterases of tree sparrows and house sparrows
Mammals	Wood mice temporarily reduced, but recovered rapidly through immigration	Shift in diet and distribution of shrews	None	Summer aphicides caused enzyme changes in rabbits; autumn molluscicides affected wood mice
Soil invertebrates	Some species decreased (e.g. *Folsomia*)	Some increased, due to changes in other species (e.g. *Isotomiella*)	None	Not investigated
Crop invertebrates				
Herbivores	Most species reduced	No effects of weed loss, as most are crop feeders	Overall reduction of 50%	Summer aphicides affected species on the crop plants
Detritivores	Little overall change	Some species increased, possibly due to reduced predation	None	None

Carnivores	Variable reductions, especially species exposed over winter or on vegetation in summer	Declines due to reduction of prey (e.g. *Agonum*) or changes in competitors (e.g. *Trechus*)	Overall reduction of 50%, lower capacity for predation on pests	Winter insecticides, especially delayed spring application and in sparse crop
Parasites	Reduction of most groups	Reduction due to decreases in numbers of hosts	Overall reduction of 50%, no consistent effect on parasitism	None identified
Plants	Seasonal reduction of growing plants, but replaced from seed-banks	None investigated	None	None investigated

Recovery

There are also many ways in which recovery from the effects of pesticides can occur. In some cases, recolonization may be almost immediate (e.g. replacement of territorial animals by non-breeding 'floaters' or those in less preferred habitats), or may be delayed until a critical event in the life cycle (reproduction, or a migration episode). However, it may also be limited by the speed at which colonists are able to move into the field. In the large expanses of contiguous cereals now grown in many areas, this constraint may prevent complete recolonization from one year to the next.

Small mammals and many invertebrates recolonize from field margins and more distant sites, on the ground or by air. For plants, replacement occurs within the field itself, from the abundant seed-banks present both in field boundary habitats and in the open fields (Fig. 16.11). Few of the animal species in cereal fields at Boxworth were multiple breeders that might recover solely by rapid reproduction within a season.

Refuges

These patterns highlight the importance of refuges within reach of cereal fields, encouraging both escape and recovery. Depending on the species, refuges may be patches of non-crop habitats, or microhabitats within the fields (underground sites for seeds, or insect larval stages). Accordingly, the resilience of the cereal ecosystem depends as heavily on the management of these areas (hedgerow structure, soil cultivation methods, etc.) as it does on the nature of the crop protection chemicals used (Wratten 1988).

Pesticide usage

Although the project was designed primarily to examine the combined effects of all pesticides used, the results also provide an indication of which particular pesticides were most damaging. This does not offer any easy generalizations applicable to all the varied forms of plant and animal wildlife. However, for most invertebrates it was the winter insecticide sprays, applied at a time when there is little protective vegetation cover, that contributed most to the population declines seen in the Full Insurance area. Although a large number of species were also exposed to summer aphicides, the effects were less consistent, because of the differences in exposure discussed above. No evidence emerged to suggest that particular herbicide applications were critical, partly because plant populations were buffered by the existence of substantial seed-banks. In the long term, those herbicides used at times when they would prevent flowering, fruiting and return of seed to the soil will be the most harmful.

One major conclusion is that the adverse effect of the Full Insurance regime was not due principally to the large number of applications, but rather to the inclusion of certain particularly damaging pesticide uses. If these were incorporated into a much less intensive pesticide programme, the effects on vulnerable parts of the wild flora and fauna might be just as great. Also, if more toxic alternative products had been used for some applications (e.g. dimethoate in place of demeton-*S*-methyl as a summer aphicide), the adverse effects might have been greater.

Another aspect is the timing of applications. For some pesticides, the planned time of use may coincide with a critical point in the annual cycle of susceptible species, but for others an effect may occur only if application is unusually late. This happened in the case of triazophos applied for control of yellow cereal fly *Opomyza florum*, which in the spring of 1986 was delayed because of unfavourable weather conditions, and was eventually applied when migration of overwintering predators from field boundaries had commenced. The chance of these mistimed events, with disproportionate impact on certain species, can be regarded as an inevitable cost of long-term prophylactic pest control.

Changes at the community level

Effects observed at Boxworth were not just on individual animals and plants, or on their populations, but could also be identified at the community level. For some groups (e.g. birds and plants), the species-composition of the community remained unchanged after the start of the Full Insurance treatment, but for both herbivorous and predatory invertebrates there were major changes (Tables 16.5 and 16.6). This reflects not only the severity of pesticide effects on some taxa, but also more complex interactions that resulted in higher populations of certain species under the Full Insurance regime. Because of these opposing patterns, it is important to investigate functional aspects such as predation and parasitism by the combined set of natural enemies, as well as to monitor numbers of individual species. This requires manipulative studies involving exclusion experiments or other techniques (Burn 1989).

As the years of high inputs accumulated in the Full Insurance area, it appeared that the overall potential for 'natural' control of cereal pests, particularly the summer aphids, was diminished. Many predators were reduced in number, by a combination of toxic effects and changes in the availability of their prey. As a result, the pests and other species which tend to be efficient colonizers became more liable to sudden, unpredictable outbreaks in the Full Insurance area. Also, when the Full Insurance programme was stopped in some fields during the continuation study, these pests benefited because they responded more quickly to the new conditions than did predators and alternative prey.

The farm landscape

Overall, the impact of a high pesticide input depends not only on the characteristics of the chemicals used, and on the species of wild animals and plants in the area, but also on several features of local geography. These include the distribution of patches of non-crop habitats around fields (influencing the availability of refuges), the scale of use of pesticide products (i.e. whether similar crops were treated at the same time over a wide area), field size (determining the speed and extent of recolonization from neighbouring habitats), and management practices, particularly in field boundaries. Widespread removal of hedgerows and enlargement of fields is likely to reduce the numbers of some species that are exposed to pesticides (e.g. most farmland birds). However, for those animals that visit larger open fields, exposure may be increased, owing to the scarcity of temporary refuge habitats.

The interface between the crop and other habitats is physically well-defined, but its functional significance varies between different groups of animals and plants. For plants, there was limited overlap of the flora of field and boundary (Fig. 16.12), there was little spread of weeds from hedges into the crop. In contrast, the zone of interchange of some mobile invertebrates was broad, with a seasonally-changing density gradient extending many metres onto the field surface. For very wide-ranging species (birds, and some invertebrates) the nature of the boundary zone is almost irrelevant, although it has been observed elsewhere that wood mice in woodland may be behaviourally isolated from those in adjacent fields (Westlake *et al.* 1980).

FUTURE STUDIES

The main part of the Boxworth Project ended in the autumn of 1988, and the continuation phase is to be completed at the harvest of 1991. However, two new projects have been initiated to extend the implications of the work done at Boxworth, for both ecological and economic aspects.

One of these, named 'SCARAB' (Seeking Confirmation About Results At Boxworth), is aimed at establishing the generality of the effects of a high pesticide programme on invertebrates, in arable rotations rather than continuous winter wheat production, and on a variety of soil types (Cooper 1990). For this study, the contrast will be between an average level of pesticide input, such as a majority of farmers might use now, and a greatly reduced input, avoiding insecticide use.

A companion project, 'TALISMAN' (Towards A Low Input System, Minimizing Agrochemicals and Nitrogen), is directed towards the development of economically viable minimum-input farming systems, on a similar range of crops and rotations. This work will include many aspects of 'Integrated' farming that reduce the need for chemical inputs (cf. Vereijken 1986; El Titi, this volume, pp. 399–411).

Both projects will be experimental, involving plots on several of MAFF's experimental husbandry farms. Together, they will continue the contribution made by the Boxworth Project to an understanding of the ecology of the cereal ecosystem.

ACKNOWLEDGMENTS

The information presented in this paper is the outcome of work by a large team of researchers. I am grateful to them all for their contributions, and to Alastair Burn, Jon Marshall and Paul Vickerman for comments on a draft of the manuscript.

REFERENCES

Aebischer, N. & Potts, G.R. (1990). Long-term changes in numbers of cereal invertebrates assessed by monitoring. *Proceedings of the 1990 British Crop Protection Conference — Pests and Diseases*, pp. 163–172. British Crop Protection Council, Farnham.

Burn, A.J. (1988a). Effects of scale on the measurement of pesticide effects on invertebrate predators and predation. *Field Methods for the Study of Environmental Effects of Pesticides.* (Ed. by M.P. Greaves, P.W. Greig-Smith & B.D. Smith), pp. 109–117. Monograph No. 40, British Crop Protection Council, Croydon.

Burn, A.J. (1988b). Assessment of the impact of pesticides on invertebrate predation in cereal crops. *Aspects of Applied Biology*, **17**, 279–288.

Burn, A.J. (1989). Long-term effects of pesticides on natural enemies of cereal crop pests. *Pesticides and Non-target Invertebrates* (Ed. by P.C. Jepson), pp. 177–193. Intercept, Wimborne, Dorset.

Burn, A.J. (1991). Interactions between cereal pests and their predators and parasites. *The Boxworth Project: Pesticides, Cereal Farming and the Environment* (Ed. by P.W. Greig-Smith, G.K. Frampton & A.R. Hardy). HMSO, London.

Cooper, D.A. (1990). Development of an experimental research programme to pursue the results of the Boxworth Project. *Proceedings of the 1990 British Crop Protection Conference — Pests and Diseases*, pp. 153–162. British Crop Protection Council, Farnham.

Crawley, M.C. (1970). Some population dynamics of the bank vole, *Clethrionomys glareolus* and the wood mouse *Apodemus sylvaticus* in mixed woodland. *Journal of Zoology, London*, **160**, 71–89.

Feare, C.J. (1984). *The Starling*. Oxford University Press, Oxford.

Green, R.E. (1979). The ecology of wood mice (*Apodemus sylvaticus*) on arable farmland. *Journal of Zoology, London*, **188**, 357–377.

Greig-Smith, P.W. (1989a). The Boxworth Project — environmental effects of cereal pesticides. *Journal of the Royal Agricultural Society of England*, **150**, 171–187.

Greig-Smith, P.W. (1989b). Effects of non-persistent insecticides on bird and mammal populations on farmland. *Ecotoxicology* (Ed. by C.H. Walker & D. Anderson), pp. 12–18. Institute of Biology, London.

Greig-Smith, P.W. & Westlake, G.E. (1988). Approaches to hazard assessment for small mammals in cereal fields. *Field Methods for the Study of Environmental Effects of Pesticides* (Ed. by M.P. Greaves, P.W. Greig-Smith & B.D. Smith), pp. 303–311. Monograph No. 40, British Crop Protection Council, Croydon.

Greig-Smith, P.W., Frampton, G.K. & Hardy, A.R. (Eds) (1991). *The Boxworth Project: Pesticides, Cereal Farming and the Environment*. HMSO, London.

Greig-Smith, P.W., Walker, C.H. & Thompson, H.M. (1990). Ecotoxicological consequences of the interactions between avian esterases and organophosphorus compounds. *Clinical and Experimental Toxicology of Anticholinesterases* (Ed. by B. Ballantyne & T.C. Marrs). Butterworths, London.

Grue, C.E., Powell, G.V.N. & McChesney, M.J. (1982). Care of nestlings by wild female starlings exposed to organophosphate pesticides. *Journal of Applied Eology*, **19**, 327–335.

Hardy, A.R. (1986). The Boxworth Project: a progress report. *Proceedings of the 1986 British Crop Protection Conference, Pests and Diseases*, 1215–1224. British Crop Protection Council, Farnham.

Hart, A.D.M., Fletcher, M.R., Greig-Smith, P.W., Hardy, A.R., Jones, S.A. & Thompson, H.M. (1990). Le Projet Boxworth: effects de regimes de pesticides contradictoires sur les oiseaux de terre arable. *Relations entre les Traitements Phytosanitaire et la Reproduction des Animaux*, pp. 225–232. Annales de l'ANPP, No. 2.

Jarvis, R.H. (1988). The Boxworth Project. *Britain since 'Silent Spring'* (Ed. by D.J.L. Harding), pp. 46–55. Institute of Biology, London.

Jepson, P.C. & Thacker, J.R.M. (1990). Analysis of the spatial component of pesticide side-effects on non-target invertebrate populations and its relevance to hazard analysis. *Functional Ecology*, **4**, 349–355.

Johnson, I.P., Hare, R. & Flowerdew, J.R. (1991). Populations and diet of small rodents and shrews in relation to pesticide usage. *The Boxworth Project: Pesticides, Cereal Farming and the Environment* (Ed. by P.W. Greig-Smith, G.K. Frampton & A.R. Hardy). HMSO, London.

Ludke, J.L., Hill, E.F. & Dieter, M.P. (1975). ChE response and related mortality among birds fed ChE inhibitors. *Archives of Environmental Contamination and Toxicology*, **3**, 1–21.

MacFadyen, A. (1961). Improved funnel-type extractor for soil arthropods. *Journal of Animal Ecology*, **22**, 65–77.

Marchant, J.H., Hudson, R., Carter, S.P. & Whittington, P. (1990). *Population Trends in British Breeding Birds*. British Trust for Ornithology/Nature Conservancy Council.

Marshall, E.J.P. (1989a). Distribution patterns of plants associated with arable field edges. *Journal of Applied Ecology*, **26**, 247–257.

Marshall, E.J.P. (1989b). Susceptibility of four hedgerow shrubs to a range of herbicides and plant growth regulators. *Annals of Applied Biology*, **115**, 469–479.

Marshall, E.J.P. (1991). Patterns of distribution of plants in the fields and their boundaries. *The Boxworth Project: Pesticides, Cereal Farming and the Environment* (Ed. by P.W. Greig-Smith, G.K. Frampton & A.R. Hardy). HMSO, London.

Marshall, E.J.P. & Smith, B.D. (1987). Field margin flora and fauna: interaction with agriculture. *Field Margins*. (Ed. by J.M. Way & P.W. Greig-Smith), pp. 23–34. Monograph No. 35, British Crop Protection Council, Croydon.

O'Connor, R.J. & Shrubb, M. (1986). *Farming and Birds*. Cambridge University Press, Cambridge.

Powell, G.V.N. (1984). Reproduction by an altricial songbird, the red-winged blackbird, in fields treated with the organophosphate insecticide fenthion. *Journal of Applied Ecology*, 21, 83–95.

Smith, G.J. (1987). Pesticide use and toxicology in relation to wildlife: organophosphorus and carbamate compounds. *US Department of the Interior, Resource Publication No. 170*. Washington, DC.

Somerville, L. & Greaves, M.P. (Eds) (1987). *Pesticide Effects on Soil Microflora*. Taylor & Francis, London.

Tarrant, K.A. & Westlake, G.E. (1988). Laboratory evaluation of the hazard to wood mice, *Apodemus sylvaticus*, from the agricultural use of methiocarb molluscicide pellets. *Bulletin of Environmental Contamination & Toxicology*, 40, 147–152.

Tarrant, K.A., Johnson, I.P., Flowerdew, J.R. & Greig-Smith, P.W. (1990). Effects of pesticide applications on small mammals in arable fields, and the recovery of their populations. *Proceedings of the 1990 British Crop Protection Conference — Pests & Diseases*, pp. 173–182. British Crop Protection Council, Farnham.

Tinbergen, J.M. (1981). Foraging decisions in starlings (*Sturnus vulgaris* L.). *Ardea*, 69, 1–67.

Vereijken, P. (1986). From conventional to integrated agriculture. *Netherlands Journal of Agricultural Science*, 34, 387–393.

Vickerman, G.P. (1988). Farm scale evaluation of the long-term effects of different pesticide regimes on the arthropod fauna of winter wheat. *Field Methods for the Study of Environmental Effects of Pesticides*. (Ed. by M.P. Greaves, P.W. Greig-Smith & B.D. Smith), pp. 127–135. Monograph No. 40, British Crop Protection Council, Croydon.

Vickerman, G.P. (1990). The effects of different pesticide regimes on the invertebrate fauna of winter wheat. *The Boxworth Project: Pesticides, Cereal Farming and the Environment*. (Ed. by P.W. Greig-Smith, G.K. Frampton & A.R. Hardy). HMSO, London.

Walker, C.H. (1983). Pesticides and birds — mechanisms of selective toxicity. *Agriculture, Ecosystems and Environment*, 9, 211–226.

Ware, G.W. (1980). Effects of pesticides on non-target organisms. *Residue Reviews*, 76, 173–201.

Watts, C.H.S. (1968). The foods eaten by wood mice (*Apodemus sylvaticus*) and bank voles (*Clethrionomys glareolus*) in Wytham woods, Berkshire. *Journal of Animal Ecology*, 37, 25–41.

Westlake, G.E., Blunden, C.A., Brown, P.M., Bunyan, P.J. Martin, A.D., Sayers, P.E., Stanley, P.I. & Tarrant, K.A. (1980). Residues and effects in mice after drilling wheat treated with chlorfenvinphos and an organomercurial fungicide. *Ecotoxicology and Environmental Safety*, 4, 1–16.

Wilson, B.J. (1986). Yield responses of winter cereals to the control of broad-leaved weeds. *Proceedings of the European Weed Research Symposium 1986, Economic Weed Control*, pp. 75–82.

Wratten, S.D. (1988). The role of field boundaries as reservoirs of beneficial insects. *Environmental Management in Agriculture: European Perspectives* (Ed. by J.R. Park), pp. 144–150. Belhaven Press, London.

17. CONSERVATION HEADLANDS: A PRACTICAL COMBINATION OF INTENSIVE CEREAL FARMING AND CONSERVATION

N.W. SOTHERTON

The Cereals and Gamebirds Research Project, The Game Conservancy Trust, Fordingbridge, Hampshire SP6 1EF, UK

INTRODUCTION

The Game Conservancy Trust has been carrying out research in the cereal ecosystem since the mid 1930s when annual surveys of the breeding densities of the grey partridge (*Perdix perdix*) began. A nationwide decline occurred from 1952 onwards, from approximately twenty-five pairs/km² to less than five pairs in the late 1980s; a decrease of over 80%. Declining populations have also been recorded in North America (USA and Canada) and both eastern and western Europe (Potts 1986).

In Britain the grey partridge is predominantly a species of the lowland arable landscape, especially cereal fields. The most important changes that have occurred on farmland coincidentally with the partridge decline have been the intensification of grain production, involving increasing field size (hedgerow removal), improved drainage, increased use of fertilizers, improved plant breeding and, especially, the increased use of pesticides. Since the 1950s, increases in both the numbers of pesticides used and increases in the areas of cereals sprayed have occurred (Sly 1986; Rands, Hudson & Sotherton 1988). As a result, much of the research effort of The Game Conservancy Trust has been directed at testing various hypotheses regarding the decline of wild grey partridges.

THE EFFECTS OF PESTICIDES

Direct toxic effects

In the early 1950s, many deaths of partridges and pheasants from the direct toxic effects of pesticides were attributed to the dinitro-ortho-cresol herbicides (DNOCs) and later to the cyclodiene insecticides (mainly dieldrin) which peaked in the early 1960s. However, since then, occur-

TABLE 17.1. Frequency of incidents of direct toxicity in gamebirds attributed to pesticides in Britain. From Sotherton (1988)

Decade	DNOCs (herbicides)	Cyclodienes (mainly dieldrin)	Other insecticides	Total
1950s	31	>85	34	>150
1960s	0	228	26	254
1970s	1	18	13	32
Total	32	>330	73	>435

rences of direct poisoning have greatly declined following restrictions in the use of these compounds (Table 17.1) (Potts 1986).

Although the amounts and numbers of pesticides used have increased since the 1950s, the rate of increase has been greatest after the period when the problems of direct toxicity were most apparent. Although fewer compounds were used, those available during the 1950s and 1960s were more toxic to gamebirds compared to those used throughout the 1970s to the present day (Table 17.2) (Potts 1986).

Sublethal effects

Sublethal or chronic effects could adversely affect birds by increasing their susceptibility to disease, lowering their fecundity or the hatchability of their eggs or by lowering the viability of their chicks. However, there has been no evidence (or it has been inconclusive) to suggest sublethal effects may be important (Potts 1986). Recent work has suggested that some pesticides may induce sublethal effects by acting as precursors which subsequently induce susceptibility in adult partridges to later applications

TABLE 17.2. Median lethal concentrations of pesticides to gamebirds (mg a.i./kg body wt). From Sotherton (1988)

1950s Cyclodienes DNOC herbicides Methyl mercury	<75
1980s Insecticides	700
Herbicides	2900
Fungicides	4300

of other pesticides (Riviere, Bach & Grolleau 1985; Johnston *et al.* 1989). This potentiation of one pesticide group by another is currently an area of concern.

It was possibly because of the failure to detect any large-scale lethal or sublethal effects of pesticides on partridges, that repeated concern expressed about the continuing adverse impact of pesticides in their decline were described as recently as 1980 as 'hysterical and unjustified' (Broadbent 1980). This judgment however ignored the possible impact of the indirect effects of pesticides.

Indirect effects

In 1984, the Cereals and Gamebirds Research Project began working on the problems associated with wild gamebird production on intensively farmed arable land. The Project arose as a result of studies on grey partridges carried out in Sussex which began in 1968 to identify the problems encountered by stocks of wild breeding birds and the factors that had contributed to the observed decline.

Previous studies (Blank, Southwood & Cross 1967; Potts 1970a) found that the key factor causing changes in a grey partridge population in Hampshire was chick mortality and had clearly linked the observed national decline with the increasingly poor levels of chick survival. Chick survival was shown to be associated with the availability of sufficient quantities of the preferred insects that are essential in the diet of young chicks (Southwood & Cross 1969; Potts 1986). Moreover, it has been shown that pesticides appeared to be a major factor reducing populations of insects preferred by chicks (Potts 1986). These insects include a group of Coleoptera (Chrysomelidae, small diurnal Carabidae and Curculionidae), larval forms of Lepidoptera and Tenthredinidae (especially species of the genus *Dolerus*) and many members of the Heteroptera (especially species of the genus *Calocoris*). Many of these preferred insects were found to be more abundant at the edges of cereal fields where grey partridge broods forage (Green 1984).

It has now been established that various types of pesticides can severely reduce numbers of these non-target species, especially in the short-term; insecticides (Vickerman 1977; Vickerman & Sunderland 1977; Sotherton, Moreby & Langley 1987) insecticidal fungicides (Sotherton & Moreby 1988; Sotherton 1989) and herbicides. The use of herbicides has probably been the most important effect, at least until recently, because of their ability to remove cereal field weeds on relatively short temporal and spatial scales; these weeds being the host plants of many of these phytophagous chick-food insects (Southwood & Cross 1969; Vickerman 1974; Sotherton 1982).

Our challenge therefore was to devise practically orientated and costed management options whereby cereal farmers could continue to farm profitably and maintain high levels of production but at the same time overcome some of the adverse effects of pesticides on farmland game and wildlife. As an independent research organization funded primarily by farmers and landowners, The Game Conservancy Trust could not easily advocate a regression in farming technology (the abandonment of pesticides) and it wanted to avoid unnecessary damage caused by insect pests, weeds or diseases. A compromise was clearly necessary and its achievement became the remit of the Cereals and Gamebirds Research Project. Uniquely, in the early years of the Project, the funding to seek this compromise came from the subscriptions of a core group of enthusiastic cereal farmers who wished to use pesticides in more judicious ways which would reduce the exposure of non-target species to the (unforeseen) adverse effects of pesticides.

APPROACHES TO REDUCE THE EXPOSURE OF NON-TARGET INVERTEBRATES TO THE ADVERSE EFFECTS OF PESTICIDES

Three approaches to solve these problems were paramount.
1   Screening pesticides for selectivity against non-target, beneficial insects and less harmful weeds; more information on the side-effects on non-target organisms was needed.
2   The exclusive use of selective compounds and some degree of pesticide exclusion on certain areas of cereal fields, especially field margins.
3   The enhancement of insect-pest natural enemy populations within integrated control programmes to reduce the use of insecticides with a consequent reduction in the detrimental effects of pesticides upon game and wildlife.
Only the results of the first two will be reviewed in this paper (see Wratten and Powell, pp. 233–257 and Altieri, pp. 259–272, both this volume, for discussion of the third point).

Pesticide screening for selectivity against non-target insects and plants

In the absence of useful data from Government or most manufacturers relating to the selectivity of pesticides approved for use in UK cereal fields in the 1980s (over eighty compounds) a limited programme of screening was carried out.

Fungicides

When The Game Conservancy's Sussex study began in 1970 (see Aebischer, pp. 305–331, this volume), none of the fields in the study area were treated with fungicide. In 1986, almost every field received at least one foliar spray. At the same time, data were made available that suggested that fungicides could be having an effect on beneficial insect populations (Colbourn & Asquith 1973; Reyes & Stevenson 1975; Vickerman 1977; Vickerman & Sotherton 1983). Also national figures for fungicide use had shown a great increase in frequency of application. Between 1974 and 1982 there was a thirty-five-fold increase in the area of winter wheat treated but only a 1·4-fold increase in the total area of crop grown (Rands, Hudson & Sotherton 1988). Screening work began in the laboratory with all the single active ingredient foliar fungicides approved for use in UK cereal fields. All compounds were tested at maximum recommended field dose rates and dilutions against two kinds of beneficial insects; hoverfly larvae (Diptera: Syrphidae) which are important predators of cereal aphids and leaf beetle larvae (Coleoptera: Chrysomelidae) which were used as representative chick food items. Screening consisted of an assessment of mortality 48 h after topical application in comparison with water-treated control larvae and larvae treated with a toxic standard insecticide. Although a crude assessment of the effects of a pesticide, this method did allow for a relatively rapid selection of candidate compounds to be isolated for further, more intensive study. Only one of the twenty-seven compounds tested displayed any biologically significant degree of insecticidal activity under laboratory conditions. This was the organophosphate compound, pyrazophos (Sotherton 1989).

Conclusions concerning the possible extent of such insecticidal properties could not be drawn from laboratory work alone. The next step was therefore to conduct large-scale field trials to simulate realistic conditions of crop growth, weather and spraying likely to be found on farmland and therefore conditions in which non-target, beneficial species would be most likely to be exposed to pyrazophos. Over 2 years, a range of cereal crops were sprayed with test compounds and the extent of the insecticidal properties under field conditions identified (Table 17.3). Further details of experimental design, methodology and statistical analysis can be found elsewhere (Sotherton, Moreby & Langley 1987; Sotherton & Moreby 1988; Sotherton 1989).

N.W. SOTHERTON

TABLE 17.3. Percentage reduction in total numbers of beneficial arthropods found on plots treated with pyrazophos relative to untreated areas in a range of cereal crops at varying growth stages in southern England, 1984 and 1985 (data gathered from ground-zone searches). From Sotherton (1989)

	Winter wheat (GS 50)	Spring barley (GS 60)	Winter barley (GS 37)	Spring wheat (GS 37)
Total predatory arthropods	65·4	73·7	76·3	74·4
Carabidae	68·6	65·1	74·0	64·9
Staphylinidae	80·2	75·0	83·4	83·7
Tachyporus spp. (Coleo.: Staph.)	80·3	78·7	88·2	85·6
Highly ranked polyphagous predators	86·5	70·0	85·6	83·3
Stenophagous predators	88·9	94·0	88·0	96·3

GS, growth stage, according to Zadoks, Chang & Konzak (1974).

It is accepted that the use of laboratory data based on an assessment of mortality after only 48 h was a crude measure of the 'safety' of each pesticide in this study. Additional information may be obtained from so-called semi-field trials, where large numbers of animals are exposed to realistic doses of pesticides. Examples of semi-field work using pyrazophos and a detailed critique of the criteria for the design, execution and analysis of field trials can be found elsewhere (Sotherton, Jepson & Pullen 1988).

Insecticides

Insecticide use in cereal fields in Britain has also greatly increased over the last decade, although this increase has not been in response to changes in the observed levels of the major summer pest species, the grain aphid (*Sitobion avenae*). Instead, trends towards insurance or prophylactic spraying of broad-spectrum aphicides have been recorded (Sotherton 1989).

However, unlike fungicide use, applications of insecticides would be expected to reduce numbers of beneficial insects. Previous work has shown that some of the compounds in use were broad-spectrum and capable of significantly reducing the numbers of beneficial insects (Vickerman & Sunderland 1977). In contrast, other compounds were thought to be relatively specific in their mode of action and therefore relatively safe or selective to beneficial, non-target arthropods, especially bees (Cole & Wilkinson 1984; Potts 1986; Sotherton 1989).

In 1986 large numbers of field-collected *Trechus quadristriatus*, a diurnal carabid beetle, were sprayed in the laboratory and several small plots in winter wheat fields were treated with solutions of four insecticides approved for use in the summer to control the grain aphid in UK cereal fields. Dimethoate caused 100% mortality and a large reduction in numbers, whereas others were far less severe in their effects (Table 17.4).

In 1986, large-scale field trials were again carried out and preliminary data suggested that the ranking of the activity of these compounds against other beneficial arthropods in the field was identical to that found in the laboratory (Fig. 17.1).

Herbicides

Farmers may be encouraged to use selective insecticides and fungicides without insecticidal properties to help conserve beneficial insects. However this is not the only scope for screening for selectivity. Approximately 60% of the preferred chick-food insects are phytophagous species feeding on

TABLE 17.4. Selectivity of cereal aphicides against non-target insects

Active ingredient (a.i.)	g a.i./ha	*Trechus quadristriatus*, percentage mortality after 48 h[1]	Heteroptera, percentage reduction after 48 h[2]
Demeton-*S*-methyl	122	16·9	68·8
Dimethoate	340	100·0	95·0
Phosalone	490	0·0	0·0
Pirimicarb	140	1·4	0·0
Water	—	0·0	3·2

[1] Laboratory study, application via Potter tower.
[2] Headland plots (100 × 12 m), percentage reduction after 48 h relative to pretreatment numbers; Hampshire, June 1987.

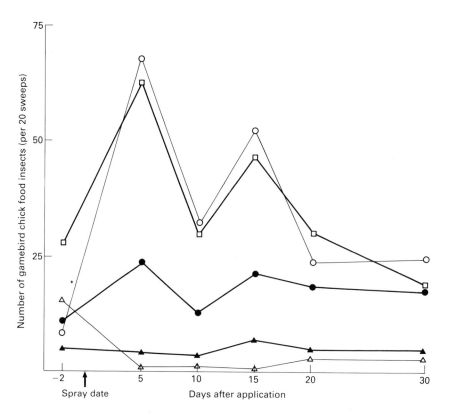

FIG. 17.1. The number of gamebird chick-food insects (per 20 sweeps) on plots of winter wheat either remaining unsprayed or treated with one of four aphicides, southern England June–July 1986. Pirimicarb (○) phosalone (●), demeton-*S*-methyl (▲), dimethoate (△) and untreated (□). From Sotherton (1989).

broad-leaved weeds. Such species as knotgrass (*Polygonum aviculare*), black bindweed (*Fallopia convolvulus*), fat hen (*Chenopodium album*), charlock (*Sinapis arvensis*) and the mayweeds (*Matricaria* spp.) are the host plants of many of these insects (Sotherton 1982; Sotherton, Boatman & Rands 1989). Some species of plants, such as *S. arvensis* and corn poppy (*Papaver rhoeas*) may also provide pollen and nectar for parasitoids and the stenophagous predators of cereal aphids. Seeds of species such as annual meadow-grass (*Poa annua*), *P. aviculare* and *F. convolvulus* are also important foods for adult partridges (Potts 1970b). Finally, many species of polyphagous predatory Coleoptera were found in weedy cereal fields, especially those found with dense understories of *P. annua* (Coombes & Sotherton 1986; N.W. Sotherton, unpublished data). Therefore herbicides have also been screened, not to detect their insecticidal properties but to establish their selectivity against 'beneficial' weeds.

Compounds were sought that removed the pernicious, unacceptable weed species that caused farmers most concern (Boatman 1987, 1989; Boatman & Wilson 1988; Boatman, Freeman & Green 1988; Bain & Boatman 1989), such as black-grass (*Alopecurus myosuroides*), wild oats (*Avena* spp.), rough meadow grass (*Poa trivialis*), common couch (*Elymus repens*), barren brome (*Bromus sterilis*) and cleavers (*Galium aparine*), but did not reduce populations of species considered to have beneficial properties.

Unlike the screening for insecticides and fungicides, herbicide work has only been carried out in small field plots (Boatman 1987). Examples of herbicide screening are given in Table 17.5. So far, residual, broad-spectrum herbicides have been excluded (for example chlortoluron, isoproturon, the sulphonyl ureas and the hydroxybenzonitriles). At present eight compounds have been found to have a degree of selectivity in removing pernicious grass weeds or *G. aparine* without also removing the required species.

Restricted use of pesticides on small areas of cereal fields

Pesticide screening has taken and will take many years to complete, and attempts to date have only begun to provide some of the necessary data (but still excluding sublethal effects and possible effects resulting from many other potential routes of exposure, e.g. ingestion of contaminated prey, or surface residual deposits). Without government input to such work or a requirement within formal pesticide registration procedures for manufacturers to provide the relevant information, farmers will not be able to use pesticides in the selective manner that ought to be possible. Recent

Table 17.5. Percentage control of useful broad-leaved weeds by various herbicides used in UK cereal fields

Herbicide	Dose (g a.i./ha)	Polygonum aviculare	Fallopia convolvulus	Matricaria spp.	Stellaria media
Chlortoluron	3500	89	94	88	
Isoproturon	2500	NS	NS	100	—
Dichlofop-methyl	1140	R	R	R	79
Fenoxaprop-ethyl	180	R	R	R	R
Tralkoxydim*	300	R	R	R	R
Tri-allate	2250	NS	NS	NS	R
Fluroxypyr					NS
March	200	—	NS	88	
April	150	—	100	57	87
Quinmerac*	500	NS	NS	NS	100
					NS

* Under development (UK clearance pending).
— Not present in trial.
NS, no significant control.
R, resistant (no activity against any dicot spp.).

legislation does however provide for the ability to refuse pesticide registration if its use cannot be shown to be an improvement on existing products. Presumably therefore the converse is true, if such products are given clearance for use, the older, broad-spectrum compounds currently being used such as dimethoate could be phased out in favour of newer, more selective compounds, or compounds that have only been given approval after more thorough, environmental-impact screening (for example some of the synthetic pyrethroids). However, farmers require answers 'yesterday' if the aims of both environmental conservation and intensive grain production are to be compatible! An approach was therefore proposed that would provide some immediate amelioration to the direct and indirect adverse effects of pesticides on non-target invertebrates by excluding unacceptable pesticides or limiting their use on relatively small areas of the crop.

Experiments were devised whereby pesticides could be selectively excluded from small areas of the crop at the field margin. To achieve this, the outermost section of the tractor-mounted pesticide spray boom (normally a width of 6 m) was either switched off when spraying around the edge of the crop, avoiding the application of certain chemicals at sensitive times of the year, or these edges or 'headlands' were sprayed separately with products screened for their selectivity and 'approved' for use. The rest of the field was sprayed with the usual complement of pesticides and only the crop margin (which, on average, accounted for 6% of the total field area) received these lower pesticide inputs. These selectively sprayed crop margins are termed 'Conservation Headlands' (Oliver-Bellasis & Sotherton 1986).

Sites and methods

From 1983 to 1987, several large blocks of cereal fields on the principal study farm in southern England were either fully sprayed or were fully sprayed except for the 6 m closest to the field boundary. Pesticide use on this outermost strip varied slightly between years as the technique was refined and pesticide screening progressed, but in all cases the aim was to avoid insecticidal compounds and herbicides affecting most species of broad-leaved weeds. The use of herbicides in the Conservation Headland aimed to selectively remove pernicious grass weeds such as black-grass, wild oats and couch without killing broad-leaved species. This has been achieved following field screening for selectivity and efficacy by using such compounds as tri-allate, dichlofop-methyl and preharvest glyphosate. The only unacceptable broad-leaved species is cleavers (*Galium aparine*). Selective control within the dicots has been less easy. A degree of selective control can be achieved however by using fluroxypyr. Excellent selective control can be achieved using a developmental compound called quinmerac awaiting formal clearance in Britain.

From 1984 onwards similar paired blocks of cereal fields were set up on farms in the eastern counties of England. Details of the experimental design may be found elsewhere (Sotherton & Robertson 1990) as may details of the methodology used to assess weeds, insects and gamebirds (Rands 1985, 1986, in press; Sotherton, Rands & Moreby 1985).

Since 1986, guidelines have been annually updated and distributed to subscribing farmers in time for autumn drilling. They contain recommendations concerning those compounds regarded as being selective enough for use on Conservation Headlands. The latest set of guidelines are summarized in Table 17.6.

Results

(a) Weeds. Selective pesticide use on cereal crop margins has resulted in increases in the numbers of many species of broad-leaved weeds both between fields (Sotherton, Rands & Moreby 1985) and within a field (Table 17.7).

TABLE 17.6. The 1989/90 summary of the guidelines for the use of pesticides in Conservation Headlands

	Autumn		Spring	
Insecticides	Yes	(avoiding drift into hedgerows)	No	
Fungicides	Yes		Yes	(except compounds containing pyrazophos)
Growth regulators	Yes		Yes	
Herbicides				
Grass weeds	Yes	(but avoid broad-spectrum residual products — use only those compounds approved for use*)		
Broad-leaved weeds	No	(except those compounds approved for use against specific problem weeds, e.g. cleaver control with fluroxypyr)		

* Tri-allate, dichlofop-methyl, difenzoquat, flamprop-*m*-isopropyl, fenoxaprop-ethyl.

TABLE 17.7. Mean densities per 0·25 m² (±1 SE) of weeds, numbers of species, biomass and estimate of cover found in headland plots of spring-sown wheat either treated with a herbicide or untreated, Hampshire, June 1988

	Headland plots without herbicide	Headland plots with herbicide	P
Matricaria spp.	5·3 ± 1·8	0·0	—
Sinapis arvensis	0·1 ± 0·04	0·0	—
Polygonum aviculare	1·0 ± 0·3	0·1 ± 0·06	<0·02
Fallopia convolvulus	0·5 ± 0·4	0·03 ± 0·01	NS
Total No. species per 0·25 m²	6·8 ± 1·0	2·1 ± 0·5	<0·01
Total biomass (g)	9·7 ± 2·3	0·9 ± 0·2	<0·001
Percentage weed cover[1]	14·2 ± 2·3	2·9 ± 0·6	<0·01

Analysis on transformed data $[\log_{10}(n + 1)]$ except [1] analysis on transformed data (% cover = Domin score $^{2·6}/4$) then (% cover by $\sqrt{}$ arc sin).

(b) Chick-food insects. Selective pesticide use also resulted in increases in the numbers of chick-food insects both between fields (Table 17.8) (Rands 1985; Sotherton 1989) and within a field (Table 17.9).

(c) Gamebird chicks. In response to the provision of this improved food supply, mean grey partridge brood sizes were, in most cases, significantly greater on blocks of cereal fields surrounded by selectively sprayed margins than on fully sprayed sites (Table 17.10) (Sotherton & Robertson 1990).

TABLE 17.8. The abundance of wild gamebird chick-food insects in sprayed and selectively sprayed headlands on the study farm in southern England, June 1983 and 1984. From Sotherton (1989)

| Prey species | 1983 mean ($\pm$1 SE) number of prey items per 50 sweeps | | | 1984 mean ($\pm$1 SE) number of prey items per 20 sweeps | | |
	Sprayed headlands ($n = 19$)	Selectively sprayed headlands ($n = 18$)	P	Sprayed headlands ($n = 46$)	Selectively sprayed headlands ($n = 42$)	P
Total chick-food items	62·2 ± 21·5	180·0 ± 45·2	<0·001	34·4 ± 6·2	60·4 ± 9·6	<0·05
Heteroptera	53·9 ± 21·4	163·2 ± 44·8	<0·02	26·5 ± 5·8	49·2 ± 8·8	<0·05
Tenthredinidae and Lepidoptera larvae	3·5 ± 1·0	5·0 ± 1·0	NS	3·7 ± 1·2	4·4 ± 0·8	<0·02
Coleoptera	4·2 ± 0·5	12·3 ± 2·8	<0·05	5·4 ± 0·9	6·5 ± 1·3	NS

Table 17.9. Mean densities (± 1 SE) per 0·5 m^2 per sampling date of gamebird chick-food insects collected by vacuum suction sampling of headland plots of spring wheat treated with herbicide or remaining untreated; May, June and July 1988.

	Headland plots without herbicide	Headland plots with herbicides	P
Total chick-food items	67·9 ± 23·4	18·9 ± 9·2	<0·001
Heteroptera	41·4 ± 16·8	8·0 ± 3·7	<0·001
Tenthredinidae and Lepidoptera larvae	1·8 ± 0·6	1·7 ± 0·6	NS
Chrysomelidae	6·4 ± 2·3	2·9 ± 1·2	<0·05

Pheasant chicks are also dependent on insects in their diet during early life (Hill 1985). Mean brood sizes of this species were also significantly higher in general where Conservation Headlands were used (Table 17.10). Significant increases in grey partridge and pheasant brood size were also found on blocks of cereal fields with selectively sprayed margins on farms in the eastern counties of England (Table 17.10) (Sotherton & Robertson 1990). Radio-telemetry studies have confirmed the benefits of Conservation Headlands to chick survival. Changes in the diet, survival, home range size and distance between successive roost sites of chicks foraging in brood-rearing areas contain Conservation Headlands or fully sprayed crop edges have been demonstrated and reported elsewhere (Rands 1986; Sotherton & Robertson 1990; Potts & Aebischer 1991). Monitoring of spring pair densities of grey partridges showed that breeding density over the entire study farm increased from 3·7 pairs/km^2 in 1979 and peaked at 11·7 pairs/km^2 in 1986 (Potts & Aebischer 1990; Rands, in press). No similar increases were observed on adjacent farms where pesticide use was unchanged.

Other forms of farmland wildlife

(a) *Lepidoptera*. The resources provided by Conservation Headlands to increase gamebird chick survival (broad-leaved weeds and insects) have also been shown to be of conservation value to other groups of the fauna and flora that inhabit the cereal ecosystem. Farmland is currently regarded as a poor wildlife habitat, especially for butterflies (Thomas 1983; Heath, Pollard & Thomas 1984). The reduced suitability of farmland for butterflies has been ascribed to changes in agricultural practices, especially the intensification of production (Anon. 1977).

Quantification of the benefits of Conservation Headlands to butterflies

TABLE 17.10. Mean grey partridge brood sizes (±1 SE) on blocks of cereal fields with sprayed and selectively sprayed headlands in Hampshire and eastern England. From Sotherton & Robertson (1990)

Study area	Year	Grey partridge, mean (±1 SE) brood size			Pheasant, mean (±1 SE) brood size		
		Sprayed headlands (n)	Selectively sprayed headlands (n)	P	Sprayed headlands (n)	Selectively sprayed headlands (n)	P
Principal study farm (Hampshire)*	1983	4·7 ± 1·1 (39)	8·4 ± 1·2 (29)	<0·01	—	—	—
	1984	7·5 ± 0·8 (34)	10·0 ± 0·6 (34)	<0·01	3·2 ± 0·5 (18)	6·9 ± 0·5 (29)	<0·001
	1985	3·3 ± 0·7 (9)	5·7 ± 0·8 (14)	<0·05	3·0 ± 1·0 (3)	4·6 ± 0·6 (8)	<0·05
	1986	5·9 ± 1·6 (17)	6·2 ± 1·0 (21)	NS	2·0 ± 0·5 (8)	5·9 ± 0·7 (10)	<0·01
	1984	4·7 ± 0·4 (71)	7·8 ± 0·6 (57)	<0·001	—	—	—
Eastern England†	1985	2·7 ± 0·4 (19)	4·0 ± 0·7 (19)	<0·05	2·6 ± 0·3 (30)	3·7 ± 0·4 (35)	<0·01
	1986	4·8 ± 0·6 (32)	8·7 ± 1·5 (6)	<0·001	3·4 ± 0·6 (14)	3·5 ± 0·7 (6)	NS

* Pooled data from each block per treatment on the farm.

† Pooled data from each block per treatment per farm.

began in 1984 when systematic butterfly censuses were undertaken. Modified 'Pollard' transects (Pollard *et al.* 1975; Dover, Sotherton & Gobbett 1990) were used to assess the relative abundance of butterflies in field margins adjacent to Conservation Headlands and those that were fully sprayed. In the 5 years that transects have been conducted, between two and four more species were observed in Conservation Headlands each year. Also, many more individuals of most species were observed in the Conservation Headlands (Table 17.11) (Rands & Sotherton 1986; Dover, Sotherton & Gobbett 1990; Dover 1991).

TABLE 17.11. Comparison of the numbers of butterflies seen in field margins adjacent to Conservation Headlands (CH) and fully sprayed headlands (FS), 1984–1988, Hampshire. $* = P < 0.05$; $** = P < 0.01$; $*** = P < 0.001$; NS = no significant difference

	1984	1985	1986	1987	1988
Pieridae					
Anthocharis cardamines	CH***[†]	NT	FS(NS)	NT	NT
Pieris napi	CH***	CH*	CH***	CH***	CH(NS)
Pieris rapae	CH(NS)	NT	CH***	NT	CH***
Gonepteryx rhamni	CH***	CH**	NT	NT	NT
Pieris brassicae	CH(NS)	FS(NS)	CH***	CH***	CH(NS)
Total Pieridae	CH***	CH*	CH***	CH***	CH***
Satyridae					
Aphantopus hyperantus	CH***	CH*	FS***	CH(NS)	FS(NS)
Maniola jurtina	CH***	CH***	CH***	CH***	CH***
Pyronia tithonus	CH***	CH***	CH**	CH***	CH(NS)
Total Satyridae	CH***	CH***	CH(NS)	CH***	CH***
Nymphaliae					
Aglais urticae	CH***	CH***	CH*	CH***	CH*
Inachis io	CH***	CH***	CH(NT)	CH***	CH(NT)
Vanessa cardui	—	CH(NS)	CH(NT)	NT	CH***
Total Nymphalidae	CH***	CH***	CH**	CH***	CH***
Hesperiidae					
Ochlodes venata	CH***	CH(NS)	FS(NS)	NT	CH*
Thymelicus spp.	CH***	NT	FS(NS)	—	NT
Total Hesperiidae	CH***	CH**	FS*	NT	CH*
Total Lepidoptera	CH***	CH***	CH***	CH***	CH***

[†] Significantly ($P < 0.001$) more observed in Conservation Headlands.
NT, no test made. — Species not recorded. NS, not significant.

Detailed observations of butterfly behaviour in the two headland treatments have shown that flight speed and duration were lower in Conservation Headlands (J.W. Dover, unpublished data), and activity patterns were also altered. Some species showed differences in behaviour between spray management regimes, although this was not demonstrated in all species studied. Small white (*Pieris rapae*) males spent almost all of their time in field margins which had fully sprayed headlands in flight with little time devoted to either feeding or resting. In field margins with Conservation Headlands, the proportion of time spent in flight was radically reduced, with a much higher proportion of time spent in feeding and resting.

The change in proportion of time spent in each behaviour category observed between the two regimes corresponded with differences in the sites where such activities were carried out. *Pieris rapae* males observed in fully sprayed field margins spent most of their time in the hedgerow area whereas the hedgerows were virtually ignored in field margins where there were Conservation Headlands.

For many species of strongly flying butterflies such as the Pieridae and the Nymphalidae, the nectar provided by weeds such as charlock (*S. arvensis*), field pansy (*Viola arvensis*) and creeping thistle (*Cirsium arvense*) provided the attraction bringing them into Conservation Headlands (Dover 1989). The host plants of some of the Pieridae (*S. arvensis* and rape, *Brassica napus*) were found in Conservation Headlands (Dover, Sotherton & Gobbett 1990) and were exploited as larval food plants. Some of the increased numbers of such species as green-veined white (*Pieris napi*), *P. rapae* and orange tip (*Anthocharis cardamines*) may also have been caused by their presence. However, the host plants of the Satyridae, Hesperiidae, Nymphalidae and Lycaenidae were not found in the Conservation Headlands. Indeed, those of the satyrid and hesperiid butterflies (Gramineae) were selectively controlled using herbicides. This was fortunate, as any eggs laid in Conservation Headlands could constitute lost reproductive effort following harvest and tillage.

For some species it was not possible to make sufficient observations in field margins with fully sprayed headlands, due to the low numbers of butterflies. However, data derived from Conservation Headlands revealed that not all species responded to the same extent as *P. rapae* in their use of headland resources. Observations of meadow brown (*Maniola jurtina*) males in 1985 showed a more even disposition of time between the two sub-areas of the field margin whilst gatekeeper (*Pyronia tithonus*) males were more strongly associated with the hedgerow area than the headland. (Dover 1991). One of the common names of this species is the hedge brown and it is well known for its association with the shrubby flora of hedgerows. This species could be less adaptable in its ability to exploit nectar sources, having a preference for hedgerow nectar, and may therefore represent species of butterfly that were less exploitive of headland resources. However, it was clearly demonstrated that its abundance was much greater in Conservation Headlands compared to those that were fully sprayed (Table 17.11; Dover, Sotherton & Gobbett 1990). It may have been that when hedgerow nectar was scarce, unsuitable or unavailable then headland nectar was used to 'top-up' or act as the major nectar source.

Another possible explanation for the increased abundance of hedgerow species could be the reduced amount of pesticide drift into field boundaries where the spray boom was switched off 6 m away from the field edge. Conservation Headlands may be acting as buffer zones protecting hedgerows from summer insecticides and broad-leaved residual herbicides throughout the year. The amount of pesticide deposition into field boundaries adjacent to Conservation Headlands and fully sprayed crop edges has been measured. Drift was decreased by as much as 50% (Cuthbertson 1988). However, even in fully sprayed headlands, only very small amounts of active ingredients were deposited into field boundaries. Current work is quantifying the sublethal effects of these pesticide fractions (of pyrethroid insecticides) on larval Lepidoptera in the laboratory. Preliminary data suggested that such low doses (ng/cm^2 or µg active ingredient per gram body weight) were adversely affecting survival, growth, longevity and fecundity of treated individuals (T. Cilgi, unpublished data).

In the longer term, trends in relative abundance have been established on the study farm over a 5-year period and comparisons made with data derived from the National Butterfly Monitoring Scheme. For two satyrid species *M. jurtina* and *P. tithonus*, populations were either maintained or increased relative to 1984 levels at the study farm. This compares with the south–south-east region of the National Butterfly Monitoring Scheme, (many of the incorporated sites being Nature Reserves) where these species showed declines over the same period (Fig. 17.2) (Dover, Sotherton & Gobbett 1990).

(b) Rare arable weeds. The absence of broad-spectrum, residual broad-leaved weed herbicides from Conservation Headlands led to an increase in numbers and frequency of occurrence of several species of uncommon rare arable weeds. In 1986 the Botanical Society of the British Isles initiated a survey of twenty-five species of arable weeds. In the same year surveys of our network of between twenty and thirty study farms across southern and eastern England found that seventeen of these species were found in Conservation Headlands.

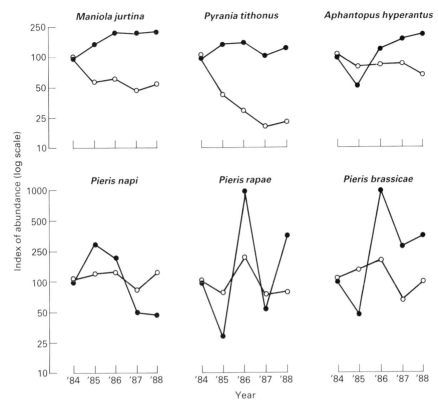

FIG. 17.2. The relative abundance of six butterfly species at a farm employing Conservation Headlands (●) and the SE region of the National Butterfly Monitoring Scheme (○) (1984–88). Index of abundance starting from an arbitrary value of 100 in 1984. From Dover, Sotherton & Gobbett (1990).

Since 1987, autecological studies have been conducted on several rare species to determine the reasons for their decline and to suggest management recommendations for their conservation on intensively-farmed arable land. One aspect of their biology that many species have in common was their distinctive within-field distributions. Many species such as rough poppy (*Papaver hybridum*) and narrow-fruited corn-salad (*Valerianella dentata*) were found at the edges of fields, the greater proportion of the total number of seedlings being found within the outermost 10 m of the crop (Wilson 1989). These edge species could therefore benefit from a change of farming practice on the crop edge especially from the implementation of Conservation Headlands. Herbicide use has been frequently cited as a major contributory factor in the change in weed communities and the decline of certain species (Fryer & Chancellor 1970). Indeed a very rare species such as shepherd's-needle (*Scandix pecten-veneris*) appeared to be susceptible to all herbicides used during the 1940s and 1950s, a problem exacerbated by its almost total lack of seed dormancy (P.J. Wilson, unpublished data). However it would be incorrect to suggest that the removal of herbicides from crop edges would be sufficient to ensure the conservation of all rare arable weeds, not without concern for other factors which have been associated with intensive grain production. For example some species such as *P. hybridum*, lamb's succory (*Arnoseris minima*), lesser snapdragon (*Misopates orontium*) and mousetail (*Myosurus minimus*) were found in significantly lower numbers and produced significantly fewer fruits when nitrogen fertilizers were added to experimental plots (Wilson, Boatman & Edwards 1991). Crop sowing date may also prove to be of importance especially for species that set seed over a prolonged period into the autumn and early winter. Species with such characteristics used to set seed on cereal stubbles left over the winter prior to spring drilling. Recent trends towards autumn-drilling and monoculture winter cereal production may also prove to be detrimental to rare weed conservation (P.J. Wilson, personal communication).

(c) Invertebrate natural enemies. Other groups of beneficial insects including the stenophagous and polyphagous natural enemies of cereal pests may also benefit from the resources provided in Conservation Headlands or from the degree of exclusion of the adverse effects of pesticides that they provide. Preliminary work has shown how some economically important species of hoverfly (Diptera: Syrphidae) such as *Episyrphus balteatus* made use of the nectar and pollen of flowering weeds in Conservation Headlands such as *S. arvensis*, *Matricaria* spp. and *C. arvense* (Cowgill 1989).

Polyphagous species such as the carabid beetles *Pterostichus melanarius* and *Agonum dorsale* were better fed (a significantly higher proportion of males, gravid females and non-gravid females had all four portions of their digestive tracts full of solid food remains) in Conservation Headlands compared to individuals from fully sprayed crop edges. It is possible that the greater prey availability (larger numbers and more diverse alternative prey items) might increase predator numbers in the longer term through improved diet and fecundity (P.A. Chiverton, unpublished data).

Agronomic consequences

The agronomic costs of implementing Conservation Headlands on cereal farms have been calculated. Losses caused by the presence of broad-leaved weeds and, in some years, higher levels of cereal aphids in the spring and summer have been calculated. Yields of grain in Conservation Headlands are between 6 and 10% less than those that have been fully sprayed. Grain moisture levels from weedy headlands were about 1% higher and weed seed contamination levels were also higher (Boatman & Sotherton 1988). Apart from such direct losses to the crop, other indirect management costs have been incorporated into the costings. These include extra costs for grain storage, slower harvesting and extra herbicide costs (purchase of selective products and additional application costs). The costs of Conservation Headlands will vary according to the type of crop grain (wheat or barley, autumn or spring sown) and its intended market (seed, milling, malting or feed). Our calculations have been based on the need to selectively spray only 6% of the total cereal area. When the costs are spread over the total acreage, the costs vary from £2·20 to £8·50 per cereal hectare (Boatman & Sotherton 1988).

Many European countries are beginning to initiate research into the use of selective pesticides on crop edges and to monitor the benefits to the flora and fauna of arable land. All these initiatives are State-funded, multidisciplinary studies. These include West Germany (Schumacher 1987; Felkl 1988; Welling 1988), Denmark (Hald 1989) Sweden (Sotherton, Boatman & Chiverton 1989) and The Netherlands (De Snoo & Canters 1990). In several states in West Germany, financial compensation is available to farmers using no herbicides or fertilizers on their crop edges for the conservation of rare weeds. In Britain, the use of Conservation Headlands has been most widely taken up on farms where there is an interest in wild gamebird production and in two of the nineteen Environmentally Sensitive Areas designated by The Ministry of Agriculture Fisheries and Food, where farmers are given financial assistance to use Conservation Headlands as an environmentally sensitive method of farming.

CONCLUSION

The Conservation Headland strategy, and the practice of reduced pesticide inputs it advocates, will fit in well with any plans for cereal extensification that may soon be forthcoming to satisfy EC demands (see also Potts, pp. 15–18, this volume). This is vital if we are to conserve species that are unique to the cereal ecosystem. For many species (e.g. grey partridge, several species of rare arable weeds) the abandonment of arable land or a change of its use to farm woodland or grassland is not the answer, in fact quite the reverse. The annual cycle of soil cultivation is the key to the survival of annual plants which are at the base of many species' food chains. The maintenance of this cycle without the pressure of high input, intensive production methods and without the disruption of food chains and depletion of vital resources is the best method of ensuring the continuance of the conservation ethic in these vast areas of land.

ACKNOWLEDGMENTS

This work was funded primarily by those UK cereal farmers subscribing to the Cereals and Gamebirds Research Project. Additional Projects were funded by the British Agrochemicals Association, the Ministry of Agriculture Fisheries and Food, the Natural Environment Research Council, the Worldwide Fund for Nature (UK) and the Nature Conservancy Council. I would like to thank all the farmers who have provided access to their farms and permission to survey or conduct experiments. Finally I am indebted to all CGRP scientists especially Dr Mike Rands, Dr Nigel Boatman, Dr John Dover, Dr Philip Wilson and Stephen Moreby.

REFERENCES

Anon. (1977). *Nature Conservation and Agriculture*. Nature Conservancy Council, London.
Bain, A.B. & Boatman, N.D. (1989). The potential of quinmerac (BAS 518H) for the selective control of cleavers (*Galium aparine*) in field boundaries. *Brighton Crop Protection Conference — Weeds 1989*, pp. 1057–1063. British Crop Protection Council, Farnham.
Blank, T.H., Southwood, T.R.E. & Cross, D.J. (1967). The ecology of the partridge I. Outline of population processes with particular reference to chick mortality and nest density. *Journal of Animal Ecology*, 36, 549–556.
Boatman, N.D. (1987). Selective grass weed control in cereal headlands to encourage game and wildlife. *British Crop Protection Conference — Weeds 1989*, pp. 277–284. British Crop Protection Council, Farnham.
Boatman, N.D. (1989). Selective weed control in field margins. *Brighton Crop Protection Conference — Weeds 1989*, pp. 785–795. British Crop Protection Council, Farnham.
Boatman, N.D. & Sotherton, N.W. (1988). The agronomic consequences and costs of managing field margins for game and wildlife conservation. *Aspects of Applied Biology*, 17, 47–56.
Boatman, N.D. & Wilson, P.J. (1988). Field edge management for game and wildlife conservation. *Aspects of Applied Biology*, 16, 53–61.
Boatman, N.D., Freeman, K.J. & Green, M.C.E. (1988). The effects of timings and dose on the control of *Galium aparine* (cleavers) and other broad-leaved weeds by fluroxypyr in cereal headlands. *Aspects of Applied Biology*, 18, 117–128.
Broadbent, L. (1980). Ecological aspects of agricultural pesticide application. *Biologist*, 27, 131–133.
Colburn, R. & Asquith, D. (1973). Tolerance of *Stethorus punctum* adults and larvae to various pesticides. *Journal of Economic Entomology*, 66, 961–962.
Cole, J.F.H. & Wilkinson, W. (1984). Selectivity of pirimicarb in cereal crops. *British Crop Protection Conference — Pests and Diseases 1984*, pp. 311–316. British Crop Protection Council, Farnham.

Coombes, D.S. & Sotherton, N.W. (1986). The dispersion of polyphagous predators from their overwintering sites into cereal fields and factors affecting their distribution in the spring and summer. *Annals of Applied Biology*, **108**, 461–474.

Cowgill, S. (1989). The role of non-crop habitats on hoverfly (Diptera: Syrphidae) foraging on arable land. *Brighton Crop Protection Conference — Weeds 1989*, pp. 1103–1108. British Crop Protection Council, Farnham.

Cuthbertson, P. (1988). The pattern and level of pesticide drift into conservation and fully sprayed arable crop headlands. *Aspects of Applied Biology*, **17**, 273–377.

De Snoo, G.R. & Canters, K.J. (1990). *Side Effects of Pesticides on Terrestrial Vertebrates.* Centre for Environmental Studies, Leiden University, Report No.35, 2nd edn, Leiden.

Dover, J.W. (1989). A method for recording and transcribing behavioural observations of butterflies. *Entomologist's Gazette*, **40**, 95–100.

Dover, J.W. (1991). The conservation of insects on arable farmland. *The Conservation of Insects and their Habitats* (Ed. by N.W. Collins & J.Thomas), pp. 293–318. Academic Press, New York.

Dover, J.W., Sotherton, N.W. & Gobbett, K. (1990). Reduced pesticide inputs on cereal field margins: the effects on butterfly abundance. *Ecological Entomology*, **15**, 17–24.

Felkl, G. (1988). First investigations on the abundance of epigeal arthropods, cereal aphids and stenophagous aphid predators in herbicide-free border strips of winter wheat fields in Hesse. *Gesunde Pflanzen*, **40**, 483–491.

Fryer, J.D. & Chancellor, R.J. (1970). Evidence of changing weed populations in arable land. *Proceedings of the 10th British Weed Control Conference*, 958–964.

Green, R.E. (1984). The feeding ecology and survival of partridge chicks (*Alectoris rufa* and *Perdix perdix*) on arable farmland in East Anglia. *Journal of Applied Ecology*, **21**, 817–830.

Hald, A.-B. (Ed.) **(1989).** *Conservation Strategies for Field Boundaries and Field Marginal Areas, Oikos Seminar*, **19**.

Heath, J., Pollard, E. & Thomas, J. (1984). *Atlas of Butterflies in Britain and Ireland.* Viking, Harmondsworth.

Hill, D.A. (1985). The feeding ecology and survival of pheasant chicks on arable farmland. *Journal of Applied Ecology*, **22**, 645–654.

Johnston, G., Collett, G., Walker, C., Dawson, A., Boyd, I. & Osborn D. (1989). Enhancement of malathion toxicity to the hybrid red-legged partridge following exposure to prochloraz. *Pesticide Biochemistry and Physiology*, **35**, 107–118.

Oliver-Bellasis, H.R. & Sotherton N.W. (1986). The Cereals and Gamebirds Research Project: an independent viewpoint. *British Crop Protection Conference — Pests and Diseases 1986*, pp. 1225–1235. British Crop Protection Council, Farnham.

Pollard, E., Elias, D.O., Skelton, M.J. & Thomas, J.A. (1975). A method of assessing the abundance of butterflies in Monks Wood National Nature Reserve in 1973. *Entomologist's Gazette*, **26**, 79–88.

Potts, G.R. (1970a). The effects of the use of herbicides in cereals on aphids and on the feeding ecology of partridges. *10th British Weed Control Conference 1970*, 299–302.

Potts, G.R. (1970b). Studies on the changing role of weeds of the genus *Polygonum* in the diet of partridges. *Journal of Applied Ecology*, **7**, 567–576.

Potts, G.R. (1986). *The Partridge: Pesticides, Predation and Conservation.* Collins, London.

Potts, G.R. & Aebischer, N.J. (1990). Control of population size in birds: the grey partridge as a case study. *Toward a More Exact Ecology* (Ed. by P.J. Grubb & J.B. Whittaker), pp. 141–161. Blackwell Scientific Publications, Oxford.

Potts, G.R. & Aebischer, N.J. (1991). Modelling the population dynamics of the grey partridge: conservation and management. *Bird Population Studies: Their Relevance to Conservation and Management* (Ed. by C.M. Perrins, J.-D. Lebreton & G.J.M.

Hirsons). Tour du Valat Symposium 1988, Oxford University Press, Oxford.

Rands, M.R.W. (1985). Pesticide use on cereals and the survival of grey partridge chicks: a field experiment. *Journal of Applied Ecology*, **22**, 49–54.

Rands, M.R.W. (1986). The survival of gamebird (Galliformes) chicks survival and breeding density. *Journal of Applied Ecology*. (In press).

Rands, M.R.W. & Sotherton, N.W. (1986). Pesticide use on cereal crops and changes in the abundance of butterflies on arable farmland. *Biological Conservation*, **36**, 71–83.

Rands, M.R.W., Hudson, P.J. & Sotherton, N.W. (1988). Gamebirds, ecology, conservation and agriculture. *The Ecology and Management of Gamebirds*. (Ed. by P.J. Hudson & M.R.W. Rands), pp. 1–17. Blackwell Scientific Publications, Oxford.

Reyes, A. & Stevenson, A. (1975). Toxicity of benomyl to the cabbage maggot *Hylemya brassicae* (Diptera: Anthomyiidae), in greenhouse tests. *Canadian Entomologist*, **107**, 685–687.

Rivière, J.L., Bach, J. & Grolleau, G. (1985). Effects of prochloraz on drug metabolism in the Japanese quail, grey partridge, chicken and pheasant. *Archives of Environmental Contamination and Toxicology*, **14**, 299–306.

Schumacher, W. (1987). Measures taken to preserve arable weeds and their associated communities in Central Europe. *Field Margins* (Ed. by J.M. Way & P.W. Greig-Smith), pp. 11–22. BCPC Monograph No. 35, London.

Sly, J.M.A. (1986). *Review of Usage of Pesticides in Agriculture, Horticulture and Animal Husbandry in England and Wales, 1980–1983*. Pesticide Usage Survey Report No. 41. Ministry of Agriculture, Fisheries and Food, Pinner.

Sotherton, N.W. (1982). Effects of herbicides on the chrysomelid beetle *Gastrophysa polygoni* (L.) in laboratory and field. *Zeitschrift für angewandte Entomologie*, **94**, 446–451.

Sotherton, N.W. (1988). The Cereals and Gamebirds Research Project: Overcoming the indirect effects of pesticides. *Britain Since Silent Spring* (Ed. by D.J.L. Harding), pp. 64–72. Institute of Biology, London 1988.

Sotherton, N.W. (1989). Farming methods to reduce the exposure of non-target arthropods to pesticides. *Pesticides and Non-target Invertebrates* (Ed. by P.C. Jepson), pp. 195–212. Intercept Ltd., Wimborne.

Sotherton, N.W. & Moreby, S.J. (1988). The effects of foliar fungicides on beneficial arthropods in wheat fields. *Entomophaga*, **33**, 87–99.

Sotherton, N.W. & Robertson, P.A. (1990). Indirect impacts of pesticides on the production of wild gamebirds in Britain. *Perdix V, Gray Partridge and Ring-necked Pheasant Workshop* (Ed. by K.E. Church, R.E. Warner & S.J. Brady). Kansas Department of Wildlife and Parks, Emporia.

Sotherton, N.W., Boatman, N.D. & Chiverton, P.A. (1989). The selective use of pesticides on cereal crop margins for game and wildlife: the British experience. *Proceedings of the 30th Swedish Plant Protection Conference*, **4**, 18–29.

Sotherton, N.W., Boatman, N.D. & Rands, M.R.W. (1989). The 'Conservation Headland' experiment in cereal ecosystems. *The Entomologist*, **108**, 135–143.

Sotherton, N.W., Jepson, P.C. & Pullen, A.J. (1988). Criteria for the design, execution and analysis of terrestrial non-target invertebrate field tests. *Field Methods for the Study of Environmental Effects of Pesticides* (Ed. by M.P. Greaves, B.P. Smith & P.W. Greig-Smith), pp. 183–190. BCPC Monograph No. 40, London.

Sotherton, N.W., Moreby, S.J. & Langley, M.G. (1987). The effects of the foliar fungicide pyrazophos on beneficial arthropods in barley fields. *Annals of Applied Biology*, **111**, 75–87.

Sotherton, N.W., Rands, M.R.W. & Moreby, S.J. (1985). Comparison of herbicide treated and untreated headlands on the survival of game and wildlife. *British Crop Protection Conference — Weeds 1985*, pp. 991–998. British Crop Protection Council, Farnham.

Southwood, T.R.E. & Cross, D.J. (1969). The ecology of the partridge III. Breeding success and the abundance of insects in natural habitats. *Journal of Animal Ecology*, **38**, 497–509.

Thomas, J.A. (1983). A 'Watch' census of common British butterflies. *Journal of Biological Education*, **17**, 333–338.

Vickerman, G.P. (1974). Some effects of grass weed control on the arthropod fauna of cereals. *12th British Weed Control Conference — 1974*, 929–939.

Vickerman, G.P. (1977). The effects of foliar fungicides on some insect pests of cereals. *1977 British Crop Protection Conference — Pests & Diseases*, 121–128.

Vickerman, G.P. & Sotherton, N.W. (1983). Effects of some foliar fungicides on the chrysomelid beetle *Gastrophysa polygoni* (L.). *Pesticide Science*, **14**, 405–411.

Vickerman, G.P. & Sunderland, K.D. (1977). Some effects of dimethoate on arthropods in winter wheat. *Journal of Applied Ecology*, **14**, 767–777.

Welling, M. (Ed.) (1988). *Crop Edges — Positive Effects for Agriculture* Paul Parey, Berlin.

Wilson, P.J. (1989). The distribution of arable weed seedbanks and the implications for the conservation of endangered species and communities. *Brighton Crop Protection Conference — Weeds 1989*, pp. 1081–1086. British Crop Protection Council, Farnham.

Wilson, P.J., Boatman, N.D. & Edwards, P.J. (1990). Strategies for the conservation of endangered arable weeds in Great Britain. *Proceedings of the EWRS Symposium — Integrated Weed Management in Cereals*, pp. 93–101.

Zadoks, J.C., Chang, T.T. & Konzak, C.F. (1974). A decimal code for the growth stages of cereals. *Weed Research*, **14**, 415–421.

18. THE LAUTENBACH PROJECT 1978–89: INTEGRATED WHEAT PRODUCTION ON A COMMERCIAL ARABLE FARM, SOUTH-WEST GERMANY

A. EL TITI

State Institute for Plant Protection, Stuttgart, Reinsburgstrasse 107,
D-7000 Stuttgart 1, Germany

INTRODUCTION

The Lautenbach Project (Steiner, El Titi & Bosch 1986) was initiated in 1978. It was a practical response to overcome the problems of intensified production methods in agriculture, declining farm-incomes and increasing endangerment of wildlife (Anon. 1978; Schroeder 1988). A major objective of these on-farm investigations was to evolve an Integrated Farming System (IFS) under commercial farming conditions and to compare its productivity, economics and environmental impacts with those of the Current Farming System (CFS). Different cereal crops were grown on the farm for seed production. These include winter wheat, spring wheat, and, to a limited extent, spring barley, oats and rye. Sugar beet and legumes (peas, beans or faba-beans) were included in the crop rotation.

A package of husbandry measures and techniques were selected to fit the requirements of the natural enemies of major pests and diseases. This package sought a practical compromise and not necessarily the maximum benefits. Hence, the Lautenbach Project was the first interdisciplinary research approach on a private farm, aiming to integrate natural regulation components into farming practices. The Project exhibited a new field of scientific research, which could perceive the possibilities and limitations of a farmer's conception of alternative technologies.

Since many results from the Lautenbach Project have already been published (El Titi 1986; Bosch 1987; El Titi & Landes 1990) this contribution will briefly review the concept and update the results.

DESIGN AND LAYOUT

The estate of Lautenbach comprises more than 240 ha of farmland. This size of farm is far above the national average in the Federal Republic of Germany of 18 ha. Two farm units of 36 ha each were set up within

Lautenbach. One of them was farmed according to the principals of IFS, whereas the other was managed conventionally. Fields of both farms were divided into six field plots each (Fig. 18.1), so that six field pairs were available for comparison. Both farming systems had the same rotation.

This rotation concept had an important technical value, because it allowed for the transition of IFS methods and techniques to the remaining field area of the estate. The crop rotation in both farming systems was: winter wheat / sugar beet / legumes / winter wheat / sugar beet / oats / winter wheat / sugar beet / spring wheat / faba-beans / winter wheat or sugar beet (alternating) / oats / winter wheat. The rotation had a balanced proportion of both winter and spring crops, which is known to suppress some potential grass weeds (e.g. black grass *Alopecurus myosuroides*). The single plot pairs were located on comparable soil types of similar topography to justify comparisons of the same crops. During the investigation period the same cultivars were cropped on fields of both systems.

The monitoring of the responses of the agro-ecosystem involved the use of a number of bioindicators. These include *Lumbricidae, Enchytraeidae,* edaphic *Arthropoda, Nematoda,* cellulose decomposition rate and soil physical properties and soil carbon and nutrient content. For some of these

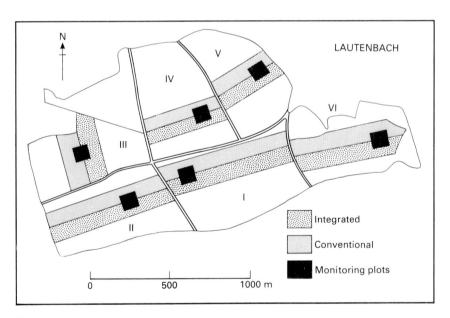

FIG. 18.1. Design and layout of field plots for integrated and conventional farming systems at Lautenbach with reference to the monitoring plots.

measurements a subplot of 1 ha, half on the IFS and half on the CFS, was set up on all field pairs (Fig. 18.1).

DIFFERENCES BETWEEN INTEGRATED AND CONVENTIONAL FARMING SYSTEMS

The established differences between the two farming systems did not change over the 12-year period of the investigation. However, various innovative technical or managerial improvements were incorporated into the IFS. The system was, therefore, a semidynamic one. The major differences are listed in Table 18.1.

TABLE 18.1. Comparison of main cultural measures used in Integrated (IFS) and Conventional (CFS) Farming Systems in the Lautenbach Project

	IFS	CFS
Crop rotation	60% cereals, 25% sugar beet, 15% legumes	60% cereals, 25% sugar beet, 15% legumes
Soil tillage	Tine loosening + rotory incorporation Non-inversion	Ploughing
Sowing		
Sugar beet	45 cm inter-row/20 cm seed space	45 cm inter-row/20 cm seed space
Cereals	Double rows 6 cm within/24 cm between	Drilling 15 cm/420 kernels/m^2
Faba-beans	45 cm inter-row, 5 cm seed space	Drilling 30 cm
Cultivars	Same variety	Same variety
Fertilization		
$Ca/K_2O/P_2O$	According to soil chemical analysis	According to soil chemical analysis
Nitrogen	According to N minerality, but reduced (25%)	According to N minerality optimal supply (recommended dose)
Control		
Weeds	Mechanical/herbicides at high incidence	Herbicides
Diseases	Fungicides at high incidence only	Fungicides
Pests	Insecticides at high thresholds only	Insecticides
Hedgerows, shelterbelts	Management considered	Not considered
Field margins	Native flora accepted	Native flora mowed
Field-edge & attractants	Management considered	Not considered

METHODS

Currently accepted assessment methods for record keeping, monitoring, extraction and analysis were as follows.

Economic data

Standard field record cards were used. They enable one to register details of all field operations, such as the machinery used, the time needed, fertilizers supplied, pesticides sprayed and so on. The analysis of the economic data of both farming systems together with more details on the methodology of the economic analysis are described by Zeddies, Jung & El Titi (1986).

Nitrogen leaching

Estimates of leachable nitrate levels in the soil were calculated by nitrate extraction of soil cores taken to 1 m depth (N-min-method after Scharpf & Wehrmann 1976). A mixed sample (12 cores each) per ha was taken on at least two occasions (the beginning of the growing season and in autumn) each year on each of the six plot pairs. Since 1983 a tractor-mounted sampling machine was used for this purpose. The cooled samples were frozen to $-18\,°C$, before being transported to laboratories.

Bioindicators

(a) Soil surface fauna. Five pitfall traps (Klinger 1984; Bosch 1987), with formaldehyde solution were used in each monitoring plot of both farming systems between 1978 and 1983. Between mid-April and October the epigeal fauna was trapped in 2-weekly intervals. The period was shortened down to trapping during June and July from 1985.

(b) Eudaphic arthropods. Sampling strategies are given by El Titi (1984), Matt (1986) and Gottfriedsen (1987). Ten soil cores were taken on three different occasions each year (spring, summer and autumn) from each monitoring plot of six field pairs using the 'split auger' sampler from the 10 cm topsoil during the 7-day extraction period (using Tullgren-extraction), only mobile mites and springtails were obtained.

(c) Nematodes. Nematodes were extracted from three soil samples per monitoring plot of both farming systems on the same dates as soil arthropods. Each sample was a mixture of soil taken by fifty pin cores at 25 cm depth and extracted by Baermann funnel method (Decker 1969).

(d) Earthworms. At least once a year earthworms were extracted from 8 × 0·125 quadrats per plot per system using a diluted formalin solution (Edwards & Lofty 1975), counted, identified and weighed to the nearest 0·001 g.

(e) Cellulose decomposition rate. Filter paper of very high purity was placed into fine mesh polyester bags (ten bags per plot per system) which were able to exclude micro-arthropods and earthworms. They were buried in 5 cm of topsoil for 6–8 weeks and their loss in weight was used as an indicator of soil microbial activity (Schroeder 1980).

(f) Vegetation diversity. Species diversity and the population density of non-crop vegetation were assessed at least once every year on the field areas of both systems before any weed control measure was carried out. In addition seed-banks were assessed on three of the six field pairs after 7 experimental years (Wahl 1988).

(g) Foliage fauna. D-Vac suction sampling and the beat-funnel methods (Steiner 1967) were used in the first 3 years. Thereafter only visual counting method on a defined number of crop plants, linear length of row crops or leaves was used to assess the density and diversity of the foliage fauna.

RESULTS

The results over the two rotation periods (2 × 6 years) showed significant differences in cereal pests and diseases, weeds, soil nutrient contents, pesticides and fertilizer consumption, soil physical properties, bioindicators in and on soil, farm income and landscape structure between the two farming systems.

Cereal diseases

The incidence of infestation of stem base disease at growth stage 31 (Zadoks, Chang & Konzak 1974) was regularly lower on the IFS plots than on the those of the CFS. This pattern was repeatedly the case prior to any fungicide sprays. Similar results were found for powdery mildew (*Erysiphe graminis*), no matter which cultivar was grown. However, leaf blotch disease (*Septoria nodorum*) was on average 5–12% higher on the IFS plots than on the CFS plots.

Cereal beetles

Both common cereal leaf beetles (*Oulema lichenis* and *O. melanopa*) occurred regularly at Lautenbach. In some years leaf damage reached high levels, but never exceeded the economic threshold (Keppler 1985). The incidence of attack, assessed as injured plants on winter wheat fields, showed either similar or significantly lower beetle activity on the IFS fields (Fig. 18.2).

Cereal gall midges

Three cecidomyiid species were known to cause damage mainly to winter wheat. These were *Sitodiplosis mosellana*, *Contarinia tritici* (yellow and orange blossom midge) and *Haplodiplosis equestris*. Their economic importance as pests correlated highly with their larval density in the soil during the previous year (Bühl & Schuette 1971). The larval density in soils of both systems assessed by a flotation technique (Wübbeler 1988) indicated significant ($P < 0.05$) differences between the farming systems (Fig. 18.3). Twice as many larvae were extracted from the CFS soils compared with samples from IFS plots.

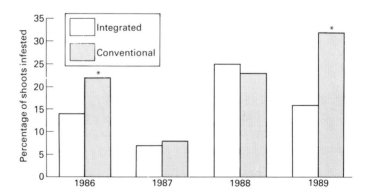

FIG. 18.2. The incidence of attack by cereal leaf beetles on winter wheat on fields in the Integrated and Conventional Farming Systems, Lautenbach, 1986–89. In the figures * $P < 0.05$; ** $P < 0.01$; *** $P < 0.001$.

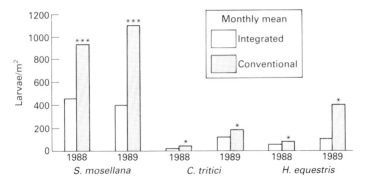

FIG. 18.3. The population density of the larvae of cereal gall-midges *Sitodiplosis mosellana*, *Contarinia tritici* and *Haplodiplosis equestris* in soils of the Integrated and Conventional Farming Systems at Lautenbach in 1988 and 1989. From Froese & Heynen (1989).

Weeds

Both annual and perennial weeds were more common on the IFS wheat fields than on the CFS ones. Analysis of the seed-bank in soils of both farming systems, particularly the vertical distribution pattern indicated significant changes (Wahl 1988); more seeds being left in the topsoil on the unploughed IFS fields. Weed emergence on these fields tended to be 1–2 weeks earlier than on the CFS. However, weed control intensity expressed as amount of herbicide active ingredient used, number of treatments, costs of control per ha and the herbicide-treated area — did not increase over the 12-year period (El Titi 1989).

Consumption of pesticides

Decisions regarding pesticide applications were made following crop monitoring. In IFS fields higher pest densities were tolerated. This contributed to reductions in the frequency of chemical control. In addition, the use of adjuvants served to reduce average doses applied to IFS fields. Figure 18.4 illustrates average pesticide usage expressed as total active ingredients per ha for herbicides, fungicides and insecticides over all crops between 1978 and 1989.

Yields and economic data

Grain yield from IFS wheat fields was usually lower than of the CFS. The average difference over the 12 years was not more than 7%. In some years

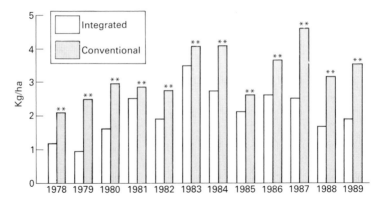

FIG. 18.4. Annual average pesticide use in the Integrated and Conventional Farming Systems, Lautenbach 1978–89, expressed as total active ingredients per ha.

(e.g. 1985), in which high infestations of fungal pathogens occurred, the IFS fields produced significantly higher yields (Fig. 18.5).

Analysis of the economic data showed the success of the integrated farming system, especially improvements in gross margin over conventionally farmed fields. This is true for winter wheat and for all crops considered together (Fig. 18.6). The same or slightly higher gross margin values were achieved with highly significant reductions in the amounts of pesticides and nitrogen fertilizers.

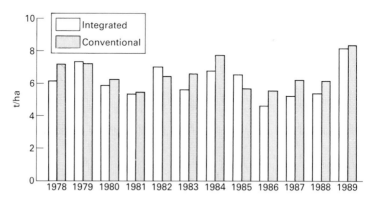

FIG. 18.5. Grain yields of winter wheat cropped in Integrated and Conventional Farming Systems, Lautenbach 1978–89.

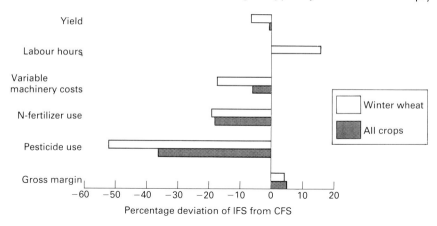

FIG. 18.6. Average percentage differences in the main economic parameters, calculated for winter wheat and for all crops in the Integrated System compared to those grown in Conventional System, Lautenbach 1978–89 (CFS = 100%).

Bioindicators

Many bioindicators responded positively to the integrated farming system. Both numbers (Fig. 18.7) and biomass of earthworms were higher in IFS than in CFS fields. Higher densities of Collembola, gamasid mites, carabid and staphylinid beetles were also found on the IFS fields (El Titi & Ipach 1989). A most spectacular result was the increase in soil humus content

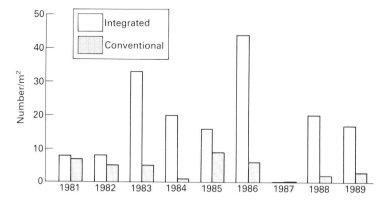

FIG. 18.7. Numbers of juveniles and adult Lumbricidae (per m²) in the Integrated and Conventional farmed fields at Lautenbach, 1981–89.

(Fig. 18.8). In soils of IFS fields the humus content was higher than in the CFS fields and far above expected levels particularly in the top layers.

The ecological infrastructure of Lautenbach has also changed especially during the last 6 years. The results obtained from the hedgerow trial on the farm (Bosch 1986) during the 10-year period (1978–1989) were most encouraging. Three hedgerows of 100 m length each were planted in 1978, 70 m apart from each other, to study the effect of hedges and vegetation diversity on crop productivity, density of both pests and beneficial arthropods as well as on the physical yields of the crops grown on the farm. A number of insect pests (e.g. cereal aphids and beet leaf miners) occurred in lower densities within the hedged area, compared with a hedge-free part of the same field. In the vicinity of hedgerows there were more beneficial arthropods, mainly Hymenopterous parasitoids, predatory Heteropteran bugs, predatory flies as well as more species of spiders. Crop yield within the plots surrounded by the hedgerows increased until 1985. The involvement of the farm manager in operations of pest and disease monitoring and yield measurements has obviously encouraged him to expand the area planted to hedgerows (from nothing in 1978 to 9 km of hedges at present). In addition, the native flora on the field boundaries is no longer mowed, effectively contributing towards more stability in the agro-ecosystem (Altieri, this volume, pp. 262–269). Significant improvements of soil physical properties, particularly soil workability and water infiltration capability, have also been recorded.

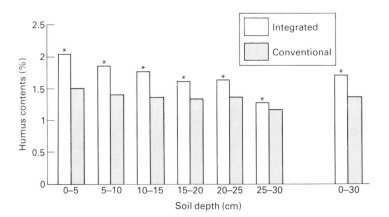

FIG. 18.8. Percentage soil humus contents of one field farmed either in the Integrated or the Conventional Farming Systems, Lautenbach, 1989.

DISCUSSION

Reductions of inputs do not necessarily mean a reduction in income, if an integrated farming approach is followed. This is particularly related to reductions in inputs of pesticides, nutrients and fuel. In the Lautenbach study minimal soil tillage has contributed to enhancement of various edaphic faunal groups, which can act as predators of pests (Karg 1962; Sharma 1971), or of fungal pathogens (Ulber 1983; Curl, Lartey & Peterson 1988), or can improve the physical properties of soils (Lee 1985). Such effects helped to reduce pesticides treatments and the use of tractor power. Disease and pest incidence was lower in the IFS fields. Reduced agrochemical use on farmland also contributed to the limitation of environmental pollution. This is of great social importance not only to farmers, but also to the whole community. This is why the strategy of the integrated farming is considered as the most promising one.

Improvement of several ecosystem components such as earthworms, carabids, soil structure, hedgerows or marginal vegetation cannot be expressed in monetary values. However, they are of substantial importance to sustain the basic value for agricultural production. They significantly contribute to regulation of both water retention and evaporation, enhance the multiplication of beneficial organisms below and above soil surface and accelerate the turnover of organic residues. Soil biota within a functioning soil ecosystem is in fact the key factor of soil fertility and productivity. This is the most precious wealth a farmer can ever possess. This wealth is also known to be of great national value for the rest of society. It is, however, most surprising that soil value and its deterioration has never been seriously quantified in farm micro-economics analysis (e.g. Costigan & Biscoe, this volume, pp. 64–66).

The Lautenbach Project has stimulated far-reaching interests among farmers, teachers, advisors and policy makers. Both on national and international levels the Lautenbach concept has been adopted in setting up of other farming systems research projects. Increasing numbers of farmers have inspected the running activities at Lautenbach, and are adopting single techniques and measures or even trying the whole concept on their own farms. On the state level many farmer clubs have now been created, which are closely supported by the continuing investigations at Lautenbach.

Official farming regulations are now changing in the state of Baden-Württemberg and throughout the European Community. There are new restrictions in the use of pesticides and nitrogen fertilizers in water conservation areas, and measures to conserve endangered wildlife species.

These serve to highlight the importance of the Lautenbach results, particularly to farmers and official extension authorities.

ACKNOWLEDGMENTS

I thank E. Joas, M. Brommer, K. Barthelemess and C. Peao for their field and sampling assistance, Dr Schmidt, Dr Bühler and Mr H. Reinfelder (State Institute for Agrochemistry) for the chemical analysis of soil, nitrate measurement and carbon estimation. Thanks are also due to the manager of Lautenbach — H. Landes — who made, by his sincere co-operation, these investigations possible, and to Dr N.W. Sotherton of The Game Conservancy Trust, UK, for valuable advice concerning this manuscript.

REFERENCES

Anon. (1978). *Umweltgutachten 1978.* Verlag W. Kohlhammer GmbH, Stuttgart & Mainz.

Bosch, J. (1986). Wirkungen von Feldhecken auf die Arthropodenfauna und die Erträge angrenzender Ackerflächen. *Mitteilungen der Biologischen Bundesanstalt*, **232**, 308.

Bosch, J. (1987). Der Einfluss einiger dominanter Ackerunkraeuter auf Nutz- und Schadarthropoden in Zuckerrueben. *Zeitschrift Pflanzenkrankheiten und Pflanzenschutz*, **94**, 398–408.

Bühl, C. & Schuette, F. (1971). *Prognose wichtiger Pflanzenschädlinge in der Landwirtschaft.* Verlag Paul Parey, Berlin & Hamburg.

Curl, E.A., Lartey, R. & Peterson, C.M. (1988). Interactions between root pathogens and soil microarthropods. *Agriculture, Ecosystem and Environment*, **24**, 249–261.

Decker, H. (1971). Die wurzelparasitäre Nematodengattung *Pratylenchoides* Winslow, 1958. *Archiv für Pflanzenschutz*, **7**, 119–130.

Edwards, C.A. & Lofty, J.R. (1975). The invertebrate fauna of the park grass plots. I. Soil fauna. *Report, Rothamsted Experimental Station*, Part 2, pp. 133–154. HMSO, London.

El Titi, A. (1984). Auswirkungen der Bodenbearbeitungsart auf die edaphischen Raubmilben (*Mesostigmata: Acarina*). *Pedobiologia*, **27**, 79–88.

El Titi, A. (1986). Integrierter Pflanzenschutz. Modellvorhaben Ackerbau. Projekt Lautenbacher Hof. *Bericht der Ladesanstalt für Pflanzenschutz*, Stuttgart.

El Titi, A. (1989). Integrierter Pflanzenschutz. Modellvorhaben Ackerbau. Projekt Lautenbacher Hof. *Bericht der Ladesanstalt für Pflanzenschutz*, Stuttgart.

El Titi, A. & Ipach, U. (1989). Soil fauna in sustainable agriculture: results of an integrated farming system at Lautenbach. *Agriculture, Ecosystem and Environment*, **27**, 561–572.

El Titi, A. & Landes, H. (1990). Integrated farming system of Lautenbach: a practical contribution toward sustainable agriculture in Europe. *Sustainable Agricultural Systems* (Ed. by C.A. Edwards, R. Lal, P. Madden, R.H. Miller & G. House), pp. 265–286. Soil and Water Conservation Society, Ankeny, Iowa.

Froese, A. & Heynen, C. (1989). Untersuchungen über die terristrische Dipterenfauna auf integriert und konventionell bewirtschafteten Ackerflächen (Lautenbach-Projekt). *Gesellschaft für Ökologie*, Verhandlung, **19**(1), 74–75.

Gottfriedsen, R. (1987). *Angewandt-zoologische Untersuchungen zum Integrierten Pflanzenschutz im Projekt Lautenbach.* Diplomarbeit, University of Tübingen.

Karg, W. (1962). Untersuchung über die Veränderung und Wechselbeziehungen der

Mikroarthropoden in kartoffelnematoden verseuchten Flächen. *Nachrichtenblatt des Deutschen Pflanzenschutzdienst.* NF, **16**, 187–193.

Keppler, R. (1985). *Ausmaß und Schadwirkungen der Verringerung photosynthetischer Blattfläche des Winterweizens durch Getreidehähnchen* (Lema melanopus) *L. und* Lema lichenis *Voet.), Coleoptera: Chrysomelidae).* Wiss. Arbeit für das Lehramt an Gymnasien, University of Tübingen.

Klinger, K. (1984). *Auswirkungen von Randstreifen an Winter weizenfeldern auf die Populationsdynamik von Getreideblattläusen und ihrer Antagonisten. Eine Untersuchung zum Integrierten Pflanzenschutz im Lautenbach-Projekt.* Diplomarbeit, University of Tübingen.

Lee, K.E. (1985). *Earthworms. Their Ecology and Relationships with Soils and Land Use.* Academic Press, Sydney.

Matt, M. (1986). *Die edaphische Collembolenfauna in unterschiedlich bewirtschafteter Feldflur im Raum Heilbronn (Baden-Württemberg). Untersuchungen im Rahmen des Lautenbach-Projekts zum integrierten Pflanzenschutz.* Diplomarbeit, University of Tübingen.

Scharpf, H.C. & Wehrmann, J. (1976). Die Bedeutung des Mineralstickstoffvorrates des Bodens zu Vegetationsbeginn für die Bemessung der N-Düngung zu Winterweizen. *Landwirtschaftliche Forschung*, **32**, 100–114.

Schroeder, D. (1980). Stroh- und Zelluloseabbau sowie Dehydrogenaseaktivität in 'biologisch' und 'konventionell' bewirtschafteten Böden. *Landwirtschaftliche Forschung, Sonderheft*, **37**, 169–175.

Schroeder, D. (1988). Agrarpolitik für Verbraucher, Bauern und Umwelt. *Pfälzer Bauer*, **23**, 28–29.

Sharma, R.D. (1971). Studies on the plant parasitic nematode *Tylenchorynchus dubius. Proefschrift Communications Agriculture, University Wageningen.*

Steiner, H. (1967). Die Anwendungsmöglichkeit der Klopfmethode bei Arbeiten über die Obstbaumfauna. *Entomophaga*, **12**, 17–20.

Ulber, B. (1983). Einfluß von *Onychiurus fimatus* Gisin (Onychiuridae, Collembola), and *Folsomia fimataria* (1) (Collembola, Isotomidae) auf *Pythium ultimum* Trow., einen Erreger des Wurzelbrandes der Zuckerrübe. *Proceedings of the VIII International Colloquium of Soil Zoology, Louvain-la-Neuve, Belgium*, pp. 261–268.

Wahl, S.A. (1988). Einfluß langjähriger pflanzenbaulicher Maßnahmen auf die Verunkrautung-Ergebnisse aus dem Lautenbach-Projekt. *Zeitschrift für Pflanzenkrankheiten und Pflanzenschutz, Sonderheft* **XI**, 109–119.

Wübbeler, H. (1988). *Die edaphische Dipterenfauna unterschiedlich bewirtschafteter Ackerflächen im Raum Heilbronn (Baden-Württemberg). Untersuchungen im Rahmen des Lautenbach-Projektes zum Integrierten Pflanzenschutz.* Diplomarbeit, University of Tübingen.

Zadoks, J.C., Chang, T.T. & Konzak, C.F. (1974). A decimal code for the growth stages of cereals. *Weed Research*, **14**, 514–521.

Zeddies, J., Jung, G. & El Titi, A. (1986). Integrierter Pflanzenschutz im Ackerbau. Das Lautenbach-Projekt. II. Ökonomische Auswirkungen. *Zeitschrift für Pflanzenkrankheiten und Pflanzenschutz*, **93**, 449–461.

19. NEW PERSPECTIVES ON AGRO-ECOLOGY: BETWEEN THEORY AND PRACTICE IN THE AGRICULTURAL ECOSYSTEM

D.W. MACDONALD AND H. SMITH

Wildlife Conservation Research Unit, Department of Zoology, South Parks Road, Oxford OX1 3PS, UK

INTRODUCTION

What is the future for agro-ecology? Having been asked to identify salient themes of this volume on *The Ecology of Temperate Cereal Fields*, we tackle this question from the perspective of ecologists motivated largely by the desire to foster the integration of conservation and agriculture. From this perspective, a backward glance at the 20th century brings to mind G.K. Chesterton's wry aphorism: the past was far from promising, the future none can tell. The background is well known: a post-war drive for self-sufficiency, fuelled by the Common Agricultural Policy, led to a triumphant spiral of production (e.g. EEC 1988). In the European Community an increase by one-third in cereal production since the early 1970s has resulted from a 50% increase in yields while the productive area has tended to decrease (see also Murphy, pp. 69–91, this volume). Much of this increase has resulted from substantial advances in crop breeding, including increases in kernel yield, better adaptation to modern farming techniques, improved yield responses to added nitrogen and improved disease and herbicide resistance (Fischbeck, pp. 31–54, this volume). From being a major cereal importer, the EC was, by the 1970s, producing substantial surpluses, but was unable to export competitively because the product was overpriced at a time of worldwide surplus and stagnant demand. The inequity of intervention buying and storing became a necessity, with surpluses worldwide reaching 34% of consumption in 1987/88. The public cost of subsidizing agriculture in the EC reached 45·3 billion ECU in 1989 (Murphy 1989). The inequities of international economics (Lal 1989) thwarted the co-operation that would have been required to redeploy this surplus to stave off famine, and so it was doomed to become waste.

Immense environmental costs, as well as overproduction, have arisen from the economic mismanagement of the biological and technological triumphs of cereal production (Potts, pp. 3–21; Fischbeck, pp. 31–54; Biscoe & Costigan, pp. 55–68; and Murphy, pp. 69–91, all this volume). However, the economic and political climate may be ripe for change. In Britain, a country in which 12 million ha is devoted to intensive agriculture, a shift in priorities appears to be statutory: agriculture ministers must endeavour to achieve a reasonable balance between agriculture, conservation and the economic and social interests of rural areas (Section 17, Agriculturae Act, HMSO 1986). Further, 'The government aims to maintain a prosperous rural economy and to conserve and enhance the rural environment' (White Paper on the Environment, HMSO 1990). What are the political options for change? Whilst the vagaries of the weather and world markets modify the extent of surplus cereal production, they seem unlikely to abolish it considering the impetus provided by cell biology and molecular genetics (Fischbeck, pp. 47–52, this volume). A more refined adjustment of supply to demand and a more environmentally friendly and sustainable farming system can only come from policy changes. Within the EC these seem likely to follow the recent pattern of redirecting financial support to reduce production, rather than risking the brutal consequences for rural economies of abandoning support. It was the first such EC measure to reduce production — the set-aside scheme — that fired an already smouldering debate about the environmental price of modern intensive farming. In the UK the scheme was designed specifically to reduce production by reducing the cropped hectarage: its potential environmental benefits were an ill-thought-out appendage.

Whilst neither the set-aside scheme nor its more environmentally friendly progeny (1990 amendments (MAAF, WOAD, DANI 1990) and the Countryside Premium Scheme, (Countryside Commission 1989)) are likely, themselves, to make a significant dent in either cereal surpluses or environmental problems, they are important because of the impetus that they have given to changes in attitude. This is reflected in more innovative policies, such as Extensification and Nitrate Sensitive Areas, and in a new impetus in funding agro-ecological research. At the pure end of the applied spectrum within the UK, the Joint Agriculture and Environment Programme (JAEP) and the AFRC Unit of Ecology and Behaviour at Oxford were the prime examples, whilst systems-based approaches were funded within the Ministry of Agriculture, Fisheries and Food (MAFF) (SCARAB — Seeking Confirmation About Results at Boxworth — and TALISMAN — Towards A Lower Input System Minimizing Agrochemicals and Nitrogen (see Greig-Smith, pp. 368–369, this volume)) and the Agriculture and Food Research Council (AFRC) Institutes (LIFE, Lower Input Farming and Environment). Similar new research initiatives have been instigated in other EC countries and elsewhere.

As we race towards the 21st century, how can ecologists best contribute to management of the agro-ecosystem? To what extent can current ecological knowledge be applied to solve new problems and to what extent is it insufficiently robust or inappropriate? We will seek answers in the papers in this volume, in ecological theory, and in case studies.

The papers in this book fall into two categories. Most are concerned with the ecology of improving crop production. They address the ways in which a better understanding of the many aspects of the ecology of crop growth can improve integrated agricultural systems, incorporating lower levels of agrochemical inputs, whilst maintaining reasonably high levels of production. The remainder represent a wider view of the ecosystem. They recognize the primacy of the need to produce crops on agricultural land, but are principally concerned with the ecology of integrating wildlife and landscape conservation.

CROP PRODUCTION ECOLOGY

These papers cover a great diversity of organisms, many of which, through difficulties of study and lack of charisma, are neglected, despite their often critical importance in the ecosystem. Chief amongst these are micro-organisms and particularly those that live in the soil. Both Jagnow, (pp. 113–137), Wu & Ocio Brookes. (pp. 95–111, both this volume) show that the 'black-box' into which soil communities are often and classified by default contains an enormous diversity of taxa and species, the specific roles of which are rarely known and a high proportion of which probably remain undescribed. But despite this taxonomic diversity, these papers are unified by familiar and recurring ecological themes.

The forceful reminders in this volume that successional processes are important *within* the quintessentially annual cycle of the cereal field is salutory to those accustomed to working with species whose life histories follow annual or longer cycles. The crop plants and their associated weed floras provide an annually renewed resource of living substrate for the phylloplane flora and fauna (Blakeman, pp. 171–191, this volume), and of both live and dead material for the heterotrophs and saprotrophs of the rhizosphere (Jagnow, pp. 117–137, this volume), the wider soil microbial biomass (Brookes, Wu & Ocio, pp. 95–111, this volume) and larger fauna (Zwart & Brussaard, pp. 139–168, this volume). Despite the difficulty of studying these micro-organisms, many parallels with 'conventional' autotrophic successions are evident. In the rhizosphere and phylloplane both the microbial biomass and numbers of species colonizing increase in a broadly predictable way as the substrate ages and its physical and biochemical complexity increases. Some species appear to facilitate colonization by others as, for example, when new substrates are made available by bacterial lysis of the epidermal cells of roots and leaves. However, the extent to which species colonize at different times and that to which they merely differ in the time at which they become dominant, as a result of different growth rates and competitive abilities, is unclear. This is likely to be critically influenced by dispersal mechanisms, some species of the rhizoplane and phylloplane being transmitted via the seed coat, others colonizing *de novo* by air, water or soilborne propagules. Both random events and small quantitative differences in environmental conditions play a critical part in determining the species composition of these seral communities. Jagnow (pp. 113–137, this volume), for example, documents how the species composition and abundance of rhizosphere-colonizing bacteria are temperature dependent and how they vary between years, between soil types, soil management practices and even between two wheat cultivars differing by only a single disomic chromosome substitution. In what appears to be an extreme example of the effects of these spatial heterogeneities — but may of course be the norm — Lambert (1987, quoted by Jagnow, p. 118, this volume) identified 236 types of maize rhizoplane bacteria by their protein banding patterns, most of which were found on only one of seventeen plants examined, 7% on two plants and none on three or more plants. But despite our woeful ignorance of the species involved, their modes of arrival and requirements for persistence, these examples appear to exhibit elements of all current models for community assembly during succession (see Lawton 1987).

Species interactions form major themes not only in the context of their role in determining the composition of seral communities but also in the context of the longer-lived inhabitants of cereal fields. Competition, interacting with the demographies of individual species, is critical in determining the yields of both weed and crop species. Herbivory can be important both directly through reductions in crop yield, and indirectly by facilitating infection by other injurious species (e.g. viral transmission by aphids, see Carter & Plumb, pp. 193–208, this volume) and by modifying the relative competitive abilities of weeds and crops. Predation and parasitism can reduce the numbers of herbivores although, for many species, the extent to which predation is regulating the herbivore populations is unclear. Thus, Aebischer (pp. 315–331, this volume) showed that ladybird populations appeared to 'track' fluctuations in aphid numbers, apparently caused by other agencies, while Jagnow (p. 119, this volume) suggests that bacteriophages can regulate bacterial populations in the rhizosphere.

Farmers struggle to achieve environmental uniformity within cereal fields. But despite their apparent success when we compare cereal fields with most other habitats, the spatial and temporal heterogeneities that Jagnow (pp. 113–137) and Brookes, Wu & Ocio (pp. 95–111) (this volume) elegantly illustrate in the seral communities of the cereal rhizosphere are also characteristic of the system on a larger spatial scale and longer time-frame. It is well known, for example, that weed distributions within fields are extremely patchy and that different species form patches on different scales (Firbank, pp. 214–216, this volume). For some species the patches persist between years despite radical changes in the environment imposed by a change in crop. For others they are ephemeral. Some of these differences are related to the stochasticities of dispersal, others to its predictability. Thus, Wilson & Brain (1990) found that blackgrass, *Alopecurus myosuroides*, distributions were consistent between years under continuous cereals and between cereals with an intervening 3-year grass break. Some distributions are related to heterogeneities in the physical environment that affect establishment (e.g. Andreasen & Streibig 1990) or in the relative efficacy of herbicide application for example. The patchy distributions of weeds in turn create patchy distributions of herbivores and omnivores. Thus, although on widely different spatial scales, the movements of partridge, *Perdix perdix*, the chrysomelid beetle, *Gastrophysa polygoni* and wood mice, *Apodemus sylvaticus*, are woven around a chain of consequences linked to the application of herbicides (Sotherton 1982; Rands 1985; Tew 1990).

An increasingly sophisticated theoretical framework is available in which to hang our knowledge of cereal field organisms and the ways they interact as individuals, species and communities. To a limited extent this framework can now cope even with the complications of stochasticities, heterogeneities and scale. But to what extent does this understanding assist us in answering the many practical questions that ecologists now face about management of agro-ecosystems? The primary objective of cereal field management remains unchanged and clear: to maintain a reasonably high level of crop production. The modern caveats with which we are concerned are also reasonably obvious: crop production must be achieved by more sustainable methods, requiring lower levels of agrochemical inputs and having either neutral or beneficial effects on non-pestilential wildlife and the wider environment. The ecological quesïons that must be answered in order to meet these objectives are concerned with how to manipulate the ecosystem so as to control species that are detrimental to crop production, to increase those that can be of benefit and to conserve the inoffensive. The first two aims usually go hand-in-hand because most of the beneficial species are so defined because they compete with, prey upon or parasitize the pestilential. Three sorts of practical approaches to these problems, requiring different levels of ecological understanding, are raised by papers in this volume. For the sake of clarity we have treated them as mutually exclusive entities although, in practice, combinations of approaches will be appropriate.

Reduced inputs through better forecasting

More precise forecasting can allow reduction of agrochemical inputs with little loss in efficacy. The principle is to trigger applications at the lower limit of abundance at which control of a particular weed or invertebrate pest becomes cost-effective (e.g. Cousens 1987). Thresholds for this trigger are calculated by combining economic models with experimentally derived yield–density curves for particular weed–crop pairs. Weed populations in one generation determine levels of weed seed production, and hence potential weed population levels in ensuing generations (depending on seed longevity). Theoretical models are therefore required to extend this empirical approach to allow forecasting of population trends, and hence potential impacts on future crop yields. Such models describe the population dynamics of the weed and the relationship between weed and crop. They can incorporate density-dependent factors such as the effects of varying weed and crop density, and density-independent factors, such as the effects of differing levels of control on both weed density and crop yield. Models are valuable tools in identifying demographic parameters of key importance in determining population size and can also be incorporated into economic threshold models. However, as Firbank (pp. 214–225, this volume), Brain & Cousens (1990) and others have pointed out, thresholds derived from both empirical experiments and from predictive mechanistic models suffer from two sorts of drawbacks.

First, they usually lack reality, inadequately accommodating spatial heterogeneities or stochastic temporal heterogeneities, such as variations in the weather (Firbank *et al.* 1990). Few take account of the effects of more than one species of weed (but see Pacala 1989 and Wilson & Wright 1990) let alone of the combined effects of weed and invertebrate pests. Second, even if they were made more complex, they lack practical realism. It is notoriously difficult for farmers to supply sufficiently accurate information. It is impossible to estimate accurately weed densities from small numbers of quadrat samples (Marshall 1988). Moreover, by the time it is possible to assess the relative germination times of the weed and crop species which are required for many population dynamics models, it is already too late to take appropriate action (Firbank, p. 214, this volume).

Finally, simply identifying the weed species at early growth stages is not a trivial problem. Considering the farmer's aversion to taking risks, these difficulties limit disproportionately the utility of the models, particularly while the costs of insurance spraying are often very low. As applied biologists we should bear in mind that elegant but impractical solutions to applied problems constitute failures.

More complicated predictive computer systems, incorporated into expert systems, have been developed for use on farms, most notably in Denmark (Baandrup & Ballegaard 1989; Kudsk 1989), in response to their legislative requirement for dramatic reductions in the use of agro-chemicals. These aim not only to aid in deciding when it is necessary to spray but also in fine-tuning the dosage. The Danish system incorporates economic tolerance thresholds for a wide range of weed species but its predictions are based on the simple principle that the activity of a particular pesticide on different weed species, at different growth stages and under different environmental conditions, can be regarded as parallel displacements of the same dose–response curve. This has led to the concept of factor-adjusted doses. The expert system selects the most appropriate pesticide for the species present, and adjusts that dose according to the susceptibility of the weeds at the time of application. Guided by economic thresholds the system frequently indicates that equally effective control can be achieved by doses that are substantially lower than those recommended by manufacturers on the basis of worst-case prophylaxis. The predictive part of this system is not based on ecological mechanisms, although the decision-making process uses biological data.

Economic thresholds are also used with some success to assess the cost effectiveness of pesticide use against invertebrate pests, although reliability is such that most use conservative critical thresholds (reviewed by Burn 1987). Many of the problems parallel those for deriving and employing weed thresholds but the difficulties are arguably even greater: the targets are usually more mobile, their life histories relatively complex (e.g. Carter & Plumb, pp. 193–208, this volume), their numbers affected by predators (e.g. Wratten & Powell, pp. 233–257, this volume) and the extent of damage that they cause can be affected by compensatory growth of the host plants. Again, the need is for continued improvement in our understanding of the population dynamics of the pests, of the relationship between damage and plant yield and of the variability in both of these, and for pragmatism in what is achievable by, and acceptable to, the farmer (Burn 1987).

An alternative approach which can give more advanced, if less precise, warnings of population levels, is to use long-term synoptic monitoring data (Woiwod, pp. 275–304, this volume). This approach is particularly appropriate for mobile taxa whose populations are determined by factors which operate over wide areas and long time frames. At one level, such data can be used to develop predictive statistical, as opposed to mechanistic, models of population trends (e.g. Aebischer, pp. 305–331, this volume; 1990). At another, suitably designed, long-term monitoring programmes can provide better understanding of the population dynamics of such species than can be obtained by short-term data collection and population dynamics modelling. The Rothamsted Insect Survey, for example, provides national and regional forecasts of aphid numbers. In combination with long-term information on numbers on the host plants, host-plant distributions and biological information on the life histories of the species involved, it also contributes to a sophisticated understanding of both density-independent and regulatory factors (see, e.g. Cammell, Tatchell & Woiwod 1989). Other examples are provided by Woiwod (pp. 289–301, this volume) and Taylor (1986). Few could disagree with their conclusion that long-term synoptic monitoring studies are critical elements in understanding the spatial and temporal dynamics of many species. This understanding can both generate novel conceptual ideas and provide the data needed to test them, thereby ensuring their utility by constraining them within the bounds of reality.

These approaches to reducing agrochemical inputs while sustaining yields, are based primarily on better understanding of, or better modelling of, populations of pest species and their relationships with the crop. They aim only to reduce the pestilential, and have no direct 'interest' in either potentially beneficial or benign species. Indeed they are inherently unlikely to benefit either group. It is almost inevitable that seasons in which spraying is advised, on the basis of thresholds, are those in which predators will be most severely affected. Predators are likely to be numerous because, for example, they are responding to high prey numbers, or to the same favourable climatic factors as their prey. For example, the brown planthopper, *Nilaparvata lugens,* a pest of rice crops, is normally regulated by spiders which cause density-dependent nymphal mortality. Kenmore *et al.* (1984) described how an application of insecticide killed the spiders and resulted in an 800-fold increase in planthopper densities.

The ecosystem approach

An alternative approach is to manipulate ecosystem function by using to best advantage both natural biological controls and cultural, but not agrochemical, control methods. Alluding to the implied understanding of the system, we refer to this as the Ecosystem Approach (see Potts &

Vickerman 1974). Such solutions are included within, but not synonymous with, organic agriculture, which also embraces the whole gamut of mechanical weed control measures. At first sight this approach would appear to require a daunting understanding of the ecosystem at community, population and individual levels. To what extent is this true?

Reductionist approaches

Much can be achieved simply by understanding the autecologies and interactions of some of the key species which limit crop growth or which promote it either directly (e.g. vesicular–arbuscular mycorrhizae, see below), or indirectly through disease and pest control. We refer to these as reductionist approaches. Thus, cultural methods of weed control depend primarily on understanding the relationships between phenologies and population dynamics of the species present and different cropping regimes (e.g. Hill, Patriouin & vander Kloet 1989). They can be made much more effective and target levels for control can be set if this information is incorporated into predictive demographic models such as those discussed above. Greater emphasis on the competitive attributes of the crop, as well as the weeds, should prove profitable to weed control (Richards 1989). Crops may, for example, be better able to compete with weeds if they are sown early relative to the timing of germination of their principle competitors, if they are grown closer together, or with other crop species (intercropping, see Bulson *et al.* 1990), or if varieties are chosen with morphologies that confer better competitive abilities (e.g. Fischbeck, p. 37, this volume). Wratten & Thomas (1990) are amongst those who focus on ways of augmenting the populations of predators of crop pests (see also Burn 1987; Wratten 1987; Wratten & Powell, pp. 233–257, this volume). Many polyphagous predators of cereal pests, and most notably some species of carabid beetles, overwinter in field boundaries and migrate into the crop in spring (Sotherton 1985). Reductions in the extent and quality of field boundary habitat, together with an enormous increase in insecticide and molluscicide use in recent years, is likely to have limited severely the effectiveness of these species as natural enemies of crop pests. Wratten & Thomas (1990) (discussed by Wratten & Powell, pp. 248–251, this volume) have shown that artificial, grass-sown earth banks within cereal fields can harbour very large numbers of carabids during the winter and enable the beetles to invade even to the centre of the crop rapidly in spring. An experimental design of blocks of different species of grasses showed that species with dense tussocky growth harboured the highest numbers of predators.

Several important caveats arise from the difficulties of extrapolating from studies of single species within complex communities. First, it is vital to test in the field predictions concerning the responses of species to perturbations. For example, from knowledge of the microbial rhizosphere fauna, Jagnow (pp. 127–130, this volume) provides testable predictions of the consequences of specific changes in agricultural systems: reduced nitrogen applications will inhibit the germination of spores of damaging parasitic fungi, will increase beneficial vesicular–arbuscular mycorrhizal populations and may decrease production of bacterial and yeast secondary metabolites that are toxic to roots. The importance of testing such predictions in field experiments is illustrated by the salutary example described by Brookes, Wu & Ocio (pp. 100–103, this volume). Their hypothesis was that increased straw incorporation would reduce leaching losses of nitrate in winter. Because straw has a high $C:N$ ratio, it seemed likely that the extra microbial biomass that develops in incorporated straw would utilize soil inorganic nitrogen at precisely the time when it is least utilized by the crop, and thus most at risk of leaching. In practice, however, soil inorganic nitrogen reserves were unaffected by adding straw; probably the new biomass was efficient in utilizing straw-derived organic nitrogen.

The second caveat arises from the sad fact that research effort lags far behind the needs of practitioners; the temptation is thus to clutch at straws of information with the risk of overinterpretation. For example, it would be all too easy to extrapolate far beyond the measured statements in which Wratten & Thomas (1990) reported their findings. Their results showed clearly that a simple habitat manipulation created a large overwintering population of predators in a situation where they were previously absent in winter, and from which they could penetrate the crop more effectively in summer. Current concepts in population regulation caution against jumping to the conclusion that this will necessarily either significantly increase the sustainable beetle population size or regulate numbers of aphids or other prey (see Wratten 1987; Wratten & Powell, pp. 237–257, this volume).

The predator-exclusion experiment within the Boxworth Project illustrates some of the potential complexities (Greig-Smith, pp. 356–358, this volume). In fields with moderate and low levels of pesticide input, aphid numbers were relatively higher in plots from which polyphagous predators were excluded than in control plots. In fields with high pesticide inputs there was no difference between aphid numbers in control and predator-exclusion plots. This suggests that predators were sufficiently abundant to regulate aphid numbers under low but not high pesticide regimes. However, even under lower pesticide regimes predators failed to have any significant effect on prey numbers in occasional outbreak years. It is unclear to what extent this effect was compounded by spraying against the pest outbreak, thereby preventing density-dependent increase in predator numbers. Aebischer (pp. 315–329, this volume) suggests, however, that in The Game Conservancy's Sussex study area, invertebrate predators tracked, rather than regulated, their prey. The complexity, and species- and circumstance-specificity, of these interactions is well illustrated by the demonstration at Boxworth that levels of predation on aphids can be higher in fields with high than in those with low pesticide inputs. This appears to occur when populations of alternative prey species of polyphagous predators are disproportionately reduced (Greig-Smith, pp. 353–355, this volume).

Two examples in this volume illustrate our third caution, which concerns the implications of measurements made on different spatial and temporal scales. There is at first sight a paradox concerning polyphagous predators of cereal aphids in the findings of the Boxworth (Greig-Smith, pp. 352–358, this volume) and The Game Conservancy (Aebischer, pp. 315–329, this volume) Projects. The Boxworth Project revealed a complex pattern in which high agrochemical inputs reduced populations of many polyphagous invertebrate predators and their prey relative to those in fields with lower inputs, while some species of both predators and prey showed contradictory patterns. However, Aebischer reported that densities, patterns of change in abundance, and overall population trends in both predators and prey were remarkably similar on five farms which differed substantially in the levels of pesticide use, in husbandry system, and in the quality and quantity of field boundary habitat where many of the predators overwinter. Some of the farms changed radically in these respects over 20 years while others were more consistent.

The paradox would be resolved if what you see depends on the scale at which you measure it. At Boxworth intensive sampling related invertebrate numbers to specified levels of agrochemical inputs over a 5-year period. The spatial and temporal scale of measurement was too fine to detect synoptically synchronous fluctuations or long-term trends. On the other hand, The Game Conservancy's small, but extensive, annual samples were not designed to detect effects on a fine scale. The Boxworth Project suggests that it is possible to achieve farm- or field-scale beneficial effects on invertebrate populations by manipulating farming methods. Aebischer's results do not refute this, but rather provide a different and all too rare perspective on the population dynamics of these species. The finding that they tend to fluctuate in synchrony irrespective of farm practice suggests that they are driven by factors acting on a broad spatial scale. Many similar examples of long-term, broad-scale synchrony in population fluctuations are reported for bird species monitored by the British Trust for Ornithology's Common Bird Census (e.g. Baillie 1990; Marchant *et al.* 1990). The fact that the mean population levels of the invertebrates are declining, also on a broad scale, is likely to be of critical importance. Short termism in research funding increases the risk of addressing detailed problems in the absence of data to show that Rome is burning. Woiwod (pp. 275–304, this volume) is surely right to plead for more long-term, synoptic studies, and the more these have an experimental component the greater is likely to be their explanatory power.

Whilst we have seen that studies of key species can be profitable, there are trade-offs between optimizing management to control or promote different species. Thus, earlier-sown winter cereals may fare better in competition with autumn-germinating weeds (Firbank, pp. 211–214, this volume) but they are also more susceptible to barley yellow dwarf virus infection carried by the autumn migration of aphid vectors (Carter & Plumb, pp. 193–208, this volume). Similarly, El Titi (pp. 405–408, this volume) demonstrates that while minimum tillage and straw incorporation conserve nitrogen and improve soil structure they are also likely to increase populations of annual and perennial weeds (see also Legere *et al.* 1990). On-farm expert systems, built to incorporate risk assessments using multi-species information, may be a profitable method of building on these reductionist approaches.

Community approaches

An alternative to assembling, whether by mechanistic or non-mechanistic means, the parts gleaned from reductionist approaches, is to tackle the whole ecosystem. Two areas of theoretical community ecology seem particularly pertinent: succession, and the relationship between diversity and stability. Having already touched on the assembly of successional communities with respect to microbial successions, we will return to this subject below (see *farm conservation ecology*, pp. 430–442).

The diversity–stability argument, amidst much redefinition and an almost complete inversion of prevailing ideas, has been seized upon in its original Eltonian form (Elton 1966) by many who advocate integrated systems in agriculture. The logic seems simple. Farmland is an artificially simplified system characterized by dramatic fluctuations in populations of pest species. The more we simplify the system, for example by abandoning rotations, the worse these problems become. Therefore by putting less effort into simplifying the system and allowing a more diverse community to persist, greater stability of its component populations will be achieved. How can this view be reconciled with the conclusion from models derived from the Lotka–Volterra predator–prey and competition equations, that the more species and richer structures of interdependence within an ecosystem, the greater the likelihood of its collapse when perturbed (May 1981)? This view suggests that the impact of major perturbations, such as applications of agrochemicals, to a system that has been allowed to become more complex, will have more major destabilizing effects than regular perturbations to a simple system. This conclusion might sound like a charter for completely organic systems and against better-targeted and reduced inputs. May (1981) explains the inverse relationship between diversity and stability by arguing that unpredictable environments are associated with *r*-selected species, able to recover rapidly from perturbation. This is not to say that their populations do not fluctuate widely, merely that they recover rapidly. Indubitably, the cereal ecosystem is populated largely by short-lived species but this would remain predominantly the case even under a more diverse, integrated system.

Perhaps these arguments have little bearing on the agro-ecosystem. May (1981) suggested that the instability of agricultural monocultures results not from their simplicity but from the lack of any significant evolutionary history between pests and pathogens. However, the annually traumatic nature of the system, and the annual or shorter life histories of the great majority of its inhabitants, suggests that selection pressures are likely to be high and evolutionary responses rapid. There are many examples of elegant evolutionary tailoring between crops and their pests. Adaptations of weed seeds to ensure their dispersal with the crop seeds are well known. For example, corncockle, *Agrostemma githago*, evolved three races with seed weights matching precisely those of wheat, rye and rape (Firbank, p. 218, this volume) and gold-of-pleasure, *Cammelina sativa*, evolved races with seeds matching in weight and shape those of the oil-seed and fibre flax crops with which it was associated (Stebbins 1950). Rapid evolution of resistance to the herbicide triazine, now identified in fifty-five species, provides a more recent example (Firbank, p. 218, this volume). Moreover, many of the relationships in cereal fields are not novel pairings of predator and prey, but probably evolved in, and still occur in, semi-natural and natural habitats. It seems likely that the explanation for major fluctuations in pest numbers in cereal fields lies in factors such as the inability of regulating predators to recover as rapidly from trauma as their prey, or the carrying capacity of predators being reduced (e.g. by loss of overwintering sites) while that of their prey is increased (e.g. by providing a superabundant food supply). Community ecological theory must surely have something important to say in this regard, but the message is not yet articulated in terms that the applied ecologist can decipher.

Community theory has few links with applied, whole-system approaches to agro-ecology. There is little evidence that one is a testing ground for the other or that realistic and useful general rules are emerging. Some systems experiments in agro-ecology appear capable of generating valuable mechanistic information, others provide useful agricultural information but scant ecological insight. It has proven to be extremely difficult to conduct large-scale experiments in the field. In illustrating this point our intention is not to denigrate attempts at such experiments, but to emphasize the magnitude of the difficulties they face. Three related categories of problems for ecological interpretation are common in these experiments.

Problems of replication are shared by all ecological experimentation in the field. The replication required to draw robust conclusions from the variance imposed by spatial and temporal heterogeneities is frequently beyond the constraints of funding and manpower. For the same reasons there is a trade-off between the size of experimental treatments and the number of times that they can be replicated. Different sizes of treatment are appropriate for measuring different processes involving different species. In the Boxworth experiment, which set out to examine the consequences for crop production and other aspects of the ecosystem, of three management systems differing primarily in their levels of pesticide use (Greig-Smith, pp. 333–371, this volume), replication was sacrificed to allow measurement of field-scale effects. Even this scale was too fine to detect treatment effects on bird and rabbit populations because their activities transgressed treatment boundaries. At the same time the inability to control for within- and between-field differences in variables such as seed-banks and soil types, has made it more difficult to draw robust conclusions from the wealth of meticulous monitoring data. Nevertheless, the detailed experiments conducted within the Boxworth framework have been extremely profitable and the experiment as a whole has pointed to many general trends that require confirmation by more specific and rigorous studies (the aim of SCARAB, Greig-Smith, pp. 368–369, this volume).

The problem of replication has been addressed in the UK on different scales in two new systems experiments on low input, integrated agriculture. LIFE (Low Input Farming and Environment), the Institute of Arable Crops Research experiment at Long Ashton, compares a conventional five-course rotation with an integrated rotation designed to optimize nitrogen conservation, minimize disease carry-over and include oil and protein crops to reduce small-grain production (Jordan, Hutcheon & Perks 1990). The rotations are compared under both 'standard farm practice' and 'low' inputs. They are replicated both within and between five fields and include all courses of the rotation each year. The Lautenbach experiment (El Titi, pp. 399–411, this volume) provides a similar, well replicated comparison of two contrasting systems. However robust the results from these single-site experiments may be, it will remain difficult to extrapolate the results with confidence to other sites. The problem of geographical replication is addressed in MAFF's post-Boxworth TALISMAN (Towards a Lower Input System Minimizing Agrochemicals and Nitrogen) project. Conventional and integrated rotations are again compared at low and high input levels but the experiment is replicated on four farms in different parts of the country. The cost of geographical replication in this experiment has been a loss of temporal replication: only two of the six courses of the rotation are compared on each farm, each year.

The second category of problem is more specific to systems experiments. In all of these experiments, the systems that are being compared differ, by definition, in many respects. In LIFE, for example, the conventional and integrated rotations differ in total length, in the crops grown, and consequently in the husbandry operations each season. It is therefore extremely difficult to associate measured differences with specific features of the system.

This problem is substantially exacerbated by a third category of problems which arise from the requirement, in many systems experiments, either to follow the pragmatic dictates of 'standard farm practice' (which is not standard but variable) or to achieve preset goals. Both requirements imply that already complex treatments, in variable environments, can be further modified during the course of the experiment by the introduction of new variables. Thus, the dictates of standard farm practice in the LIFE experiment mean that agrochemical use, for example, can be varied between years, as it would be on normal farms (these features are more rigorously controlled within the Lautenbach experiment). The husbandry methods for the low input treatments in this experiment are modified annually to optimize the likelihood of achieving a preset goal of a 50% reduction in pesticide use, while sustaining yields at 80% of those attainable from intensive methods and maintaining or improving gross margins.

The features that we have described as problems are precisely those that make these experiments invaluable in providing robust information about the potential agro-economic viability of particular systems, and about the gross consequences for selected environmental or ecological attributes. These features become problems at the level of extrapolation to different systems and different sites because they made it very difficult to derive insights into cause, mechanism and effect in the ecological processes underlying observed changes.

Despite these difficulties, it is vital that the predictability and variability of systems are measured in the face of different scales of environmental heterogeneities. The most profitable approach to ecological understanding of the mechanisms underlying community-wide responses to manipulation may lie in simpler, well-replicated experiments, that are hypothesis-testing rather than goal-orientated (we return to this theme in *Research for the future*, below). In practice, a level of inadequacy of replication on both spatial and temporal scales, must be accepted in the majority of field experiments. Because of these inherent difficulties in studying complex systems by manipulative field experiments, multi-pronged approaches are often likely to be the most productive. For example, synoptic and long-term monitoring programmes, can often provide powerful precursors (raising many questions) and adjuncts (providing wider spatial and temporal perspectives) to manipulative field experiments.

Augmented biological control

Natural biological controls, such as those used in integrated pest managements systems, can be artificially augmented using a diversity of approaches. For some of these ecological understanding is only peripheral. Thus, for example, genetic engineering to confer on crop plants traits such as resistance to viral pathogens, or insecticidal or insect repellent properties, requires ecological input only at the stage of risk assessment. This entails evaluating the potential of the engineered organisms for persistence, spread and invasion, and the likelihood of transfer of the engineered trait to other species (e.g. Crawley 1989).

Artificially augmenting the populations of species that are either antagonistic to crop pests and pathogens or directly beneficial to crop growth, requires more direct ecological input. Many such possibilities have been discussed in this volume. Jagnow (pp. 121–127, this volume) gives examples of species or taxa that have been shown to have direct positive effects on plant growth and for which inoculation trials have been attempted. For example, symbiotic associations between certain Zygomycete fungi and plant roots, known as vesicular–arbuscular mycorrhizae (VAM), occur on many plant species and improve their phosphorous uptake. Inoculation of wheat roots with the spores of efficient VAM-forming fungi resulted in an increased length of VAM-infected roots and a significant increase in the proportion of the root system that was healthy. In maize roots, simultaneous infection with VAM-forming fungi and *Azospirillum* bacteria resulted in a 20% increase in the amount of VAM infection (and higher phosphorus levels than those infected with VAM alone). Jagnow also reports growth promotion by inoculation with naturally occurring rhizosphere bacteria known to produce plant growth substances. Some *Azospirillum* species, for example, produce indole acetic acid and giberellins. Seed inoculation enhances early development in wheat, the effects of which persist until harvest, and, in sorgum, improves drought tolerance by increasing rooting depth.

Advances are also being made in augmenting populations of species that are antagonistic to crop pathogens. Competition for foliar nutrients by saprophytic yeasts has been shown to reduce populations of necrotrophic fungal pathogens and inoculation can significantly enhance this effect. Similarly, strains of the bacterium *Erwinia herbicola* that induce the formation of ice-nuclei on cereal leaves, leading to frost damage, can be outcompeted with naturally occurring isolates which lack ice-nucleation activity (Blakeman, pp. 180–181, this volume). In 1982 a genetically engineered ice-nucleation-minus strain of a *Pseudomonas* species was the first genetically engineered micro-organism to be released in the field (Crawley 1989). Examples of augmenting, by inoculation, species parasitic on root and leaf pathogenic fungi are given by Jagnow (pp. 125–126) and Blakeman (pp. 179–180, both this volume). For example, inoculation of the cereal rhizosphere with a *Bacillus* sp. controlled an infection of the cereal rust *Puccinia recondita* through lysis of its uredospores.

All of these examples require understanding of the autecology of the potential biocontrol agents, of the target species and of the many other species with which they might interact. It seems likely that artificially elevated population levels will often be short-lived (and therefore potentially of dubious cost-effectiveness) as density-dependent control mechanisms come into play. Even agents with very precise host specificity, and even short-term perturbations, can potentially have longer-term repercussions, particularly for adversely affected, non-target species with small populations. Changes in the target population may itself have unforseen ramifications. These approaches have considerable potential benefits as well as problems. Their requirement for manipulative experimentation, together with an increasingly productive link between theoreticians and applied biologists (Kareiva 1989), offers the opportunity to address simultaneously both applied problems and the mechanisms underlying webs of inter-species interactions.

FARM CONSERVATION ECOLOGY

The new emphasis in crop production ecology is for more integrated and environmentally sustainable production methods. Although some of these are likely to benefit agriculturally benign species, one cannot assume that changed farming methods will automatically reverse the deleterious effects that many species have suffered under post-war agriculture. Understanding such species and their communities, and the consequent ability to manage them, is the remit of farm conservation ecology. This subject requires a new, wider view of the agricultural ecosystem which must encompass the relative simplicity of large blocks of cropped fields and the enormous complexity of the many interstitial habitats that comprise narrow corridors and small, distant islands. The agricultural ecosystem encompasses all the species that live in one, or the other, or both of these habitats, and which do so on varying scales of time and space (Macdonald & Smith 1990).

Many general ecological concepts should help us to understand this system: for example, succession is crucially relevant in the context of set-aside land, and species interactions and patchiness are the stuff of community ecology. Whatever political balance is struck for the future of farmland, ecologists may expect to be dealing with secondary succession, involving small populations persisting within, or colonizing, a spatially and temporally heterogeneous landscape. Insofar as this involves, for many species, oases of suitable habitat amidst a desert of unsuitable habitat it seems to fall within the scope of the Theory of Island Biogeography (MacArthur & Wilson 1967). Colonization and succession require organisms getting from one place to another, and highlight the importance of dispersal, especially regarding the fate of fallowed land (Macdonald & Smith 1990). Conservation ecologists are no strangers to these topics (Pimm & Gilpin 1989). On a larger scale, Wilcox & Murphy (1985) conclude, 'habitat fragmentation is the most serious threat to biological diversity and is the primary cause of the present extinction crisis'. However, European farmland is inherently a fragmented mosaic in which the conservation aim must be to manage heterogeneity. Farmland provides a template for exploring how the geometry of fragmentation affects species' survival. Patchiness has effects at the level of metapopulations (Levin 1974), populations (Den Boer 1968), community structure (Leigh 1990), spatial organization (Macdonald & Carr 1989), and foraging behaviour (Stephens & Krebs 1986). Furthermore, the link between the movements and spacing of individuals in a patchy world, and the dynamics of their populations is the topic of theory that seeks to unite behaviour and ecology (Hassell & May 1985).

The papers in this book demonstrate that farmland is a good place to gather data with which to test theories. How useful are the generalizations that theory offers the practitioner? We will consider briefly what farmland ecologists can learn from concepts arising from theories concerned with small and fragmented populations.

Concepts

The Minimum Viable Population (MVP) is conventionally defined as the smallest population having a $\geqslant 95\%$ chance of persisting for 200 years (Shaffer 1981; Soulé 1987). Although conservation on farmland generally concerns much shorter time scales the concept of MVP is relevant because farmland is fragmented and prone to seasonal and inter-annual trauma, with the consequence that farmland ecologists are often faced with nurturing (or destroying) small populations.

Small populations are more vulnerable to extinction for three reasons: demographic stochasticity, environmental stochasticity and genetic effects such as inbreeding and loss of heterozygosity through drift. Therefore they are likely to disappear more quickly, and more often, than large populations (e.g. Lande 1987). Environmental stochasticity, which amounts to being in the wrong place at the wrong time, is crucial because neither rapid population growth nor being abundant guarantee protection against density-independent catastrophe (Goodman 1987). The key to survival in the face of environmental mishap lies in: (i) risk averaging, by being distributed on a wider scale than that of environmental variation; and (ii) reliance on repopulation, following local extinctions, by immigrants from enclaves where local population growth rate is positive. Therefore, MVP focuses attention on the importance of understanding a species' ability to recolonize patches from which it has become extinct. This understanding is likely to be crucial to species' management (Bunce & Howard 1990; Macdonald & Smith 1990).

The threat to populations due to environmental variation was inadequately emphasized during much of the 1980s when genetic risks were widely considered as paramount (MacArthur 1972). In answer to the question of how many individuals were needed as a minimum to forestall genetic impoverishment, the figure 500 came to assume almost biblical sanctity. As Lande (1988) pointed out, extinction now beckons species whose conservation was misguidedly based on adherence to the 500-credo when it would have been much better if people had paid less heed to generalizations and more to specific details of their biology. The prevailing consensus is that environmental, demographic and genetic threats to small populations interact, and that interaction is affected by life-history variables such as body size and age structure (e.g. Pimm & Gilpin 1989). However, untested models suggest that environmental variability may preserve species or threaten them, depending on circumstances, so aside from asserting that stochasticity is important, generalizations are again elusive (Chesson 1988).

MVP theory indicates that conservation value does not necessarily increase with size of protected area. This is illustrated by Ehrlich & Murphy's (1987) discovery that for the checkerspot butterfly the biggest reserve was far from being the best. This reserve had a uniform, eastern aspect. However, a mosaic of northern and southern slopes was necessary to survive the different conditions of wet and dry years. Survival of the checkerspot depends on some subpopulations producing colonists to replace other subpopulations that have died out. MVP provides a useful generalization in showing why overall survival may depend on the topographic heterogeneity of reserves, not their individual or collective area. Predictions about the consequences of the pattern of fragmentation are amenable to experimental test, and especially so on farmland (e.g. Lefkovitch & Fahrig 1985; Gilpin 1988).

The second theory to which we turn for useful generalizations is the Theory of Island Biogeography (TIB). The nub of the theory is that the equilibrium number of species is lower on small islands because extinction rates are higher, and also lower on isolated islands because immigrant rates are lower (MacArthur & Wilson 1967). A variety of authors (e.g. Diamond 1975; Wilson & Willis 1975; Diamond & May 1981; IUCN 1980) have used TIB as a basis for planning mainland conservation. Special importance has been attached to corridors, which are predicted to elevate immigration rate, thereby lowering extinction rate and raising the equilibrium level, and to fragmentation, which is predicted to result in loss of species because of (i) stochasticity, (ii) loss of enclaves and (iii) loss of habitat diversity. Indeed Diamond wrote, in 1975, that islands and land-bridges 'may furnish a model for what could happen when a fraction of an expanse of habitat is set-aside as a reserve'.

Technically, TIB has enjoyed some support (e.g. Diamond 1975), but its application to 'islands' of habitat on the mainland has been criticized (e.g. Simberloff & Abele 1976; Abele & Connor 1979; Margules & Usher 1981). Nonetheless, advocates of the theory have produced schemes for how nature reserves might be planned (Diamond 1975), and it is now commonplace to hear island biogeography invoked as a theoretical framework within which to consider farmland conservation.

The most obvious tenet of TIB is that, because of their lower extinction rates, larger reserves are better than smaller ones. There are cases where this conclusion appears to hold. For example, Foreman, Galli & Leck (1976) found that the number of bird species increased with the size of woodland, as did the number of phytophagous insect species with the size of stand of Juniper (Ward & Lakhani 1977) and the number of species of higher plants with the areas of nature reserves (Usher 1979). However, in addition to the chequerspot butterfly (see above), there are numerous cases where two smaller reserves support more species than one large one (e.g. Miller & Harris 1977; Connor & McCoy 1979; Higgs 1981). Higgs & Usher (1980) argue that the relative virtue of one large versus two small reserves depends on proportion of species they share: a single large reserve being preferable only if the two smaller ones share many species.

A second prediction of TIB is that separate reserves will support more species if they are close together, due to increased immigration. Again, there are cases where this prediction holds (e.g. Fritz 1979; Lynch & Whigham 1984). However, these and other predictions can be swamped by the effects of environmental stochasticity (predicted by MVP), and ignore many confounding considerations, especially concerning scale (e.g. more distant reserves have less in common, therefore higher combined diversity). Disturbed environments, such as farmland, are especially prone to confounding forces: Helliwell (1976) found small isolated woods had the greatest conservation value in terms of numbers and scarcity of higher plants, perhaps because larger woods had attracted more intensive management. Furthermore it is always easy to find exceptions to rules. For example, Diamond (1975) proposed that circular reserves are preferable in that they minimize dispersal distances within the reserve. For species such as blackcap, *Sylvia atricapilla*, however, which is dependent on woodland margins, numbers correlate with the square root of woodland area (Moore & Hooper 1975; see also Kareiva 1984; Lovejoy *et al.* 1986).

Our point is not to criticize TIB, which has prompted many interesting speculations about the optimal design of reserves; on the contrary, it is that while theories may succeed in identifying an abstract framework that directs the problem-solver's thinking, they cannot be expected to accommodate all the complications of reality. This point is illustrated by the following studies of farmland ecology stimulated by TIB.

Opdam, Rijsdijk & Hustings (1985) and van Dorp & Opdam (1987) report on long-term studies of the distribution of birds amongst isolated woodlands on Dutch farmland. Having censused 235 such copses, of between 0·1 and 39 ha, they found that, in accordance with TIB, woodland size was the best indicator (53–72% of variance) of bird species diversity and probability of occurrence for twenty-six of thirty-two species (but made no difference at all to some, such as chaffinch, *Fringilla coelebs*, woodpigeon, *Columba palumbus*, and blackbird, *Turdus merula*, which were found everywhere). The effect of patch size is complicated because a small woodland is not necessarily a representative subsample of a large one (see Howe 1984). For example, some deciduous woodlands larger than 10 ha supported species otherwise restricted to coniferous woods (e.g. goldcrests, *Regulus regulus*), simply because larger deciduous woodland had a greater chance of including a few conifers.

The effects of isolation, the second TIB parameter, were very varied. Isolation only affected forest-interior birds. Furthermore, although the number of forest-interior species significantly increased with density of these connecting elements (hedges and wooded banks; see also Bull, Mead & Williamson 1976), for no single species was frequency of occurrence significantly affected by the density of connecting elements. Nuthatches, *Sitta europaea*, illustrated the consequences of strong site tenacity and low dispersal capacity: they occurred in all copses larger than 6 ha, but only in smaller copses if they were near a wood; the probability of nuthatches occupying a copse increased with the density of hedges.

Studies of other vertebrates also indicate that the applicability of TIB on farmland varies from species to species. White-footed mice, *Peromyscus leucopus*, avoid open areas (Wegner & Merriam 1979). Fahrig & Merriam (1985) removed these mice from selected spinneys, choosing some that where linked to others by hedges and others that were isolated. Their subsequent study of recolonization showed that the population growth rate was lower in isolated woodlots, indicating that corridors did have an effect on immigration. However, maximum population size was unaffected by the size of woodlots (although overwinter survival, and hence probability of local extinction, may have been). On the other hand, a very similar species, the wood mouse, *Apodemus sylvaticus*, may scarcely venture into a hedgerow at all for the several months of the year spent in cereal fields (Tew 1990). Equally, van Apeldoorn (1987) described circumstances in which tall maize crops can be vital for the dispersal of bank voles, *Clethrionomys glareolus*. Prestt (1971) reports that hedgerows are crucial for adders on farmland. Returning to Opdam's birds, what he actually found was that each species had different requirements: tree creepers, *Certhia familiaris*, needed woodlands of larger than 1 ha, great spotted woodpeckers, *Dendrocopus major*, needed more than 3 ha; the presence of marsh tits, *Parus palustris*, was influenced by the isolation of copses, whereas wrens, *Troglodytes troglodytes*, and robins, *Erithacus rubecula*, increased with access to grassy banks. In short, most farmland bird species were unaffected by an island effect on the scale relevant to farmland conservation.

For the time being, the general lesson offered by conservation theory to the agro-ecologist is that the persistence of comparatively rare species depends on the large-scale heterogeneity of their environment. In summary, Goodman's (1987) list of the implications of this for nature reserves can be adapted for our purposes, as follows:

1   many populations within refuge patches will inevitably be too small to survive environmental variance indefinitely;

2   it is therefore desirable to have an archipelago of dispersed refugia;

3   special steps may be necessary to foster given populations when they are at a low ebb, e.g. resource augmentation, predator and disease control; and

4   the characteristics of enclaves of robust population growth are likely to be species specific, so any given reserve is unlikely to cater for more than a limited number of uncommon species.

This brief account of only two topics leads us to three conclusions about the conceptual framework of conservation ecology on farmland:

1   ecological theory in general, and conservation theory in particular, provide stimulating ideas to direct the thinking and research of agro-ecologists;

2   the agro-ecosystem, because of its mosaic qualities, annual calendar and manipulability, offers great opportunities for testing ecological theory; and

3   because farmland is a mosaic, theories concerned with patches and scale are crucially important.

Indeed, when more than one type of organism is considered the whole problem of scale becomes overwhelming: the grassy bank that constitutes just one patch in the fox's territory may be the universe for the vole it hunts there. However, in turning to ecological theory for help we should be mindful of its function. In that the role of abstract theory is to produce generalizations whose robustness lies in their simplification of a complicated reality, it is unsurprising if exceptions to a rule outnumber adherents to it. It is also not surprising if the predictions of different theories appear to be at odds. There are two equally important solutions to this mismatch of theory and practice.

First, where the assumptions of theory are manifestly false, more aposite theories can be developed. For example, on the farmland mosaic, population models assuming homogeneity of resource dispersion for mathematical simplicity are probably invalid (see, e.g. Hastings 1978; Ehrlich & Murphy 1987; Kareiva 1987). Similarly, through its emphasis on environmental stochasticity, MVP theory directs attention to the concept of the metapopulation, which is a population of subpopulations which may go extinct locally and recolonize (Levin 1974). However, it is not obvious how current metapopulation models can be usefully applied, considering their clearly unrealistic assumptions (e.g. all patches are equally accessible, all individuals are equal). In the case of applying the Theory of Island Biogeography, inappropriate assumptions include ignoring habitat heterogeneity, dismissal of competition, and concern with only one criterion for conservation: maximization of species richness (e.g. Gilbert 1980; Margules, Higgs & Rafe 1982). However, while it is prudent to be sceptical of such assumptions, it would be incorrect to conclude that no relevant theory is available. On the contrary, although some of them might require a little tailoring to the wider agro-ecosystem, there is a mountainous backlog of predictions awaiting the attention of empiricists (e.g. Pacala 1989). As emphasized by Kareiva (1989) in a paper that should be read by all empiricists, better dialogue is needed to catalyse the reaction of theory and practice. Indeed, we conclude that, paradoxically, the proliferation of increasingly sophisticated and potentially more useful theories is contributing to them not being used; the time, and the algebraic skills, required to sift interesting predictions from the plethora of theoretical publications is increasingly daunting. A long-term solution might lie in the education of more numerate ecologists, but meanwhile theoreticians can help guide empiricists to test their models by pointing clearly towards measurable parameters.

Second, the key species or communities can be identified and, guided by generalizations, their specific ecologies can be explored. The simple conclusion to be drawn from the varied reactions of different species to patch size, corridors, etc., is that the requirements of white-footed mice differ from those of wood mice, and both differ from bank voles, whose requirements differ from those of adders, wrens and nuthatches. Inevitably, TIB provides inadequate explanation of the dispersion of at least some of these organisms on farmland, and cannot be expected to provide a landscape design that is a panacea for them all. This is not to say the theory is wrong: it is hard to see how a proposition as straightforward as that isolation diminishes immigration rate when all else is equal could be wrong. Indeed, the theory has succeeded in that it identified the importance of processes such as local extinctions and immigration. However, in nature it is so rare for 'all else to be equal' that generalizations tend to be qualified out of utility. The assumption that the countryside is analagous to islands is too simple when reality is that habitats differ on a continuum of utility, and that continuum has different parameters for each species. In a similar context, Simberloff & Cox (1987) observed that corridors might have 'cons' as well as 'pros' and concluded 'Generalizations made from theoretical considerations cannot be universally applied'. Specific problems usually require specific answers, but what level of specificity characterizes the best research strategy to solve these problems?

Research for the future

How do the foregoing conclusions bear on research strategy? Research strategy can be viewed on a spectrum from the purely theoretical to the goal-directed operational. The research tackled by ecologists in Britain is increasingly determined by the short-term structure of funding. This climate inevitably fosters theoretical research at the expense of empiricism. Theories provide the framework for ecological thinking, but they have so far proven too abstract to be of much utility in predicting the outcome of agricultural change. This is partly because, in the context of the wider cereal ecosystem, theorists have only rarely focused on specific, environment-dependent problems, and partly because it is not in the nature of generalizations to answer specific questions. At the other end of the spectrum, goal-directed systems research may provide answers without understanding. We suggest that both approaches are unfulfilled when empirical research on fundamental questions and on autecology lags behind. It is our perception that for reasons of funding and fashion there is just such a lag in research on the wider agro-ecosystem. The momentum with which theory should suffuse ecology is thwarted if there are inadequate data with which to test it. Yet ecologists are manifestly not yet able to predict the consequences of agricultural change by piecing together empirical information. The store of empirical data, even for species and communities that are of direct economic importance to agriculture, is small. For benign species, most of which are not sufficiently uncommon to have warranted conservation research, almost nothing is known of their autecology on farmland. We have concluded, all too tritely, that specific problems require specific solutions and yet it is clearly not feasible to research the autecology of every species. What type of empiricism is feasible, and what level of specificity do we have in mind?

The urgent need is to understand the relationship between key species and communities. On what criteria are species to be selected from the myriad candidates for study? The obvious answer is those species that are either keystone, pestilential, emblematic or rare. Another important category is those whose features facilitate the testing of theory. Conservation on farmland is constrained by either coexisting in a matrix of intensive agriculture or by the need to integrate into a low input system, where conservation of benign species might be easier but concern about the pestilential is likely to be greater. Within these constraints, questions for the future include how to provide conditions suitable for species that live in the crop for part or all of their lives, how to minimize the effects on such species of occasional pesticide trauma, how to design, and populate predictably with acceptable species, new areas of uncropped habitats, and how to repopulate depauperate pockets of semi-natural habitats. It seems to us that such questions can most productively be tackled in programmes designed to generate the greatest possible understanding of the mechanisms at individual, population and species levels as well as of their outcome at community level. The need to predict and understand responses to change invites long-term manipulative experimentation, which must be replicated to accommodate the heterogeneity characteristic of farmland and most especially of the mosaic of non-cropped habitats. We will illustrate this briefly with the two examples with which we are most familiar. They concern the adjacent, but dissimilar, communities of headland and margin of cereal fields.

The grey partridge, *Perdix perdix*, is both an economically important game species and an emblem species of cereal fields. Monitoring of its numbers by an earlier incarnation of The Game Conservancy revealed a substantial decline beginning in the 1950s. On both commercial and conservation grounds, it was important to reverse this trend and therefore desirable to understand it. This understanding grew from autecological studies which demonstrated that the key factor in the decline was chick survival (Blank, Southwood & Cross 1967), and that chick survival depended upon the availability of insect prey (Southwood & Cross 1969). These results formed the basis of the whole ecosystem approach advocated by Potts & Vickerman (1974) which generated research on diverse aspects of the food chain in which the partridge is a link (Wratten & Powell, pp. 233–257; Sotherton, pp. 373–397, both this volume), and culminated in proof that the demise of the partridge stemmed largely from the effects of agro-chemicals on the broad-leaved weeds that otherwise sustain the phytophagous insects that comprised 60% of the prey of their chicks (Potts 1986).

Four features of what has become the Cereals and Gamebirds Research Project are pertinent to our quest for the best strategy for research within the wider agro-ecosystem. First, hypotheses built on autecological study were tested by natural-scale experimentation: by contrasting the effects of different spraying regimes it was demonstrated that omission of chemicals killing broad-leaved weeds increased the numbers of broad-leaved plants (Sotherton, Rands & Moreby 1985), increased the abundance of the relevant insects and the size of partridge broods (Rands 1985), and ultimately the spring density of breeding adults (Potts & Aebischer 1990). Second, although the motivating applied problem defined the arena in which they worked, the scientists in this team sought fundamental understanding of the processes underlying the autecology of their subjects. For example, the role of predators in the population processes of partridges has relevance to the wider issue of predation as a limiting factor of vertebrate numbers (e.g. Potts & Aebischer 1990), can be tackled through field experimentation (the Salisbury Plain Experiment, Potts & Aebischer 1991), and is a major and complicated controversy in practical conservation. Third, from intensive work on partridges at the hub, research trickled out along spokes to extensive work on other components of the cereal headland community, such as wood mice (Tew 1990) and butterflies (Rands & Sotherton 1986) which in turn stimulated intensive autecological work on new key species such as the small white, *Pieris rapae* (Sotherton, pp. 386–390, this volume). Finally, the biological information was synthesized with economic understanding (Boatman & Sotherton 1988), to provide a practical solution to the original problem as embodied in Table 17.6 of Sotherton's chapter in this volume.

The second example is of our own work at the University of Oxford's farm at Wytham, where we are addressing the practical question of how to manage the boundaries of arable fields in order to improve their value as wildlife habitat. This question was put to us both by farmers and their conservation advisers (principally from Farming and Wildlife Advisory Groups), who could see the potential for conservation of this vast web of uncropped land but were faced with field margins that are species-poor and present weed-control problems. These problems appear to have arisen from erosion in width of margins and enrichment of their soils by indiscriminate fertilizer application. The usual response to the resulting depauperate communities, dominated by 'aggressive' perennials, has been to apply broad-spectrum herbicides. Predominantly annual communities result, dominated by pernicious weeds species such as Barren Brome, *Bromus sterilis*, and Cleavers, *Galium aparine* (Marshall 1989).

How might these trends be reversed? In order to reinstate field margins (those at Wytham were 0.5 m wide) they must first be expanded in width by fallowing a strip of cultivated land. Annual weed control might be combined with benefit to a diversity of wildlife by establishing species-rich, perennial, grassy swards: on fallowed strips this could be achieved either by allowing secondary succession or by side-stepping succession by sowing an appropriate seed mixture. Such swards are likely to require management, the impact of which, on both plant and animal communities, will depend on its timing and frequency and on whether or not any mown clippings are removed. These sorts of ecological considerations led us to devise ten contrasting management regimes that might be practicable for farmers, and to design an experiment to quantify their effects on the flora and fauna of existing field margins and on the development of successional communities on fallowed extensions to these margins (Smith & Macdonald 1989). In addition to broad-scale monitoring of plants, selected invertebrate groups and small mammals, the impact of the treatments on the demographies of selected species or taxa of key agricultural or conservation importance (for example Vanessid and Satyrid butterflies and annual grass weeds (Watt, Smith & Macdonald 1990)) were studied in greater detail. As well as helping us to answer specific practical questions, the experiment allowed us to study successional processes (encompassing invasion, establishment, persistence) and the impact of manipulating the course of succession by management, on fallowed agricultural soils. These processes are all of direct relevance to the wider debate about the future of agricultural land. Although many detailed studies were made of old field successions in America (e.g. Deever 1950; Beckwith 1954) these are unlikely to generate accurate predictions about equivalent events in modern agricultural systems. Soil fertilities are likely to be much higher, sources of propagules more depleted and, in the context of field margins, we are additionally concerned with the persistence of small populations adjacent to intensively managed monocultures.

The design of our experiment (the ten treatments are replicated in eight randomized blocks; each block occupies a single field) allows us to attach confidence to repeatability of the results and to measure the extent and scale of heterogeneities within them. The data allow us not only to test generalized models for succession under these conditions but also give direct empirical generalizations that are of applied value. For example, the seed-bank, much vaunted for its potential in habitat restoration, bore little quantitative relationship with the colonizing flora, contained few species typical of species-rich, perennial grassland swards and only one species that was not present in the extant field flora (see also Graham & Hutchings 1988). A high proportion of species occurring in the old field boundaries, immediately adjacent to the new field margins, colonized very rapidly (79% in 2 years). Only two species from more distant seed sources colonized during the same period, apparently because high productivity resulted in rapid sward closure and removed establishment opportunities.

For the same reason, the peak in abundance of annuals was extremely short lived (the number of annual species in permanent quadrats declined by 45% between the first and second years) whilst annual grass weed densities in sown swards were very significantly lower than in naturally regenerating swards only 1 year after establishment (Watt, Smith & Macdonald 1990). While we attempt to test these and other results in contrasting sites, the design provides: (i) continuing opportunities for comparing with theoretical expectations the observed effects of patchiness and stochasticities; and (ii) insight into the influences of this specialized habitat on the autecologies of keystone, pestilential and emblematic species.

If, despite its great potential, recent ecological theory has largely failed to influence those faced with predicting the effects of changes to the agricultural ecosystem, could a better empirical understanding of that ecosystem help to stimulate a more accessible and useful body of theory? It seems both likely and desirable that a better quantitative understanding of many of the features peculiar to this system should stimulate both conceptual thought and empirically based predictive models. It should also provide the data needed to test them and thereby avoid the criticism that they operate too freely beyond the bounds of reality (Hansson 1988, Krebs 1988). After all, a depth of biological knowledge has surely been a prerequisite for elegant mathematical truths in ecology. However, whilst some questions might be assailed successfully via a more appropriate theoretical framework, others seem likely to remain the province of empiricism and autecology. Our experience of successional processes on fallowed arable land appears to be a case in point (Smith & Macdonald 1989): better theoretical understanding of the assembly rules for successional communities and of the effects on rates and processes of high, or of progressively declining, soil nutrient levels, might enable prediction, for given starting conditions, of certain features of the seral communities over time. For example, the prediction might be that after 5 years the community would be species-poor and dominated by rhizomatous, herbaceous perennials. Theory is unlikely, however, to predict what the species will be: the generalized expectation would be fulfilled equally by a reed-bed, a patch of creeping thistle or a sward of couch grass. There are doubtless theoreticians to whom such fine-grain distinctions are not merely intractable, but also uninteresting. However, to applied ecologists these species imply very different physical structures and different complements of associated plants and animals. Within our fallowed field margins, areas dominated by all three of these species developed around a single field. Not only are stochasticity and heterogeneity as interesting as predictability, but they are crucial to the ecologist's 'customers' who advise on, farm, or merely enjoy the countryside: couch so far has no advocates, but creeping thistles, although noxious arable weeds also provide an important nectar source for many insects, while reed-beds are aesthetically pleasing and can support breeding populations of 'desirable' species of birds.

In advocating autecological manipulative fieldwork we have in mind two goals. The first is to understand the ecological processes, and the second is to apply that understanding to conservation problems. If the former was the only goal then studies in artificial and simplified systems would be the most effective strategy; it is the need to comprehend processes at a realistic scale and under realistic constraints that ultimately forces the research into the field.

CONCLUSIONS

The aim of this paper has been to draw on the contributions to this volume, on the literature and on our own experience, to decide what are the most fruitful ways of researching the ecological processes relevant to a new perspective on cereal production and to wildlife conservation in the wider agricultural landscape. The cereal ecosystem may now be amongst the best understood in the world, yet it is obvious that ecologists have made only small inroads into understanding its ramifications and have hardly started down the road to understanding the wider agricultural landscape. In order to make best use of scarce research resources we should direct our attentions to the research needed to underpin likely practical management problems. In that sense, we should be seeking specific and pragmatic answers to specific questions, but at a level at which the goal is to foster understanding as a weapon against a perverse future. For crop-production ecologists the broad goals of the practitioner remain unchanged — to sustain production — although new constraints are emerging within which this must be achieved. For the farm-conservation ecologist the applied objectives have always been less clear because value judgements about what is desirable differ, and frequently fail to take account of what is likely to be possible. One farmer may want to manage uncropped, interstitial habitats for wild flowers, birds and butterflies, another his field edges for partridges, while another may be concerned only to forestall proliferation of pernicious weeds on set-aside fallow. The ardent conservationist might nominate such nebulous goals as maximizing diversity or nurturing rarity while the pragmatist might aim for more sustainable populations of many attractive, relatively common but hitherto declining species, that the countryside-using public wish to enjoy.

These choices remind us that in conservation issues the scientist is faced with taking the plunge off the mountain of data and into the chasm of values. The cereal ecosystem is home to partridges and their predators; ecologists aspire to understanding how the numbers of one vary with numbers of the other, but society must decide what balance of the two it wishes to foster. Only some of the factors that might affect that decision are quantifiable, and many of those cannot sensibly be compared because they are measured in units that cannot be converted into any common currency. Conservation ecologists may seek to direct opinion by making it informed, but they are ultimately constrained by the customers' present and likely future needs and values. We have tried to rationalize the type of research that is likely to provide a basis for solving practical problems in practical ways. While one sort of bigotry presumes that useful research is uninteresting, another comforts itself in the belief that pure research is 'necessarily divorced from the real world and experimental scrutiny' (Karieva 1989). In fact, both ends of the spectrum are invigorated where they unite. Many of the practical questions with which this volume has been concerned should be of immense fundamental interest to ecologists, of whatever persuasion, because so little is understood of either the important generalities or the equally important specificities that underlie them.

ACKNOWLEDGMENTS

Our own research has been supported by the Nature Conservancy Council, NERC, JAEP and the Ernest Cook Trust. We gratefully acknowledge comments on an earlier draft of this paper from Drs S. Baillie, P. Doncaster, J. Ginsberg, S. Nee, T. Tew and the editors.

REFERENCES

Abele, L.G. & Connor, E.F. (1979). Application of island biogeography theory to refuge design: making the right decision for the wrong reasons. *Proceedings of the First Conference on Scientific Research in National Parks* (Ed. by R.M. Linn), pp. 89–94. USDI National Park Service Transactions and Proceedings, No. 5.

Andreasen, CHR. & Streibig, J.C. (1990). Impact of soil factors on weeds in Danish cereal crops. *Integrated Weed Management in Cereals*, pp. 53–60. EWRS Symposium Proceedings, Helsinki.

Apeldoorn, R. van. (1987). *Annual Report of the Rijksinstituut voor natuurbeheer, Leersum, Netherlands*, pp. 69–70.

Baandrup, M. & Ballegaard, T. (1989). Three years' field experience with an advisory computer system applying factor-adjusted doses. *Proceedings of the Brighton Crop*

Protection Conference — Weeds — 1989, pp. 555–560. BCPC Publications, Farnham, Surrey.

Baillie, S.R. (1990). Integrated population monitoring of breeding birds in Britain and Ireland. *Ibis*, 132, 151–166.

Beckwith, S.L. (1954). Ecological succession on abandoned farm lands and its relationship to wildlife management. *Ecological Monographs*, 24, 349–376.

Blank, T.H., Southwood, T.R.E. & Cross, D.J. (1967). The ecology of the partridge I. Outline of population processes with particular reference to chick mortality and nest density. *Journal of Animal Ecology*, 36, 549–556.

Boatman, N.D. & Sotherton, N.W. (1988). The agronomic consequences and costs of managing field margins for game and wildlife conservation. *Aspects of Applied Biology*, 17, 47–56.

Brain, P. & Cousens, R. (1990). The effect of weed distribution on predictions of crop yield. *Journal of Ecology*, 27, 735–742.

Bull, A.L., Mead, C.J. & Williamson, K. (1976). Bird-life on a Norfolk farm in relation to agricultural changes. *Bird Study*, 23, 163–182.

Bulson, H.A.J., Snaydon, R.W. & Stopes, C.E. (1990). Intercropping of autumn-sown field beans and wheat: effects on weeds under organic farming conditions. *Crop Protection in Low Input and Organic Agriculture* (Ed. by R. Unwin). BCPC Publications, Farnham.

Bunce, R.G.H. & Howard, D.C. (1990). *Species Dispersal in Agricultural Habitats*. Belhaven Press, London.

Burn, A.J. (1987). Cereal crops. *Integrated Pest Management* (Ed. by A.J. Burn, T.H. Coaker & P.C. Jepson) pp. 209–256. Academic Press, New York.

Cammell, M.E., Tatchell, G.M. & Woiwood, I.P. (1989). Spatial patterns of *Aphis fabae* in Britain. *Journal of Applied Ecology*, 55, 463–472.

Chesson, P. (1988). A general model of the role of environmental variability in communities of competing species. *Community Ecology* (Ed. by A. Hastings), pp. 68–83. Springer Verlag, Berlin.

Connor, E.F. & McCoy, E.D. (1979). The statistics and biology of the species–area relationship. *American Naturalist*, 113, 791–833.

Countryside Commission (1989). *The Countryside Premium Scheme for Set-aside Land*. Countryside Commission, Cambridge.

Cousens, R. (1987). Theory and reality of weed thresholds. *Plant Protection Quarterly*, 2, 13–20.

Crawley, M. J. (1989). Biocontrol and biotechnology. *Brighton Crop Protection Conference — Weeds — 1989*: pp. 969–978. BCPC Publications, Farnham.

Deever, C. (1950). Causes of succession on old fields of the Piedmont, North Carolina. *Ecological Monographs*, 20, 229–286.

Den Boer, P.J. (1968). Spreading of risk and stabilisation of animal numbers. *Acta Biother.* 18, 165–194.

Diamond, J. (1975). The island dilemma: lessons of modern biogeographic studies for the design of natural reserves. *Biological Conservation*, 7, 129–146.

Diamond, J.M. & May, R.M. (1981). Island biogeography and the design of natural reserves. *Theoretical Ecology: Principles and Applications* (Ed. by R.M. May), pp. 163–186. Blackwell Scientific Publications, Oxford.

Dorp, D.van & Opdam, P. (1987). Effects of patch size, isolation and regional abundance on forest bird communities. *Landscape Ecology*, 1, 59–73.

EC (1988). *The Agricultural Situation in the Community, 1987 Report*. Office for Official Publications of the European Communities, Brussels.

Ehrlich, P.R. & Murphy, D.D. (1987). Conservation lessons from long-term studies of checkerspot butterflies. *Conservation Biology*, 1, 122–131.

Elton, C.S. (1966). *The Pattern of Animal Communities.* Methuen, London.

Fahrig, L. & Merriam, G. (1985). Habitat patch connectivity and population survival. *Ecology*, 66, 1762–1768.

Firbank. L.G., Cousens, R., Mortimer, A.M. & Smith R.G.R. (1990). Effects of soil-type on crop yield–weed density relationships between winter wheat and *Bromus sterilis. Journal of Applied Ecology*, 27, 308–318.

Foreman, R.T.T., Galli, A.E. & Leck, D.F. (1976). Forest size and avian diversity in New Jersey woodlots with some land use implications. *Oecologia*, 26, 1–8.

Fritz, R.S. (1979). Consequences of insular population structure: Distribution and extinction of spruce grouse populations. *Oecologia*, 42, 57–65.

Gilbert, F.S. (1980). The equilibrium theory of island biogeography: fact or fiction? *Journal of Biogeography*, 7, 209–235.

Gilpin, M.E. (1988). Extinction of metapopulations in correlated environments. *Life in a Patchy Environment* (Ed. B. Shorrocks). Oxford University Press, Oxford.

Goodman, D. (1987). How do any species persist? Lessons for conservation biology. *Conservation Biology*, 1, 59–62.

Graham, D.J. & Hutchings, M.J. (1988). Estimation of the seed bank of a chalk grassland ley established on former arable land. *Journal of Applied Ecology*, 20, 241–252.

Hansson, L. (1988). Empiricism and modelling in small rodent research: how to partition efforts. *Oikos*, 52, 150–155.

Hassell, M. & May, R.M. (1985). From individual behaviour to population dynamics. *Behavioural Ecology* (Ed. by R. Sibley & R. Smith), pp. 3–32. Blackwell Scientific Publications, Oxford.

Hastings, A. (1978). Spatial heterogeneity and the stability of predator–prey systems: predator-mediated coexistence. *Theoretical Population Biology*, 14, 380–395.

Helliwell, D.R. (1976). The effects of size and isolation on the conservation value of wooded sites in Britain. *Journal of Biogeography*, 3, 407–416.

Higgs, A.J. (1981). Island biogeography theory and nature reserve design. *Journal of Biogeography*, 8, 117–124.

Higgs, A.J. & Usher, M.B. (1980). Should nature reserves be large or small? *Nature*, 285, 568–569.

Hill, N.M., Patriouin, D.G. and Kloet, S.P. vander (1989). Weed seed banks and vegetation at the beginning and end of the first cycle of a four-year crop rotation with minimal weed control. *Journal of Applied Ecology*, 26, 233–246.

HMSO (1986). *Agriculture Act.* HMSO, London.

HMSO (1990). *White Paper on the Environment. This Common Inheritance.* (Cm 1200). HMSO, London.

Howe, R.W. (1984). Local dynamics of bird assemblages in Australia and North America. *Ecology*, 65, 1585–1601.

IUCN (1980). *World Conservation Strategy.* Report prepared by the International Union for the Conservation of Nature and Natural Resources (IUCN), 1196 Gland, Switzerland, IUCN.

Jordan, V.W.L., Hutcheon, J.A. & Perks, D.A. (1990). Approaches to the development of low input farming systems. *Crop Production in Organic and Low Input Agriculture.* BCPC Monograph No. 45, BCPC Publications, Farnham.

Karieva, P. (1984). Predator–prey dynamics in spatially structured populations: manipulating dispersal in a coccinellid–aphid interaction. *Lecture Notes in Biomathematics*, 54, 368–389.

Kareiva, P. (1987). Habitat fragmentation and the stability of predator–prey interactions. *Nature*, 321, 388–391.

Kareiva, P. (1989). Renewing the dialogue between theory and experiments in population ecology. *Perspectives in Theoretical Ecology* (Ed. by J. Roughgarden, R.M. May & S.A. Levin), pp. 68–88. Princeton University Press, Princeton, New Jersey.

Kenmore, P.E., Carino, F.O., Perez, C.A., Dyck, V.A. & Gutierrez, A.P. (1984). Population regulation of the rice brown planthopper (*Nilaparvata lugens* Stal) within rice fields in the Philippines. *Journal of Plant Protection in the Tropics*, **1**, 19–37.

Krebs, C.J. (1988). The experimental approach to rodent population dynamics. *Oikos*, **52**, 143–149.

Kudsk, P. (1989). Experiences with reduced herbicide doses in Denmark and the development of the concept of factor-adjusted doses. *Proceedings of the Brighton Crop Protection Conference — Weeds — 1989*, pp. 545–554. BCPC Publications, Farnham.

Lal, D. (1989). The limits of international co-operation. *Twentieth Wincott Memorial Lecture*. IEA Occasional Paper No. 83, Institute of Economic Affairs, London.

Lande, R. (1987). Extinction thresholds in demographic models of territorial populations. *American Naturalist*, **130**, 624–635.

Lande, R. (1988). Genetics and demography in biological conservation. *Science*, **241**, 1455–1460.

Lawton, J.H. (1987). Are there assembly rules for successional communities? In *Colonisation, Succession and Stability* (Ed. by A.J. Gray, M.J.Crawley & P.J. Edwards), pp. 225–244. Blackwell Scientific Publications, Oxford.

Lefkovitch, L. & Fahrig, L. (1985). Spatial characteristics of habitat patches and population survival. *Ecological Models*, **30**, 297–308.

Legere, A., Samson, N., Lemieux, C. & Rioux, R. (1990). Effects of weed management level and reduced tillage on weed populations and barley yields. *Integrated Weed Management in Cereals. Proceedings of the EWRS Symposium 1990*, pp. 111–118.

Leigh Jr., E.G. (1990). Community diversity and environmental stability. *Trends in Ecology Evolution*, **5**, 340–344.

Levin, S.A. (1974). Dispersion and population interactions. *American Naturalist*, **104**, 413–423.

Lovejoy, T.E., Bierregaard Jr., R.O., Rylands, A.B., Malcome, J.R., Quintela, C.E., Harper, L.H., Brown Jr., K.S., Powell, A.H., Powell, G.V.N., Schubart, H.O.R. & Hays, M.B. (1986). Edge and other isolation effects on Amazon Forest fragments. *Conservation Biology: The Science of Scarcity and Diversity* (Ed. by M.E. Soulé) pp. 257–285 Sinauer, Sunderland.

Lynch, J.F. & Whigham, D.F. (1984). Effects of forest fragmentation on breeding bird communities in Maryland, USA. *Biological Conservation*, **28**, 287–324.

MacArthur, R.H. (1972). *Geographical Ecology: Patterns in the Distribution of Species*. Harper & Row, New York.

MacArthur, R.H. & Wilson, E.O. (1967). *The Theory of Island Biogeography*. Princeton University Press, Princeton.

Macdonald, D.W. & Carr, G.M. (1989). Food security and the rewards of tolerance. *Comparative Socioecology: The Behavioural Ecology of Humans and other Mammals.* (Ed. by V. Standen & R.A. Foley), pp. 75–99. Special Publication No. 8 of the British Ecological Society. Blackwell Scientific Publications, Oxford.

Macdonald, D.W. & Smith, H. (1990). Dispersal, dispersion and conservation in the agricultural ecosystem. *Species Dispersal in Agricultural Habitats* (Ed. by R.G.H. Bunce & D.C. Howard), pp. 18–64. Belhaven Press, London.

MAFF, WOAD, DANI (1990). *Set-aside*. MAFF, WOAD, DANI Publication SA1 (Rev. 2).

Marchant, J.H., Hudson, R., Carter, S.P. and Whittington, P. (1990). *Population Trends in British Breeding Birds*. BTO, Tring.

Margules, C. & Usher, M.B. (1981). Criteria used in assessing wildlife conservation potential: a review. *Biological Conservation*, **21**, 79–109.

Margules, C., Higgs, A.J. & Rafe, R.W. (1982). Modern biogeographic theory: are there any lessons for nature reserve design? *Biological Conservation*, **24**, 115–128.

Marshall, E.J.P. (1988). Field-scale estimates of grass weed populations in arable land. *Weed Research*, **28**, 191–198.

Marshall, E.J.P. (1989). Distribution patterns of plants associated with arable field edges. *Journal of Applied Ecology*, **26**, 247–258.

May, R.M. (1981). Patterns in multispecies communities. *Theoretical Ecology* (Ed. by R.M. May), pp. 197–227. Blackwell Scientific Publications, Oxford.

Miller, R.I. & Harris, L.D. (1977). Isolation and extirpations in wildlife reserves. *Biological Conservation*, **12**, 311–315.

Moore, N.W. & Hooper, M.D. (1975). On the number of bird species in British woods. *Biological Conservation*, **8**, 239–250.

Murphy, M. (1989). Economic implications of supply and environmental control of UK–EC agriculture. *Proceedings of the Brighton Crop Protection Conference — Weeds — 1989*, pp. 533–542. BCPC Publications, Farnham.

Opdam, P., Rijsdijk, G. & Hustings, F. (1985). Bird communities in small woods in an agricultural landscape: effects of area and isolation. *Biological Conservation*, **34**, 333–352.

Pacala, S.W. (1989). Plant population dynamic theory. *Perspectives in Theoretical Ecology* (Ed. by J. Roughgarden, R.M. May & S.A. Levin), pp. 54–67. Princeton University Press, Princeton.

Pimm, S.L. & Gilpin, M.E. (1989). Theoretical issues in conservation biology. *Perspectives in Theoretical Ecology* (Ed. by J. Roughgarden, R.M. May & S.A. Levin), pp. 287–305. Princeton University Press, Princeton.

Potts, G.R. (1986). *The Partridge, Pesticides, Predation and Conservation.* Collins, London.

Potts, G.R. & Aebischer, N.J. (1990). The control of population size in birds: the grey partridge as a case study. *Toward a More Exact Ecology* (Ed. by P.J. Grubb & J.B. Whittaker). Blackwell Scientific Publications, London.

Potts, G.R. & Aebischer, N.J. (1991). Modelling the population dynamics of the grey partridge: conservation and management. *Bird Population Studies: Relevance to Conservation and Management.* (Ed. by C.M. Perrins, J.D. Lebreton & G.J.M. Hirons, pp. 373–390. Oxford University Press, Oxford.

Potts, G.R. & Vickerman, G.P. (1974). Studies on the cereal ecosystem. *Advances in Ecological Research*, **8**, 107–197.

Prestt, I. (1971). An ecological study of the viper *Vipera berus* in southern Britain. *Journal of Zoology, London*, **164**, 373–418.

Rands, M.W.R. (1985). Pesticide use on cereals and the survival of grey partridge chicks: a field experiment. *Journal of Applied Ecology*, **22**, 49–54.

Rands, M.W.R. & Sotherton, N.W. (1986). Pesticide use on cereal crops and changes in the abundance of butterflies on arable farmland. *Biological Conservation*, **36**, 71–83.

Richards, M.C. (1989). Crop competitiveness as an aid to weed control. *Proceedings of the Brighton Crop Protection Conference — Weeds — 1989*, pp. 573–578. BCPC Publications, Farnham.

Shaffer, M.L. (1981). Minimum population sizes for species conservation. *Bioscience*, **31**, 131–134.

Simberloff, D.S. & Abele, L.G. (1976). Island biogeography theory and conservation in practice. *Science*, **191**, 285–286.

Simberloff, D.S & Cox, J. (1987). Consequences and costs of conservation corridors. *Conservation Biology*, **1**, 63–71.

Smith, H. & Macdonald, D.W. (1989). Secondary succession on extended arable fields: its manipulation for wildlife benefit and weed control. *Brighton Crop Protection*

Conference — *Weeds* — *1989*, pp. 1063–1068. BCPC Publications, Farnham.

Sotherton, N.W. (1982). Observations on the biology and ecology of the chrysomelid beetle *Gastrophysa polygoni* in cereals. *Ecological Entomology*, **7**, 197–206.

Sotherton, N.W. (1985). The distribution and abundance of predatory Coleoptera overwintering in field boundaries. *Annals of Applied Biology*, **106**, 17–21.

Sotherton, N.W., Rands, M.R.W. & Moreby, S.J. (1985). Comparison of herbicide treated and untreated headlands on the survival of game and wildlife. *Proceedings of 1985 British Crop Protection Conference* — *Weeds*, 991–998. BCPC Publications, Farnham.

Soulé, M.E. (1987). *Viable Populations for Conservation*. Cambridge University Press, Cambridge.

Southwood, T.R.E. & Cross, D.J. (1969). The ecology of the partridge III. Breeding success and the abundance of insects in natural habitats. *Journal of Animal Ecology*, **38**, 497–509.

Stebbins, G.L. (1950). *Variation and Evolution in Plants*. Columbia University Press, New York.

Stephens, D.W. & Krebs, J.R. (1986). *Foraging Theory*. Princeton University Press, Princeton.

Taylor, L.R. (1986). Synoptic dynamics, migration and the Rothamsted Insect Survey. *Journal of Animal Ecology*, **55**, 1–38.

Tew, T.E. (1990). *The behavioural ecology of the wood mouse in the cereal field ecosystem*. PhD thesis, Oxford University.

Usher, M.B. (1979). Changes in the species–area relationships of higher plants on nature reserves. *Journal of Applied Ecology*, **16**, 213–215.

Ward, L.K. & Lakhani, K.H. (1977). The conservation of juniper: the fauna of food-plant island sites in southern England. *Journal of Applied Ecology*, **14**, 121–135.

Watt, T.A., Smith, H. & Macdonald, D.W. (1990). The control of annual grass weeds in fallowed field margins managed to encourage wildlife. *Integrated Weed Management in Cereals. Proceedings of the EWRS Symposium 1990*, pp. 187–196.

Wegner, J.F. & Merriam, H.G. (1979). Movements by birds and small mammals between a wood and adjoining farmland habitats. *Journal of Applied Ecology*, **16**, 349–358.

Wilcox, B.A. & Murphy, D.D. (1985). Conservation strategy: the effects of fragmentation on extinction. *American Naturalist*, **125**, 879–887.

Wilson, B.P. & Brain, P. (1990). Weed monitoring on a whole farm — patchiness and the stability of the distribution of *Alopecurus myosuroides* over a ten year period. *Integrated Weed Management in Cereals. Proceedings of the EWRS Symposium 1990*. pp. 45–52.

Wilson, E.O. & Willis, E.O. (1975). Applied biogeography. *Ecology and Evolution of Communities* (Ed. by M.L. Cody & J.M. Diamond), pp. 522–534. Harvard University Press, Cambridge.

Wilson, B.J. & Wright, K.J. (1990). Predicting the growth and competitive effects of annual weeds in wheat. *Weed Research*, **30**, 201–211.

Wratten, S.D. (1987). The effectiveness of native natural enemies. *Integrated Pest Management* (Ed. by A.J. Burn, T.H. Coaker & P.C. Jepson), pp. 89–112. Academic Press, New York.

Wratten, S.D. & Thomas, M.B. (1990). Environmental manipulation for the encouragement of natural enemies of pests. *Crop Protection in Low Input and Organic Agriculture* (Ed. by R. Unwin). BCPC Publications, Farnham.

AUTHOR INDEX

Figures in italics refer to pages where full references appear.

449

SPECIES INDEX

SUBJECT INDEX

466